Laboratory Manual
for
Comparative Veterinary
Anatomy and Physiology

Join us on the web at

agriculture.delmar.com

Laboratory Manual for Comparative Veterinary Anatomy and Physiology

Phillip E. Cochran, MS, DVM
Department Chair, Program Veterinarian
Veterinary Technology Program
Portland Community College
Portland, Oregon

DELMAR
CENGAGE Learning

AUSTRALIA CANADA MEXICO SINGAPORE SPAIN UNITED KINGDOM UNITED STATES

Laboratory Manual for Comparative Veterinary Anatomy and Physiology
Phillip E. Cochran, MS, DVM

Career Education Strategic Business Unit: Vice President:

Dawn Gerrain

Director of Editorial:

Sherry Gomoll

Developmental Editor:

Andrea Edwards

Director of Production:

Wendy A. Troeger

Production Manager:

Carolyn Miller

Director of Marketing:

Donna J. Lewis

Channel Manager:

Nigar Hale

Cover Images:

GettyImages/Photo Disc

For product information and technology assistance, contact us at
Cengage Learning Customer & Sales Support, 1-800-354-9706

For permission to use material from this text or product, submit all requests online at **cengage.com/permissions**
Further permissions questions can be emailed to
permissionrequest@cengage.com

ExamView® and ExamView Pro® are registered trademarks of FSCreations, Inc. Windows is a registered trademark of the Microsoft Corporation used herein under license. Macintosh and Power Macintosh are registered trademarks of Apple Computer, Inc. Used herein under license.

© 2007 Cengage Learning. All Rights Reserved. Cengage Learning WebTutor™ is a trademark of Cengage Learning.

Library of Congress Control Number: 2003047389

ISBN-13: 978-0-7668-6185-5

ISBN-10: 0-7668-6185-6

Delmar Cengage Learning
5 Maxwell Drive
Clifton Park, NY 12065-2919
USA

Cengage Learning products are represented in Canada by Nelson Education, Ltd.

For your lifelong learning solutions, visit **delmar.cengage.com**

Visit our corporate website at **www.cengage.com**

Notice to the Reader

Printed in Canada
5 6 7 8 9 10 11 12 11 10 09 08

TABLE OF CONTENTS

PREFACE

This text is a laboratory manual for a detailed study of comparative veterinary anatomy and physiology. It was written because there is no adequate laboratory manual on this subject currently in publication for students of veterinary technology or pre-veterinary medicine. There are a few laboratory manuals currently in publication that describe methods for the dissection of the cat; however, these guides are written as laboratory manuals for human anatomy and physiology courses, and therefore the cat's anatomical nomenclature is humanized. For example, the *pectoralis profundus* muscle is labeled the *pectoralis minor* muscle in these laboratory guides.

In addition to the need for an anatomically correct dissection manual for the cat, there is a need for a laboratory manual that compares the anatomies of the major species encountered in the practice of veterinary medicine. Because this manual is primarily aimed at students in veterinary technology programs, or in pre-veterinary medicine anatomy and physiology courses, it does not approach either the scope or the depth of courses in veterinary histology or anatomy taught in a college of veterinary medicine. However, it is a valuable resource book for veterinary medicine in that it is perhaps the only feline dissection manual that uses accurate veterinary anatomical nomenclature.

At the beginning of each chapter there is a list of the major objectives students will learn. In addition, there is a list of materials needed to complete the chapter. This suggested list will help instructors purchase materials, anatomical models, instruments, and supplies for students to use during the study of the organ system that the chapter covers.

This laboratory manual is more than just a dissection guide; it also contains information to enhance the study of anatomy and physiology. In addition to text that explains both the anatomy and physiology of an organ system, there is at least one physiology exercise in each chapter, beginning with Chapter 7, The Skeletal System. These exercises are designed to demonstrate a physiological principle that relates to the study of the specific organ system covered in the chapter. Some of these exercises also serve as a preview to a course in clinical laboratory medicine.

In each chapter there are words that are in colored bold print, black bold print, or italics. The purpose of this differentiation is to establish a hierarchy of importance with regards to structures and functions of special emphasis that students should know. The levels of special emphasis are listed as follows:

- **Colored bold print:** the most important names of functions, structures, and physiological principles are listed in colored bold print. If a structure is listed in colored bold print, students should identify this structure and know it. Individual instructors may add structures they think are also important or eliminate structures they think are of minimal importance. If the same structure is to be located again later in the exercise, then it is given in italics, unless it needs some special emphasis, in which case it appears in black bold print.

- **Black bold print:** Words in black bold print are names of structures, functions, and principles that need special emphasis, but are not as crucial to know as words in colored bold print. Structures included in optional dissections are listed in black bold print.

- *Italics:* Words in italics identify structures, functions, and principles that need emphasis, but not as great as words in black bold print. Also, structures that were noted for identification previously in an exercise with colored bold or black bold print are listed in italics when noted again. In this case, students should identify and know these structures.

In the chapters on microscopic anatomy and histology, areas for students to draw what they see through their microscopes are included. These areas are represented by large white circles, which of course is what would be seen when looking through a microscope. The illustrations included in the chapter are not intended to serve as substitutes for these drawings; rather, these are idealized drawings of tissue to represent the morphology of cells and tissues being studied. What students view through their microscopes may not have the exact same appearance as the illustrations because of differences in slides or histological sectioning. Students need to learn what their own slides look like because this is the material on which they will be tested.

It was important to add a section in each chapter, where applicable, on a topic of clinical significance. By the nature of the material, students may occasionally feel that the information they are asked to learn is more than they will ever use, and this section illustrates the clinical relevancy of the information presented. Also, the Veterinary Vignettes sections are true stories that the author, Dr. Cochran, experienced when he owned his own practice, worked as a relief veterinarian, or encountered ill or injured animals during his teaching years. The names of the owners and animals are fictional as a courtesy to protect their identity. These stories are designed to help hold students' interest in the subject matter, and again to show the clinical relevance of the material presented. The anatomical and pathophysiological terminology used in these stories would not be used in similar stories for the non-veterinary educated public. However, because the purpose of this book is to help students learn and understand veterinary anatomical and medical terms, their use in this text is applicable.

This book is designed to accommodate all courses in anatomy and physiology at the veterinary technician or pre-veterinary level of education. Each instructor may choose the depth in which he or she covers each topic, and the order in which it is covered. The physiology lab exercises teach students some very important physiological principles, and they can be a lot of fun. The intent of this book is to present the material in a sufficiently flexible manner to accommodate all laboratory courses in anatomy and physiology.

I would like to dedicate this book to my family: my parents, wife, son, and daughter. This dedication would not be complete, however, without a special thanks to the College of Veterinary Medicine and its professors at Colorado State University. I have always appreciated the quality of my veterinary education, and I try to emulate it in my teaching today.

ACKNOWLEDGMENTS

Delmar and the author wish to thank the following individuals who devoted their time and professional experience reviewing this manuscript:

Carole Maltby, DVM
Maple Woods Community College
Kansas City, MO

Jody Rockett, DVM
College of Southern Idaho
Twin Falls, ID

Darwin Yoder, DVM
Sul Ross State University
Alpine, TX

Brenda Woodard, DVM
Northwestern State University
Natchitoches, LA

ABOUT THE AUTHOR

Phillip E. Cochran, M.S., D.V.M., is a graduate of Oregon State University with a B.S. in Zoology, an M.S. in Genetics, and a minor in Biochemistry, and he received his D.V.M. from Colorado State University. He is currently Department Chair, Attending Veterinarian, and Instructor for the Veterinary Technology Program at Portland Community College, Portland, Oregon, as well as owner of his own business, Cochran Veterinary Services.

Dr. Cochran is a published author and has several veterinary appliances registered with the U.S. Patent office. He is listed in the *Who's Who in Veterinary Science and Medicine (1991)* and in *Strathmore's Who's Who in America, Centennial Edition (2000)*. He is a member of the American Veterinary Medical Association, the Association of Veterinary Technician Educators, the Oregon Veterinary Medical Association, the Portland Veterinary Medical Association, and the Washington County Veterinary Medical Association.

CHAPTER 1

THE TERMINOLOGY OF ANATOMY

OBJECTIVES:

- begin to learn the terminology of anatomy in order to understand the descriptions of body parts and their relative positions to one another
- be able to locate these various anatomical areas on an animal
- understand the terms used to describe body planes and sections in order to understand histological sections of tissues

MATERIALS:

- cat or dog skeleton, or model of an animal
- dog or cat
- muzzle, if necessary

Introduction

In the study of any field of science it is necessary to understand the language, or specialized terminology, that is used. This is especially true in anatomy, where terms of direction, position, and movement are used to describe both the position of organs in relation to one another and the action of muscles. In veterinary medicine, anatomical terminology varies significantly from that of human medicine because we are concerned with quadruped animals rather than biped humans.

In recent years, veterinary anatomists have attempted to develop a system of terms that better describes animals and eliminates some of the confusing terminology common to human and animal medicine. The best example has been the virtual elimination of the words *anterior* and *posterior*. *Anterior,* in human anatomy, means toward the front of the body (where the mammary glands are in males and females), and *posterior* means toward the back of the body (where you can feel your backbone). In animals, the area where the mammary glands are located is on the *ventral* part of the body, and the backbone is on the *dorsal* aspect of the body. *Anterior* in veterinary medicine used to mean toward the front or head end of the body, but now the term used is *cranial,* and it has been for years. To eliminate the possibility of confusion, *cranial* has almost completely replaced the term *anterior.* However, two areas of the body to which the word *anterior* still refers are the anterior chamber of the eye and the anterior pituitary gland. In addition, in radiography, veterinarians often still use A-P (anterior-posterior) when ordering radiographs of the limbs, rather than cranial-caudal or dorso-palmar. For this reason, technicians should know what these terms mean.

This chapter introduces students to the most important terms used in the study of **gross anatomy,** which is the study of body structures visible to the naked eye. The following is a list of words in veterinary anatomical terminology used to describe direction or position relative to body parts.

Directional Terms

adjacent: Next to, adjoining, or close to. For example, the tongue is adjacent to the teeth.

cranial: Pertaining to the **cranium** or head end of the body or denoting a position more toward the cranium or head end of the body than some other reference point (body part). For example, the head is cranial to the tail. **Craniad,** or **cranially** means in the direction of the cranium or head end of the body.

caudal: Pertaining to the tail end of the body or denoting a position more toward the tail or rear of the body than some other reference point (body part). For example, the tail is caudal to the head. **Caudad** or **caudally** means in the direction of the caudal or tail end of the body (Figure 1.1).

cephalic: Pertaining to the head. This term is not used as frequently in veterinary medicine as is *cranial*. For example, the top of the head is cephalic to the neck.

rostral: Pertaining to the nose end of the head or toward the nose. For example, the nose is rostral to the eyes.

dorsal: Pertaining to the back area of the quadruped (an animal with four legs) or denoting a position more toward the back (upward) than some other reference point (body part). For example, the backbone is dorsal to the belly. **Dorsum** is a noun that refers to the back area of the body.

ventral: Pertaining to the belly or underside of a quadruped or denoting a position more toward the belly (downward) than some other reference point (body part). For example, the kidneys are ventral to the backbone. **Ventrum** is a noun that refers to the belly area of the body.

lateral: Denoting a position farther away from the median plane of the body or of a structure, on the side or toward the side away from the median plane, or pertaining to the side of the body or of a structure. For example, the lateral surface of the leg is the outside surface.

medial: Denoting a position closer to the *median plane* of the body or of a structure, toward the middle or median plane, or pertaining to the middle or a position closer to the median plane of the body or of a structure. For example, the medial surface of the leg is the inside surface.

oblique: At an angle or pertaining to an angle. For example, the vein crossed obliquely from the upper left side down to the lower right side.

superficial: Near the surface; not deep. For example, the skin is superficial to the underlying muscle.

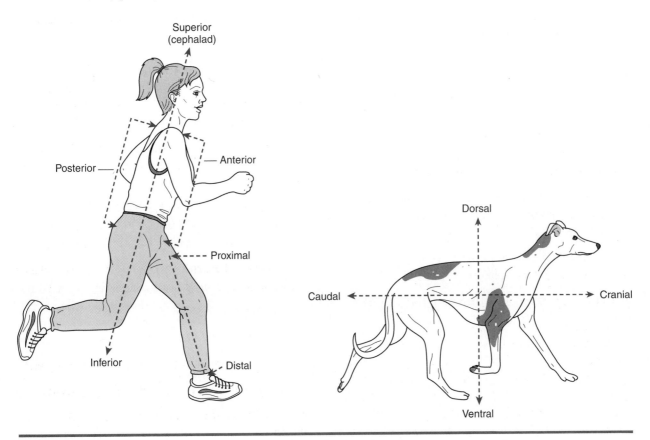

Figure 1.1: Anatomical terminology describing body direction: human vs. animal.

deep: Situated far beneath the surface; not superficial. For example, the bones are deep to the skin.

peripheral: Pertaining to or situated near the outer areas or surface of the body or a structure. For example, the subcutaneous fat is peripheral to the muscles.

proximal: Nearest to the center of the body relative to another body part, or a location on a body part relative to another more distant location. For example, the femur is proximal to the tibia, and the upper part of the humerus is the proximal part.

distal: Farthest from the center of the body relative to another body part, or a location on a body part relative to another closer location. For example, the tibia is distal to the femur, and the lower part of the humerus is the distal part.

superior: Above, directed above, or pertaining to that which is above. For example, the *nasal canal* is superior to the mouth.

inferior: Below, underneath, directed below, or pertaining to that which is below. For example, the mouth is inferior to the nasal canal.

Positional Terms

caudal: Pertaining to the back side of the leg above the carpus and tarsus.

cranial: Can also be a positional term that means the front side of the leg above the carpus and tarsus.

dorsal: Can also be a positional term and applies to the side of the leg opposite the palmar and plantar sides or, in other words, the front side of the leg from the carpus and tarsus distally (down).

palmar (volar): The caudal surface of the front leg from the carpus to the phalanges. This includes the bottom surface of the front foot (see Figure 1.5).

plantar: The caudal surface of the hind leg from the hock to the phalanges. This includes the bottom surface of the hind foot.

pronation: The act of turning the body or arm (leg) so the ventral aspect of the body or the palm is down.

prone: To lie face down, in ventral recumbency.

recumbent: Lying down. A modifying term is used to describe the surface on which an animal is lying. For example, **dorsal recumbency** means the animal is lying on its back, face up. **Ventral recumbency** is an animal lying on its ventral surface, or its belly. If the animal is lying on its side, it is in **lateral recumbency**. On the right side, it is called *right lateral recumbency* and on the left side, *left lateral recumbency.*

supination: The act of turning the body or arm (leg) so the ventral aspect of the body or the palm is up.

supine: To lie face up, in dorsal recumbency.

Anatomical Terms

abdominal: Pertaining to the abdomen.

antebrachium: The distal area of the front legs of an animal, below the elbow joint.

appendicular: Related to the limbs and their attachments to the axis of the body.

axial: Related to the head, neck, and trunk or torso, the axis of the body.

axillary: Pertaining to the armpit area or on the medial aspect of where the front leg meets the torso.

brachial: Pertaining to the proximal area of the front legs of an animal, above the elbow joint.

cervical: Pertaining to the neck area, the cervical vertebrae (the first 7 vertebrae in the dog and cat) or the region around these vertebrae, and to the cervix in the female's reproductive system.

coccygeal: Pertaining to the tail or vertebrae of the tail, the coccygeal vertebrae.

cranium: The part of the skull that encases the brain.

crural: Pertaining to the rear legs of an animal.

digital: Pertaining to the area of the foot where the animal's toes or toe bones, the phalanges, are located.

frontal: Pertaining to the forehead, or the area of the head above the eyes where the frontal bone and frontal sinuses are located.

inguinal: Pertaining to the groin or the medial aspect of the rear leg where it is attached to the torso of the body.

lumbar: Pertaining to the lumbar vertebrae (the part of the backbone between the thoracic vertebrae and the sacrum) or region around these vertebrae.

mammary: Pertaining to the mammary glands (the milk-producing glands).

nasal: Pertaining to the nose.

oral: Pertaining to the mouth.

orbital: Pertaining to the bony eye socket (orbit).

patellar: Pertaining to the patella or knee cap.

pelvic: Pertaining to the pelvis or hip bones (which are made up of four bones of each side of the pelvis: the ilium, ischium, pubic bone, and acetabular bone, which are joined in the middle by a symphysis of the right and left pubic bones).

perineal: Pertaining to the region between or surrounding the anus and the external genitalia.

peritoneal: Pertaining to the cavity inside the abdomen and the membrane (the peritoneum) that lines this cavity. The peritoneal cavity is the space inside the abdomen between the organs and the body wall.

pleural: Pertaining to the cavity inside the chest and the membranes that line this cavity. The pleural cavity is the space inside the chest between the lungs and heart and the inside of the chest wall.

popliteal: Pertaining to the caudal area or back of the true knee, the stifle joint, where the popliteal lymph node is located.

pubic: Pertaining to the bone located between the animal's rear legs, known as the pubic bone, which is part of the pelvis.

quadrants: Arbitrary divisions of the abdominal cavity into four equal sections using the mid-abdominal transverse and median planes as the dividing lines.

sacral: Pertaining to the sacrum, the fused vertebrae by which the pelvis is attached to the backbone.

scapular: Pertaining to the scapula or shoulder blade area.

sternal: Pertaining to the region of the sternum, or breastbone.

stifle: The true knee, the femorotibial joint.

thoracic: Pertaining to the thorax or chest, the thoracic vertebrae (the part of the backbone between the cervical and lumbar vertebrae), or the region around these vertebrae.

umbilical: Pertaining to the umbilicus or navel (bellybutton).

vertebral: Pertaining to the vertebrae or spinal column.

Anatomical Terms for the Mouth and Teeth (Figure 1.2)

buccal: Surface of the tooth that is next to the cheek. As a directional term, *buccal* means toward the cheek.

contact: Surface of the tooth that is adjacent to the next tooth.

labial: Surface of the incisor teeth that is next to the lips.

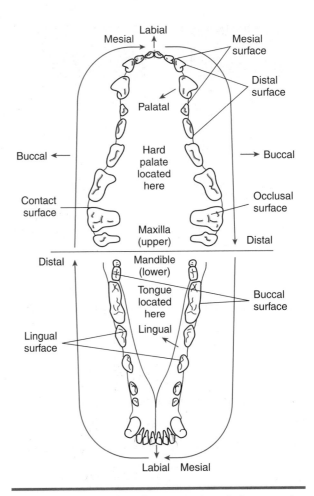

Figure 1.2: Positional terms pertaining to the teeth.

palatal: Surface of the upper teeth that is next to the hard palate.

distal (teeth position): Surfaces of the incisor teeth that are away from the middle or median plane of the mouth. The lateral surfaces of the incisor teeth and the caudal surfaces of the canine, premolar, and molar teeth.

lingual: Surface of the lower teeth that is next to the tongue. As an adjective, *lingual* means pertaining to the tongue.

mesial: Surfaces of the incisor teeth that are toward the middle or median plane of the mouth, and the rostral surface of the canine, premolar, and molar teeth.

occlusal: Surface of the tooth that makes *contact* with the opposing tooth. It is the chewing or biting surface. As a directional term, *occlusal* means toward the chewing or biting surface.

Terms of Movement

abduction: Movement of a limb away from the median line or middle of the body (Figure 1.3).

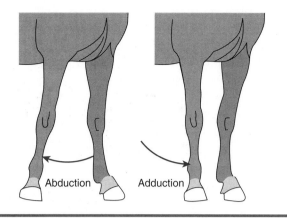
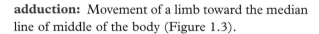

Figure 1.3: Abduction vs. adduction.

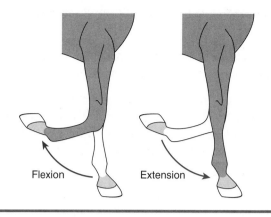

Figure 1.4: Flexion vs. extension.

adduction: Movement of a limb toward the median line of middle of the body (Figure 1.3).

eversion: Turning outward or inside out.

extension: Straightening or the act of straightening, as with a joint (Figure 1.4).

flexion: Bending or the act of bending, as with a joint (Figure 1.4).

rotation: Turning about an axis.

Planes of the Body

dorsal plane: Divides the body dorsally and ventrally, not necessarily in equal divisions (Figure 1.5).

median plane: Divides the body into left and right halves, equally.

paramedian plane: Parallel to the median plane and also divides the body into left and right, but not equally.

sagittal plane: The same as a paramedian plane. A *midsagittal plane* is the same as the median plane.

transverse plane: Divides the body cranially and caudally, not necessarily in equal divisions. It also divides the leg into upper and lower parts, not necessarily in equal divisions.

CLINICAL SIGNIFICANCE

Veterinarians use positional and directional terms every day, as when they ask a technician to "put the dog in right lateral recumbency." These anatomical,

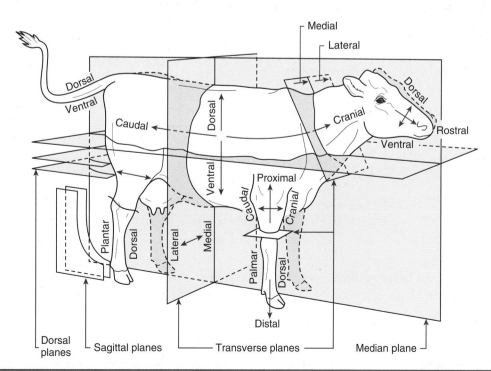

Figure 1.5: Planes of the body.

directional, and positional terms are the nomenclature used in veterinary radiography to describe views taken of the animal's body. If an animal is lying on its back in dorsal recumbency on an x-ray table, an x-ray beam would enter the ventrum of the animal and exit its dorsum prior to striking the cassette and film to produce a radiographic image. This view is called a ventral-dorsal view, or V-D view. The view is named according to how the x-ray beam penetrates the animal's body, in this case from ventral to dorsal.

If we flip the animal over and put it in ventral recumbency, the view is a D-V or dorsal-ventral. An x-ray entering the lateral side traversing to the medial side would be a lateral-medial, but for simplicity's sake it is just called a lateral or lateral view. An x-ray beam penetrating the front part of the leg and exiting the back part above the carpus or tarsus (hock) is called a cranial-caudal. It is also still commonly called an anterior-posterior by many veterinarians. Below the carpus (and tarsus) and

including the carpus (and tarsus), the view is called a dorsal-palmar (or dorsopalmar) in the front leg and dorsal-plantar (or dorsoplantar) in the rear leg.

In large animals, oblique views of the lower leg are also done in addition to the dorsopalmar (D-P) and lateral. This is so a veterinarian can essentially view 360° around the leg radiographically. Therefore, a view taken at a 45° angle with the x-ray machine between the front and the lateral side of the front leg and the film on the inside back part of the front leg would be a **dorsolateral palmaromedial oblique,** *or a DLPM oblique (Figure 1.6). The fourth view of the lower front leg normally taken would be a* **dorsomedial palmarolateral oblique.** *Can you picture these views in your mind, including where the x-ray beam enters and exits the leg and where the machine and film are placed in each of the four views?*

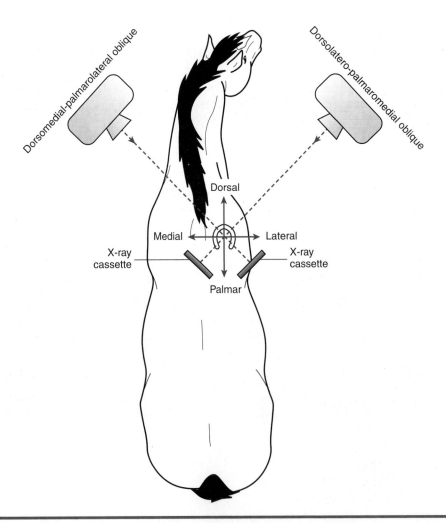

Figure 1.6: Radiographic anatomical directional terms for oblique views.

USING ANATOMICAL TERMS TO DESCRIBE BODY STRUCTURES OR AREAS

Using an animal model or skeleton, and Figure 1.7, locate the following:

1. cranium

2. tail

3. cranial vs. caudal areas

4. dorsum

5. ventrum

6. dorsal vs. ventral areas

7. proximal aspect of the humerus (upper bone of the front leg)

8. distal part of the humerus

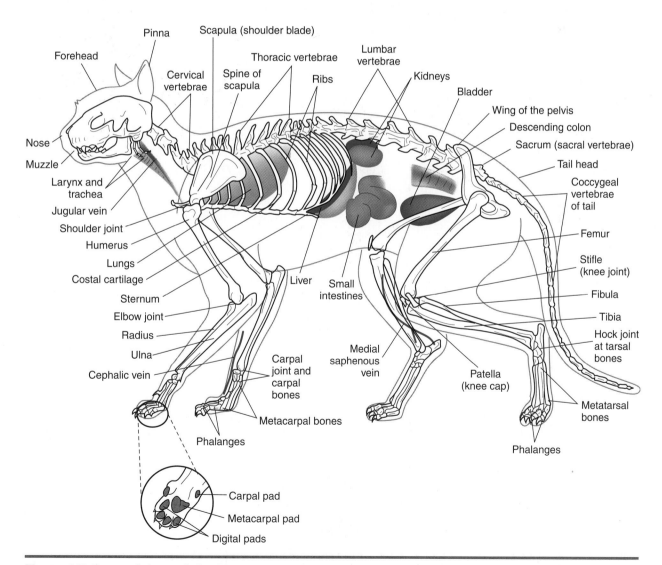

Figure 1.7: Parts of the cat's body.

9. carpus

10. elbow joint; describe the relationship of the elbow joint and the carpus using the terms *proximal* and *distal*.

11. lateral and medial sides of the leg

12. tarsus or hock

13. cranial and caudal aspects of the leg

14. dorsal and palmar/plantar aspects of the front and rear legs

---

EXERCISE 1.2

TOPOGRAPHICAL ANATOMY USING A DOG OR CAT

On a live dog or cat, perform the following:

1. Abduct, then adduct a leg.

2. Flex the stifle joint, and flex the hock joint. Note that they flex together at the same time and flex in opposite directions. Now extend both of these joints.

3. Place the animal in the following positions:
   a. dorsal recumbency
   b. ventral recumbency
   c. left or right lateral recumbency

4. Find the following areas on the animal's body:
   a. cervical region
   b. thoracic region
   c. lumbar region
   d. sacral area
   e. coccygeal region
   f. perineum
   g. popliteal area
   h. axilla
   i. inguinal area
   j. brachium
   k. antebrachium
   l. umbilicus

5. In the mouth, find the following:
   a. buccal surface of the teeth
   b. lingual surface of the teeth
   c. contact surface of the teeth
   d. occlusal surface of the teeth
   e. mesial surface of the teeth
   f. distal surface of the teeth
   g. palatal surface of the teeth
   h. labial surface of the incisor teeth

SUMMARY

In this chapter you have begun to learn the language of anatomy and medicine. This information will be particularly useful in chapters to come, especially when studying the dissection of the cat and other selected organs. The words used more than any others are: *cranial, caudal, dorsal, ventral, medial, lateral, proximal,* and *distal.* These words give direction on all the planes of the body. Even if you have had a course in medical terminology, reviewing the definitions in this chapter will be most helpful.

CHAPTER 2

THE USE OF THE MICROSCOPE

OBJECTIVES:

- identify the parts of the microscope, their purpose, and how they work
- describe and demonstrate the proper care of the microscope
- demonstrate proper focusing technique
- estimate the size of objects in a field
- return to a specific site on a prepared slide using the vernier scales on a mechanical stage

MATERIALS:

- compound light microscope
- immersion oil
- lens paper
- lens cleaner
- prepared blood smear slide
- prepared slide of grid, ruled in millimeters (grid slide)

Introduction

Most students enrolled in courses on anatomy and physiology have already completed one or more general biology courses. For students of veterinary technology and for graduate technicians, the microscope is one of the most important diagnostic instruments used in their profession. A diagnostic-quality microscope is necessary to read fecal, blood smear, tissue smear, and urine sediment slides. The technician's work is to produce an accurate reading of the previously listed specimens. Without a quality microscope and a working knowledge of its use, this is not possible. Therefore, knowledge of a microscope's proper use, care, and safety considerations is critical. For some students, much of this chapter's information on the use of the microscope will be review. However, it may have been years since others took a biology course; for this reason the following material is provided.

Care of the Compound Microscope

The following is a list of operating procedures and safety rules for using a microscope. These should *always* be followed when transporting, cleaning, using, and storing your microscope.

1. *Always* use two hands to carry and transport a microscope; hold it in an upright position with one hand on the *arm* and the other under the *base*. Avoid jarring the microscope when setting it down.

2. Use *only* special grit-free lens paper for cleaning microscope lenses. Do this before using the microscope and before putting it away.

3. Always use a coverslip with wet-mount preparations so the liquid does not get on the *objective lens*.

4. Before putting the microscope in the storage cabinet, be sure to remove the slide from the **stage,** rotate to the lowest-power objective lens, and replace the dust cover (if available). If the stage is movable rather than fixed, move it to its lowest point.

5. *Always* begin focusing with the scanning objective lens; this is the lowest-power objective lens, usually the 4X objective. Use the *coarse adjustment knob* only with the lowest power lens. *Never* focus using the high magnification or the oil immersion objective lens by moving the objective lens and slide toward each other; this may cause breakage of the glass slide or, far worse, scratch the lens. Once you have focused the microscope with the lowest power objective, then you may move a higher power objective into place.

6. When using oil, use the focusing method listed previously or start with the objective lens touching the oil, and use the *fine adjustment knob* to bring the slide's structure into view (preferably by moving the slide away from the objective).

7. *Always* clean the oil completely from the oil immersion objective lens and the microscope stage before switching to the other lenses.

EXERCISE 2.1

IDENTIFICATION OF THE PARTS OF THE MICROSCOPE

Using Figure 2.1 and your microscope, identify the following parts.

1. **ocular (eyepiece):** Microscopes have one or two eyepieces. These are called *monocular* or *binocular* microscopes, respectively. Each eyepiece, or ocular, consists of a tube containing the ocular lens that fits into a tube at the top of the *head* or *body tube* of the microscope. The ocular lens is usually a 10X or 10-power magnification lens (some are 15X). In other words, it increases the apparent size of the object by 10 times or 10 diameters. Usually one of the oculars contains a pointer, which can be rotated to indicate a specific area of concern to the viewer. If your microscope's oculars have no pointer, ask the instructor if one can be added, or perhaps another microscope has a pointer in both oculars and can be traded.

2. **objective lens:** Scanning, low power, high power, and oil immersion (usually 4X, 10X, 40X, and 100X, respectively). To calculate the total magnification using each lens, multiply the power of the eyepiece by the power of the objective lens.

$$10X \times 40X = 400X$$

3. **base:** The bottom support for the microscope on which it rests on the table. Some microscopes have an inclination joint, which allows them to be tilted backward for viewing dry preparations.

4. **head (body tube):** Supports both the oculars and the rotating *nosepiece* containing the objective lenses. Through a system of mirrors within, the body tube projects the image from the objective lens to the eyepiece.

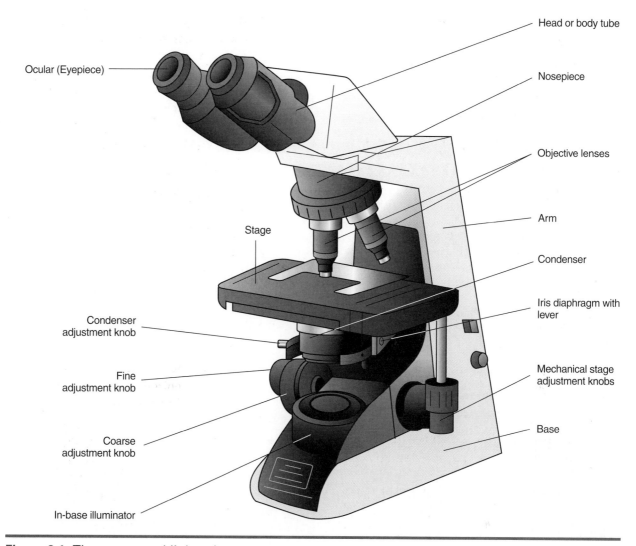

Head or body tube

Ocular (Eyepiece)

Nosepiece

Objective lenses

Stage

Arm

Condenser

Condenser
adjustment knob

Iris diaphragm with
lever

Fine
adjustment knob

Mechanical stage
adjustment knobs

Coarse
adjustment knob

Base

In-base illuminator

Figure 2.1: The compound light microscope.

5. **arm:** Vertical portion of the microscope connecting the base to the head.

6. **coarse adjustment knob:** Used to make large adjustments in the distance between the specimen and the objective lens to focus on the specimen.

7. **fine adjustment knob:** Used to make small, or fine, adjustments in the distance between the specimen and the objective lens to focus on the specimen. It is used for precise focusing once the coarse focusing has been done.

8. **condenser:** Contains a lens that focuses light on the specimen. It may have a height-adjustment knob that raises and lowers the condenser (and lens) to vary the light delivery. Generally, the best position for the condenser is close to the bottom surface of the stage.

9. **stage:** This is the platform upon which the slide rests during viewing. There are two types of stages, a mechanical stage and one equipped with spring clips to hold the slide in place. Most diagnostic-quality microscopes have a mechanical stage to permit precise movements of the specimen.

10. **nosepiece:** This contains three or four objective lenses, which are changed by rotating the desired lens into proper position over the specimen. The nosepiece should be used to rotate the lens rather than grabbing the lenses themselves.

11. **iris diaphragm with lever:** As the arm, or lever, attached to the condenser is moved, the iris diaphragm opens and closes to permit a variable amount of light to pass through the condenser. It provides the best possible contrast for viewing a specimen.

12. **filter:** A blue filter is placed between the illuminator and the condenser to change the color of light from a yellow-white to a cooler blue-white. This is easier on the viewer's eyes, and the colors of prepared specimens appear more vivid.

13. **in-base illuminator:** Also known as the *substage light* (or a mirror in low-tech microscopes), this directs light toward the condenser.

14. **rheostat:** Also known as the *light control*, this knob allows for variable degrees of light output by the in-base illuminator.

EXERCISE 2.2

USE OF THE COMPOUND MICROSCOPE

The following are the steps for focusing your microscope using the 4X, 10X, and 40X objective lenses.

1. Determine if the microscope you are using has a movable stage (i.e., it moves up and down, toward and away from the objectives) or if the objectives move (toward and away from a fixed stage).

2. Plug in the microscope's cord and turn it on. Adjust the rheostat or light control to the middle of its range.

3. Close the iris diaphragm completely, then open it halfway.

4. Raise the condenser to the fully up position.

5. Position the fine adjustment knob about halfway between the limits of its movement.

6. Make sure all the objectives are screwed in tightly.

7. Rotate the 4X objective into the viewing position (you should feel it click into place). It is best to start with the 4X objective when focusing on an object.

8. Obtain a microscope slide and place it snugly in the mechanical stage forks with the specimen side up. Center it exactly over the condenser.

9. Prior to looking through the eyepieces, observe the microscope *from the side* and turn the coarse adjustment knob so the objective is positioned as close to the slide as possible. The specimen should be brought into focus by turning the coarse adjustment knob to move the objective and stage apart. You should not attempt to adjust the coarse adjustment knob toward the specimen when looking through the eyepiece; breakage of the slide or coverslip may occur.

10. Look into the eyepieces and close your left eye. Bring the specimen into focus using the coarse adjustment first, then the fine adjustment. The slide may have to be moved to bring it into the field of view.

Steps 10-12 assume you are using a standard microscope having a rotational focusing left eyepiece.

11. With both eyes open, adjust the interpupillary difference between the left and right oculars so that one circular field of view is visible with both eyes open. Make note of the distance so it may be easily set the next time you use the microscope.

12. Now close your right eye and rotate the left-eyepiece focus adjustment *(not the fine adjustment knob)* to obtain a sharp focus for the left eye alone.

13. With both eyes open, adjust the iris diaphragm to obtain the sharpest image possible and adjust the illuminator for comfortable and clear viewing brightness. The microscope is now adjusted precisely for your eyes and should provide the sharpest image possible.

14. Before changing to a higher-power objective lens, move the part of the specimen you are studying to the *exact center* of the field of view.

15. Change to the 10X objective. If you centered your specimen properly, the part you chose to study should be in view. Modern microscopes are **parfocal,** meaning only slight fine-adjustment corrections will be necessary when changing lenses. However, brightness and diaphragm adjustment may also be needed to produce a crisp image.

16. Now change to the 40X objective lens. If the part of the specimen you chose to study is not in view, rather than searching around for it, go back to the 10X objective, re-center, and change back to the 40X objective. Again, slight corrections may be needed in the fine adjustment, the diaphragm, and the illumination.

Method of focusing using the oil immersion lens.
The use of an oil immersion lens requires a drop of immersion oil to be in contact with both the slide and the lens itself. If the oil is not in proper contact, light is refracted (bent) in such a way that some is lost and the image detail becomes blurred. As the level of magnification increases, the working distance (space between the lens and the specimen) decreases, and more light is required for accurate viewing. With an oil-immersion objective lens the working distance is very short, and thus, care must be taken when positioning the lens. Otherwise, broken slides, cracked coverslips, or even damage to the lens surface may occur. Follow these steps to view a specimen with an oil immersion lens.

1. Obtain a prepared slide of blood.

2. Start with the scanning lens, progress to the high-power lens, and focus as previously described. Find a part of the slide you would like to observe and center it.

3. Rotate the high-power lens aside, and place one drop of immersion oil on the slide over the light opening.

4. Rotate the oil immersion objective into position. The lens should touch the oil and be just short of touching the slide. Do not rotate the other objective lenses over the immersion oil.

5. Look through the eyepieces and move the iris diaphragm to adjust the light. It also may be necessary to adjust the condenser up and down. If the light is too bright, it will be difficult to distinguish the cells from the background.

6. Focus upward with the fine adjustment until the blood cells come clearly into view. Note the red blood cells, without nuclei, and the white blood cells, with nuclei.

7. Remove the slide when you are done and clean it with lens cleaner and lens paper.

8. Wipe the oil from the oil immersion objective lens, and swing the scanning power objective back into position.

In order to prevent damage and misalignment of internal parts, always follow the following steps when preparing a microscope for storage.

1. Remove any slide from the microscope's stage.

2. Clean the instrument, including the lenses, to remove any fluid from the specimens or any oil left in contact with the objectives.

3. Open the iris diaphragm.

4. Raise the condenser to its highest position.

5. Move the scanning objective into viewing position, if not already done.

6. Center the mechanical stage so it has minimum projection on either side of the fixed stage.

7. If the stage is movable, lower it to its lowest position.

8. Wrap the electric cord around the base.

9. If available, place the dust cover over the instrument.

10. Using both hands, as previously described, return the microscope to its proper storage place.

EXERCISE 2.3

DETERMINING THE SIZE OF THE MICROSCOPE FIELD

During the previous exercise, you undoubtedly noticed that the size of the microscope field decreased as the magnification increased. It is useful to know how to determine the diameter of each of the microscope's fields. This information will allow you to make a fairly accurate estimation of the size of the objects in any field of view. For example, if you know the field of view is 4 mm, and the object takes up approximately ¼ of the field, the object is 1 mm in size.

Microscope specimens are measured in micrometers or microns (μm) and millimeters (mm). A millimeter is 1/1,000 of a meter (10^{-3} m) and a micrometer or micron is 1/1,000,000 of a meter (10^{-6} m). Using your microscope, use the following steps to determine the size of a field.

1. Obtain a grid slide (a slide prepared with graph paper ruled in millimeters). Each grid section is 1-mm square. Using the lowest power objective (4×, scanning objective), bring the grid lines into focus.

2. Move the slide so that one grid line touches the left edge of the field, then count the number of squares across the diameter of the field at its widest point (Figure 2.2). On the right side, if only part of a square is visible, estimate how much of the square can be seen and calculate the decimal equivalent to the portion of the square visible (i.e., if half the square can be seen, 0.5 mm is visible).

Compound-
lens in occula
& in objectius

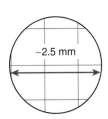

~2.5 mm

4x =4.5mm
10x = 2 mm
40x =0.5 mm
100x =

Figure 2.2: The grid scale.

3. Repeat the procedure using the low power objective (10× objective), and record your observations in the chart below. The high power and oil immersion data will need to be calculated using the formula in step 4 below.

Summary Chart for Microscope # ___11___

	Objective Lens Magnification	Total Magnification	Field Size
Scanning	4x	40x	4.5mm
Low power	10x	100x	2 mm
High power	40x	400x	0.5mm
Oil immersion	100x	1000x	0.2mm

4. Complete the chart by computing the approximate diameter of the high power and oil immersion fields using the following formula (note the following abbreviations: HPF=high power field, LPF=low power field, HP=high power, LP=low power)

$$\text{diameter of HPF} = \frac{\text{diameter of LPF} \times \text{total magnification of the LP objective}}{\text{total magnification of the HP objective}}$$

Therefore, if the diameter of the LPF is 2 mm and the total magnification of the LPF is 100 (10 × 10), and if the total magnification of the HPF is 400 (10 × 40), then

$$\text{diameter of HPF} = \frac{2 \text{ mm} \times 100}{400} = 0.50 \text{ mm}$$

5. Find a white blood cell on the blood smear slide under the oil immersion lens and estimate its size. The white blood cells are the cells that are nucleated.

USING THE VERNIER SCALES ON THE MECHANICAL STAGE

Occasionally, during the course of using the microscope, you may need to return to an exact point on a specimen slide or to record where a certain cell or lesion can be found. Remembering *landmarks* on the slide may be one way of doing this, but it is not precise. A more exact way is to use the verniers to re-locate structures. **Verniers** are scales located on both the x and y axes of the mechanical stage. A vernier consists of two parallel, graduated, sliding scales, one long and one short. The smaller scale is 9 mm long and is divided into 10 subdivisions (0 to 10). The larger scale is several centimeters long and is graduated in millimeters (e.g., 0 to 100 mm). Read the following steps to learn how to pinpoint locations using the **vernier scale.**

1. First, center the object of concern in the field of view.

2. Read the vernier scales on the x and y axes. To do this, locate the point at which the 0 line of the short scale meets the large scale. In Figure 2.3, the 0 line is located between 42 and 43 on the large scale. Now find the line on the small scale that coincides exactly with a line on the long scale. Then, count on the small scale the number of spaces between 0 and the point of coincidence. This number is your decimal point. In Figure 2.3, the line of coincidence is the sixth line on the short scale; thus, it is 0.6, and thus, the final reading for the location would be 42.6 on the x or y axis, depending on which you were reading.

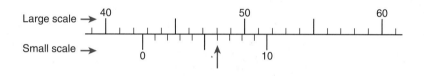

Figure 2.3: The small and large vernier scales.

3. Using your blood slide again, locate an unusual-looking white blood cell and center it in the field of view.

4. Using the technique described previously, record the location of this cell on the x and y axes in the space provided below (note: units are in mm).

    x axis = ___135.6___   y axis = ___38.5___

5. Without removing the slide, move the mechanical stage so you are looking at the periphery of the blood smear or coverslip.

6. Return to the cell found in step #3 by moving the mechanical stage controls until the verniers are adjusted to the number recorded in step #4. The cell you observed previously should be somewhere within the microscope's field of view.

SUMMARY

Because the microscope is one of the most often used and expensive pieces of equipment in a veterinary hospital, knowledge of its care and use is vital to both the veterinarian and veterinary technician. In this chapter you learned about the parts of the microscope and their use. One of the most underused skills is logging a location on a slide and returning to it using the vernier scales, and for this reason it was explained in this chapter.

CHAPTER 3

CELLULAR ANATOMY AND MORPHOLOGY

OBJECTIVES:

- define, understand, and differentiate between a cellular organelle and an inclusion body
- identify the component organelles in a cell on a cell model and a diagram
- understand the structure and function of the cell organelles
- identify the major cellular organelles using prepared slides

MATERIALS:

- three-dimensional model of an animal cell, complete with cellular organelles
- compound microscope
- prepared slide of a giant multipolar neuron
- prepared slide of a Golgi complex

Introduction

The **cell** is the basic unit of organization of an animal's body. The cell may be a single-celled life form, or in mass, cells combine to make up organized tissue within a more complex animal. Cells are extremely complex and are both the structural and functional unit of the tissues they form. Cells have the ability to maintain their boundaries, metabolize and digest nutrients, dispose of wastes, grow and reproduce, move, and respond to stimulus. Each cell must perform certain functions to sustain its life. Cells are highly diverse; differences in size, shape, and internal composition reflect their specific function in the body.

The Anatomy of the Cell

Generally, a cell is composed of three major parts: the *plasma membrane,* the *nucleus,* and the *cytoplasm* (Figure 3.1). All of these are readily observable with a light microscope. Within the cytoplasm are the *organelles,* which are either too small or require a special stain to be seen with ordinary light microscopy. However, since the advent of the electron microscope, even the smallest organelles have been identified. **Organelles,** by definition, are highly organized, living, subcellular structures, and each has a characteristic shape and function. They are, in essence, the internal working parts of the cell. They differ from **inclusion bodies,** which are non-active masses within a cell.

Identification of the Parts of a Cell

Using a cell model and Figure 3.2, identify the following parts of a cell.

1. **plasma membrane:** The **plasma membrane** is composed of bilayer phospholipids and globular protein molecules. This arrangement of compounds is called the *fluid mosaic model* of the plasma membrane, and the proteins are said to be floating in a double layer of phospholipids (see Figure 3.2). Some of the externally facing proteins and lipids have sugar (carbohydrate)

side chains attached to them that are important in cellular interactions. There are also occasional cholesterol molecules dispersed in the fluid phospholipid bilayer to help stabilize it. The proteins in the plasma membrane are also responsible for its *antigenicity.*

The plasma membrane is a flexible, elastic, protective barrier that separates the cell's internal components from the external environment. The plasma membrane has selective permeability and thus plays

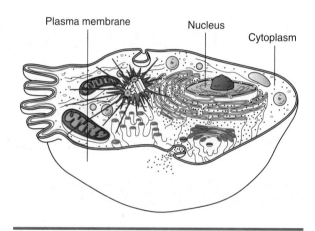

Figure 3.1: Structural components of a cell.

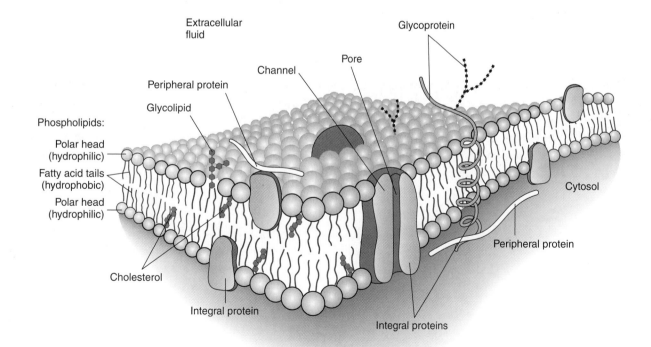

Figure 3.2: Structure of the plasma membrane.

an active role in determining what enters and leaves the cell, and in what quantity. Transport through the plasma membrane can occur in two basic ways, actively and passively. In **active transport,** the cell must provide energy in the form of **adenosine triphosphate (ATP)** to power the movement through the membrane. In **passive transport,** the movement of the substance is driven by concentration or pressure differences. The plasma membrane is also able to maintain a *resting potential*, which is absolutely necessary for cells that are excitable, such as nerve or heart cells. The resting potential of a membrane is associated with the type and quantity of ions inside the cell and on the outside. These ions impart a potential difference across the membrane; for example, a nerve cell membrane's resting potential is about –85mV. These characteristics play a vital role in cell signaling and cell-to-cell interactions.

In addition, cells that are tightly joined together to form functional units of tissue connect to each other at **cell junctions.** These contact points between plasma membranes may serve one of three functions: (1) to form fluid-tight seals between cells, (2) to anchor cells together or to anchor them to extracellular material, or (3) to act as channels that allow ions and molecules to pass from cell to cell within a tissue.

2. **nucleus:** The **nucleus** is a spherical or oval structure and is the most prominent, visible feature inside the cell when viewed with a light microscope because of its contrasting, dark appearance. Nuclei usually stain a dark blue or purple (Figure 3.3). Most cells have one or more nuclei. For example, a red blood cell has a nucleus during its developmental stage, but the nucleus becomes pyknotic and disappears when the cell becomes mature. In contrast, skeletal muscle cells have numerous nuclei, or in other words, they are *multinucleated.*

The nucleus is the control center of the cell and is necessary for cell reproduction. It is technically an organelle, but it is considered separately because of its numerous and

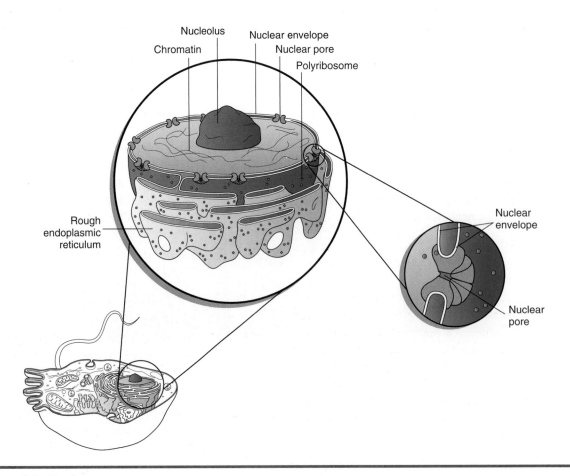

Figure 3.3: Nucleus.

diverse functions. On the exterior of the nucleus is the **nuclear envelope.** It is similar to the plasma membrane in that it is a bilayered (has two layers) lipid structure. The outer membrane of the nuclear envelope is continuous with the rough **endoplasmic reticulum (ER),** a membranous cytoplasmic structure that is the site of protein synthesis, and is very similar in structure. The nuclear envelope is also perforated by numerous channels called **nuclear pores.** Each pore is made up of proteins arranged in a circle, forming a central channel. This channel is approximately 10 times larger than the pore within a protein in the plasma membrane. Nuclear pores control the movement of substances between the nucleus and the cytoplasm. Small molecules can pass via diffusion, but larger molecules, such as RNA and other proteins, must use active transport for passage. This permits the selective transport of proteins from the cytoplasm into the nucleus and of ribonucleic acid (RNA) molecules produced in the nucleus into the cytoplasm.

Another visible feature of the cell is the **nucleolus,** which is located within the nucleus. There may be one or more nucleoli per nucleus. It is spherical and is composed of a cluster of proteins, deoxyribonucleic acid (DNA), and RNA. It is not enclosed by a membrane. The nucleolus is where *ribosomes,* which play an important role in protein synthesis, are assembled. Thus, nucleoli are quite prominent in cells that produce large amounts of proteins, such as muscle and liver cells. The nucleoli disperse during cell division and reorganize when the new cells are formed.

The bulk of the nucleus is made up of **chromatin,** which may appear either diffuse or granular when viewed by a light microscope. Chromatin is the cell's genetic material, or **chromosomes,** and is found in a dispersed, unorganized arrangement in the non-dividing cell. The chromosomes and their component **genes** control cellular structure and most cellular activities. The genes, of course, are the hereditary material, and they ultimately control, among other things, the structure and appearance of the animal.

3. **cytoplasm:** Cytoplasm is composed of two components: (1) *cytosol* and (2) *organelles.*

The **cytosol** is the fluid portion of the cytoplasm and makes up 55% of the total cell volume. It varies in consistency and composition from one part of a cell to another and is composed of 75 to 90% water, with the balance being made of various ions, glucose, amino acids, fatty acids, proteins, lipids, ATP, and waste products. Also located in the cytosol are various organic molecules that aggregate into masses that are stored and used as required for metabolism. The cytosol is the site of many chemical reactions and is the place where energy is produced and captured to drive the cellular activities necessary for life. The bulk of the cytoplasm stains pink under a microscope because of the basic proteins that compose the cytosol.

4. **organelles:** The word **organelles** literally means "small organs." These structures are the machinery of the cell and are highly specialized to carry out specific functions (Figure 3.4).

a. **ribosomes:** Ribosomes are densestaining, spherical bodies composed of **ribosomal RNA (rRNA)** and protein. They are the actual site of protein synthesis, or the place where a new amino acid is added to the chain of amino acids in the protein being created. Structurally, a ribosome consists of two subunits, one about half the size of the other. These are formed separately in the nucleolus, then they exit the nucleus to unite within the cytosol. They may unite on the nuclear membrane or on the endoplasmic reticulum. These are called *membrane-bound ribosomes.* These ribosomes synthesize proteins destined for insertion within the plasma membrane or for export from the cell. Others may be free in the cytosol, unattached to any structure in the cytoplasm; these are called *free ribosomes.* Primarily, free ribosomes synthesize proteins used inside the cell. Ribosomes are also located within the *mitochondria,* where they synthesize mitochondrial proteins. Sometimes 10 to 20 ribosomes join together in string-like arrangements called *polyribosomes.*

b. **endoplasmic reticulum (ER):** The ER is a network of membranes that form flattened sacs or tubules called *cisterns* that extend from the nuclear envelope (as mentioned previously). The ER is so

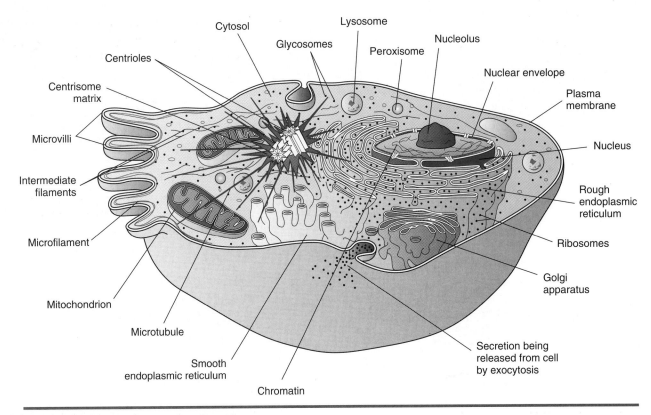

Figure 3.4: The cell and its organelles.

extensive throughout the cytosol that it constitutes more than half of the membranous surfaces found within the cytoplasm of most cells.

There are two forms of endoplasmic reticulum in cells, **rough ER** and **smooth ER.** They differ in structure and function. The membranous rough ER is continuous with the nuclear envelope and is studded with ribosomes. Proteins synthesized enter the rough ER's cisterns for processing and sorting (Figure 3.5). As stated previously, these proteins will be exported from the cell or be used in the plasma membrane. In certain cells, enzymes within the cisterns attach the proteins to carbohydrates to form *glycoproteins.* In other cells, the enzymes attach proteins to phospholipids to form plasma membranes or the membranes of other organelles. Thus, the rough ER is responsible for synthesizing secretory proteins and membrane molecules.

The *smooth ER* extends off of the rough ER at multiple sites to form another network of membranous tubules. It lacks the ribosomes of the rough ER, but it does contain unique enzymes that make it more functionally diverse than the rough ER. Smooth ER is able to synthesize phospholipids for membrane surfaces, and also synthesizes fats and steroids, such as estrogen and testosterone. In the hepatic cells of the liver, the enzymes of the smooth ER help release glucose into the bloodstream, and they inactivate or detoxify drugs and other potentially harmful substances, such as alcohol. In muscle cells, calcium ions are released from a form of smooth ER, called the *sarcoplasmic reticulum,* which triggers the contraction process.

c. **Golgi complex:** The Golgi complex consists of 3 to 20 flattened, membranous sacs stacked on top of one another with both ends dilated (Figure 3.6). The areas within these sacs are also called *cisterns* (like the similar areas found in the ER). The overall shape of the Golgi complex is cup-like because of the curvature of the cisterns. Most cells only have one Golgi complex, although some may have several.

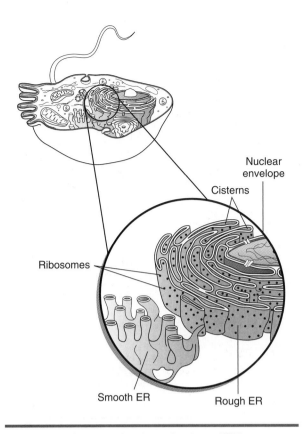

Figure 3.5: Endoplasmic reticulum (ER).

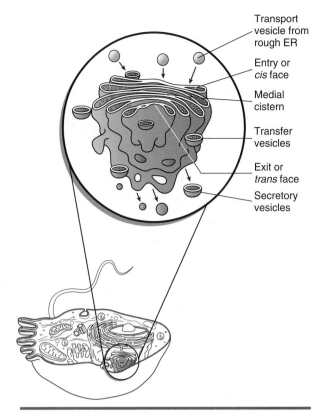

Figure 3.6: Golgi complex.

In cells that secrete many proteins into the extracellular fluid, the Golgi complex is large and extensive. Its function is to modify, sort, package, and transport products received from the rough ER and form *secretory vesicles* that discharge processed proteins, via a process called *exocytosis,* into the extracellular fluid. The Golgi complex also replaces or modifies the existing plasma membrane and forms *lysosomes* and *peroxisomes.*

d. **lysosomes: Lysosomes** are membrane-enclosed vesicles that form in the Golgi complex (Figure 3.7). Inside the lysosomes are as many as 40 different kinds of digestive or hydrolytic enzymes capable of breaking down a variety of molecules. The enzymes tend to work best in an acid pH, and thus the lysosomal membrane actively transports hydrogen ions (H^+) into the lysosomes. The pH inside a lysosome is about 5.0, compared to the pH of the cytosol, which is neutral or 7.0. The lysosomal membrane also allows the final products of digestion, such as sugars and amino acids, to be transported into the

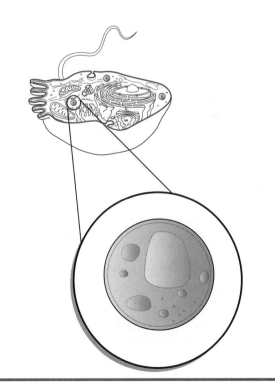

Figure 3.7: Lysosome.

cytosol. The functions of the lysosomes are to (1) digest substances that enter the cell via *endocytosis,* (2) digest worn-out organelles, a process called *autophagy,* (3) digest the entire cell if it dies, known as *autolysis,* and (4) occasionally conduct extracellular digestion, such as when spermatozoa release enzymes to help them penetrate the surface of an ovum.

e. **peroxisomes: Peroxisome** vesicles are smaller than lysosomes and contain one or more enzymes that can oxidize (by removal of a hydrogen atom) various organic substances. Peroxisomes detoxify harmful substances and are therefore abundant in kidney and liver cells. In addition to being produced by the Golgi complex, these organelles can self-replicate.

f. **mitochondria:** The **mitochondria** are the powerhouses of the cell because of their ability to produce *ATP.* A cell may have as few as 100 of them or as many as several thousand. Cells that have high physiologic activity, such as those in the muscles, liver, and kidneys, have many mitochondria because of their need for energy in the form of ATP. Mitochondria

are found within the cytoplasm where the need for energy is greatest, such as between the contractile proteins of muscle cells. Structurally, a mitochondrion is oval, like a small sausage. It is bounded by two membranes similar to the plasma membrane. The outer mitochondrion layer is smooth, but the inner membrane is arranged in a series of folds called **cristae.** The center of the mitochondrion, bounded by the inner membrane and cristae, is fluid-filled and called the *matrix.* The cristae's structure provide a large inner surface area in the mitochondria for the chemical reactions in the aerobic phase of cellular respiration (Figure 3.8). The enzymes that catalyze these reactions are located on the cristae and in the matrix. To reiterate, the main product of these reactions is ATP.

Mitochondria also can self-replicate, and often occurs during times of increased cellular demand. Each mitochondrion has multiple identical copies of circular DNA. The genes formed by this DNA, along with the genes of the nucleus, control the production of proteins that build the mitochondrial components. Because there are also ribosomes inside the mitochondrial matrix, some protein synthesis also occurs within the inner membrane. In most cells the genes of the nucleus contain two sets of DNA, one each from the male and female parent. However, mitochondrial genes are inherited only from the female parent. This is because the head of the sperm (the only part that penetrates and fertilizes the ovum) normally lacks most organelles, such as mitochondria, ribosomes, endoplasmic reticulum, and Golgi complexes.

g. **centrioles:** The paired bodies of centrioles are cylindrical and are located close to the nucleus in all animal cells capable of reproducing themselves. They lie at right angles to each other. Internally, each cylinder is composed of nine triplets of microtubules, called a *9 + 0 array,* because there are nine sets of microtubules in a circle without any in the center. The centrioles direct the formation of the mitotic spindle during cell division. They also form the basis for cell projections called *cilia* and *flagella.* A

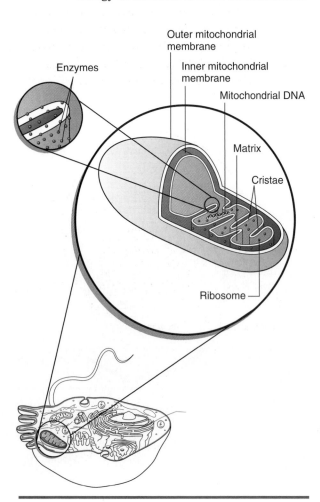

Outer mitochondrial membrane

Enzymes

Inner mitochondrial membrane

Mitochondrial DNA

Matrix

Cristae

Ribosome

Figure 3.8: Mitochondria.

centrosome is the pericentriolar area plus the centrioles (Figure 3.9).

h. **cytoskeletal elements:** Inside the cell, within the cytosol, an internal scaffolding is formed that supports and moves substances within the cell. This structure is composed of several elements, including **microtubules,** which are slender tubules formed of proteins called *tubulins*. These tubulins have the ability to aggregate and disintegrate spontaneously. The centrosome serves as the initiation site for the assembly of the microtubules. They grow outward from the centrosome toward the periphery of the cell. The microtubules aid in intracellular transport of organelles, such as secretory vesicles, and form the spindle for the migration of chromosomes during cell division. They also act in the process of transporting molecules down the length of an elongated cell, such as a neuron, and they help determine and maintain cell shape by providing rigidity to the cytosol. The **intermediate filaments,** another component of the cell's internal scaffolding, are stable proteinaceous filaments that act as internal guy wires to resist mechanical (pulling) forces that act on the cells. Yet another component, **microfilaments,** are ribbon or cordlike elements within a cell formed from contractile proteins. Because they have the ability to shorten and lengthen (by relaxation), they are critical elements of cells that are mobile or cells that have the ability to contract, such as muscle cells. A cross-linked network of microfilaments braces and strengthens the internal face of the plasma membrane.

5. **inclusions (inclusion bodies):** Within the cytoplasm, various other substances and structures exist; these are stored foods (glycogen granules and lipid droplets), pigment granules, crystals of various types, water vacuoles, and ingested foreign materials. These are not active parts of the machinery of the cell; rather, they are passive masses and are therefore referred to as *inclusions*.

6. **cilia** and **flagella: Cilia** and **flagella** are motile, cell-surface projections composed of a 9 + 2 array of microtubule doublets (nine in a circle with two in the center) (Figure 3.10). Cilia are short and move fluids and debris over a cell's surface, whereas flagella are long and can propel the entire cell.

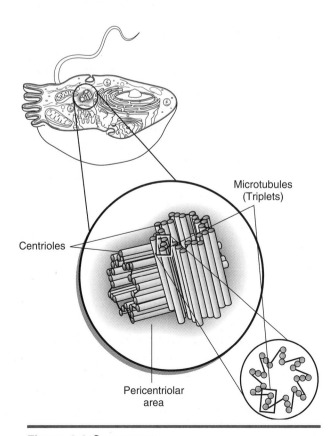

Figure 3.9 Centrosome.

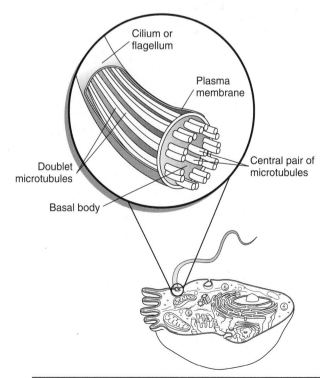

Figure 3.10: Cilia and flagella.

EXERCISE 3.1

IDENTIFICATION OF THE PARTS OF THE CELL

Using a model of a cell plus Figures 3.1 through 3.10, locate the following parts of a cell:

1. plasma membrane
2. nucleus
3. nuclear membrane
4. nuclear pores
5. nucleolus
6. chromatin
7. cytoplasm
8. cytoskeletal elements
9. ribosomes
10. rough endoplasmic reticulum
11. smooth endoplasmic reticulum
12. Golgi complex
13. lysosome
14. peroxisome
15. mitochondria
16. cristae
17. centrosome/centrioles
18. cilia
19. flagella
20. inclusion body

EXERCISE 3.2

IDENTIFICATION OF CELL PARTS USING A LIGHT MICROSCOPE

Follow the steps to complete the following procedure.

1. First, examine the giant multipolar neuron smear slide using the 10X objective lens on your light microscope. Locate a dark-blue, large motor neuron. Center the neuron in the field of view and change to the 40X objective.

   Note that various parts of the slide take on different colors. We use various stains to help identify different cells or different parts of a cell according to color. A common stain is the *hematoxylin* and *eosin combination stain*. The hematoxylin tends to stain acidic portions of the cell dark blue or purple. These parts of the cell are called *basophilic* because they *like* or bond with basic stains. Because acid and base chemicals tend to bond together, cell parts high in RNA and DNA, such as the nucleus and nucleoli, attract basic stains and stain dark blue or purple. Using the same reasoning, eosin tends to stain basic portions of the cell pink or red. These stained parts are called *acidophilic* areas and include the more basic proteins within the cell, which are located within the cytoplasm. Other stains are more specific and may stain blue only the RNA for example, as is the case with the cell you have on your microscope.

   Note that at the cell boundary there is a distinct separation between the blue and the pink background. This, of course, is where the *plasma membrane* is located, but the details of this membrane are not visible using the light microscope. There are multiple cytoplasmic extensions, called **cell processes,** branching off the neuron's *cell body.* It is for this reason it is called a *multipolar neuron.* Some of these processes were cut off during the sectioning of the tissue. Each cell, if sliced perfectly down the middle, would have visible one very long process, called an **axon,** and numerous shorter ones, called **dendrites.** At the center of the cell body is where the *nucleus* is located. Because the stain used for this slide stains the RNA darker, the nucleus (with mostly DNA) has a halo appearance around the RNA-rich nucleolus. Therefore, the dark dot in the center of the nucleus is the *nucleolus.* The *nuclear membrane* is either not visible or barely visible (Figure 3.11).

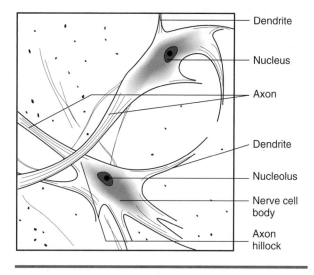

Figure 3.11: Nerve cell body with processes.

In the space below, draw and label the parts of the neuron.

Think of the boundary of the circle below as the boundary of the microscope's field of view.

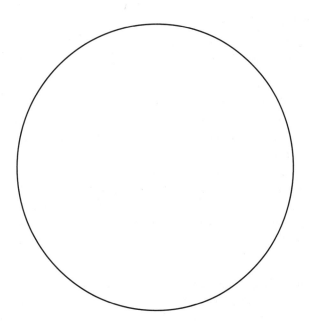

2. Next, take out the slide labeled *Golgi complex*. In the space below, draw and label the Golgi complex and any other part of the cell that is visible.

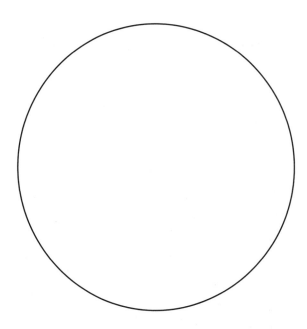

CLINICAL SIGNIFICANCE

There are numerous diseases specifically attributed to faulty cellular structure and functions. Glycogen storage disease *is a congenital absence of lysosomes, which permits accumulation of glycogen in cells.* Vitamin A intoxication *occurs when lysosomal enzymes are released to the exterior of cells, resulting in breakdown of the material between the cells.* Cystic fibrosis *is an inherited disorder that results in the production of a defective protein that fails to reach the plasma membrane. Because of its absence, the plasma membrane is unable to excrete chloride ions from the cell. This results in a thick mucus build-up outside the cells. In the lungs this causes difficulty in breathing, and elsewhere it results in improper secretion of pancreatic enzymes for digestion.*

SUMMARY

Because all pathological processes are ultimately the reflection of cellular dysfunction, understanding the inner workings of the cell is crucial to understanding how and why cells and tissues break down and cause disease. To understand pathology, it is first necessary to understand normal physiology. Hence, we study the structure and function of the cell and its organelles. This chapter began by explaining the structure of the cell's plasma membrane, then it discussed the nucleus and its internal structures. Finally, it addressed the structure and function of each individual organelle.

CHAPTER 4

GENERAL PRINCIPLES OF HISTOLOGY

OBJECTIVES:

- define and understand the principle of histology
- understand the methodology of preparing a histological section
- know how a histological section relates to the gross structure of an organ
- identify common artifacts occurring in histological sections
- understand how to section and prepare a piece of tissue for fixation so it can be sent to a laboratory for histopathological examination

MATERIALS:

- a three-dimension model of a glomerulus and Bowman's capsule
- one apple, one lemon, and one crabapple (or small apple)

Introduction

Histology is the study of *tissue;* more specifically, it is the study of the microscopic anatomy of tissues. Thus, it complements the study of gross anatomy and provides the structural basis for the study of organ physiology. A **tissue** is a group of cells that are of similar structure and are united in the performance of a particular function. Tissues are organized into **organs,** such as the intestines, heart, lungs, and kidneys. Most organs contain numerous types of tissues, and the arrangement of the tissues determines the organ's structure and function.

Preparation of Histological Sections

Histological sections are very thin, usually from 0.5 to 10 micrometers (μm), or microns, thick. The tissue must be preserved to prevent **autolysis.** This can be done by fixing the tissue in a preservative or by freezing it. Frozen tissues can be cut and observed immediately, but for a more permanent preparation, the tissue must be infiltrated with a supporting medium, cut into sections, and then mounted on a slide. Tissues infiltrated with plastic can be as thin as 0.5 μm and show superior detail. Paraffin also can be used as the supporting medium and produces excellent preparations sectioned as thin as 2 or 3 μm. Slides are stained after mounting to increase visibility and to differentiate between the various cellular and intracellular components.

Figure 4.1 outlines the steps involved in producing a stained histological slide via the paraffin procedure. This same process may be used to produce a histological section of normal tissue for study or a histopathological section of diseased tissue for analysis and diagnosis. The study of the pathology of tissues at the microscope level is called **histopathology.** Tissue sections are obtained from animals, usually during a surgical procedure or during a **post-mortem examination** or **necropsy** procedure. The pieces of tissue are placed into a fixative, such as buffered formalin or Bouin's fixative. These solutions preserve the normal morphology and facilitate further processing. To remove the water from the cells, after **fixation** the specimen is transferred through a series of alcohols with successive increases in concentration up to 100%. Then, the alcohol is removed by placing the specimen in xylene or a xylene substitute; this process is called **clearing.** It is necessary because paraffin and alcohol do not mix. After clearing, the specimen is placed in a paraffin bath heated to just above the melting point of the paraffin. Through this process of **infiltration,** the paraffin completely replaces the xylene. The sample is then transferred to an embedded mold of fresh paraffin, which is allowed to harden. The mold is removed and the excess paraffin is cut away.

A special instrument called a *microtome* is used to make thin slices of tissue. The paraffin is attached to the microtome adjacent to the microtome knife blade, and with each revolution of the instrument's handle, a slice of tissue of a desired thickness is made. Because the paraffin tends to stick to itself, a ribbon of tissue sections is produced. The sections are floated in a warm water bath, which softens the paraffin and flattens the section, eliminating wrinkles. Once flattened, the desired section is placed on a slide and allowed to dry. Next, the paraffin is removed with xylene or another solvent, and the specimen is rehydrated. It is then stained, dehydrated, cleared (made transparent) with xylene, and covered with a resinous mounting medium. Finally, a cover slip is placed on top.

Hematoxylin and eosin (H&E) stains were mentioned in the previous chapter. The hematoxylin stains *acidic* areas of a sample blue or purple (the basophilic areas), and eosin stains *basic* areas red (the acidophilic areas). Other stains that act similarly to hematoxylin are methylene blue, toluidine blue, and basic fuchsin (all are basic stains). Other acid stains, similar to the eosin stain, are orange G, phloxine, and aniline blue. There are other widely used stain combinations that are useful for identifying certain elements of tissues. Trichrome procedures, such as Mallory trichrome stain, specifically stain the collagenic fibers of connective tissue. Orcein and Weigert's resorcin fuchsin are stains that color elastic fibers. Some silver stains enable viewing of reticular fibers and nervous tissue components, such as neurons, myelin, and cells of the neuroglia. Finally, stains such as Wright's and Giemsa's (Romanovsky stains) are used for differentiating the various cells found in blood and bone marrow.

Interpretation of Histological Sections

It is necessary to have an idea of the structure of the organ sectioned to comprehend the microscopic anatomy of the specimen under the microscope. First, you need to know or learn what type of organ it is, its shape, whether it is solid or has a hollow interior, and how the cut was made through the organ to produce the sample. It is helpful to know whether it was a cross section (x.s.), a longitudinal section (l.s.), or an oblique slice through the organ. Also, knowing if the section shows the entire organ or just a piece of it is beneficial.

Most prepared slides are labeled to indicate the particular orientation of the section. This is very important when considering symmetrical organs or those with a hollow core, such as the urinary bladder or a section of small intestine. These organs have radial symmetry and their appearance is affected by the direction of the cut. On the other hand, the appearance of an asymmetrical organ, such as the liver or spleen, is unaffected by the direction of the cut.

It is also important to know where the section of tissue was taken from in organs that have dramatically different areas of tissue. The kidney's cortex, for example, is the area where the glomeruli, Bowman's capsule, and the convoluted tubules of the nephron are located; a sample section from the outer part of the kidney would include the cortex and not the medulla.

Therefore, it is necessary to consider the three-dimensional structure of organs and their components when examining histological preparations. This enables viewers to understand the appearance of the cells in terms of their size, shape, and arrangement. Because cells themselves are three-dimensional and vary in size and shape, the way they appear on a slide is dependent on their shape as well as how they were cut. Figures 4.2 and 4.3 show a variety of organs, the ways they might be sectioned, and the ways the cells and tissues taken from them might appear.

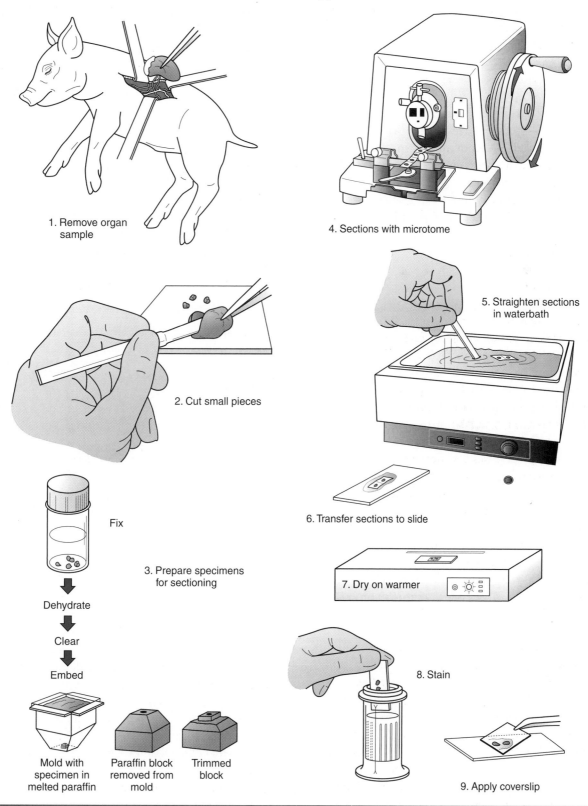

1. Remove organ sample

2. Cut small pieces

3. Prepare specimens for sectioning

Fix

Dehydrate

Clear

Embed

Mold with specimen in melted paraffin

Paraffin block removed from mold

Trimmed block

4. Sections with microtome

5. Straighten sections in waterbath

6. Transfer sections to slide

7. Dry on warmer

8. Stain

9. Apply coverslip

Figure 4.1: Steps in producing a histology slide.

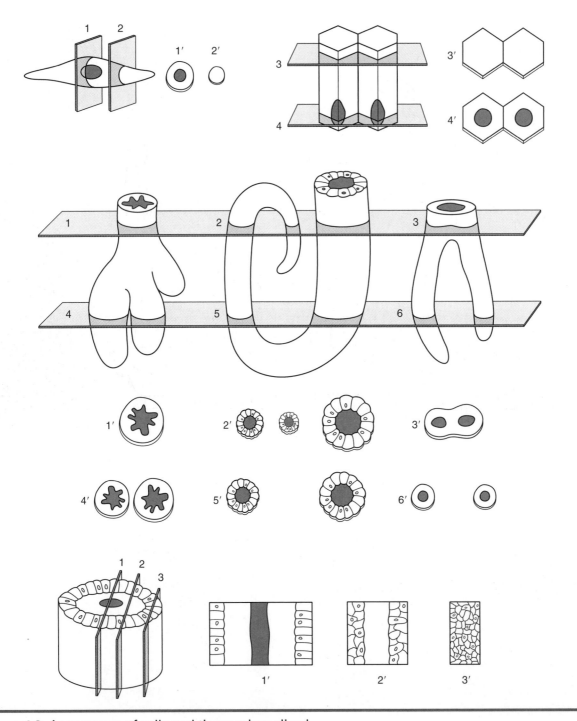

Figure 4.2: Appearance of cells and tissue when sliced.

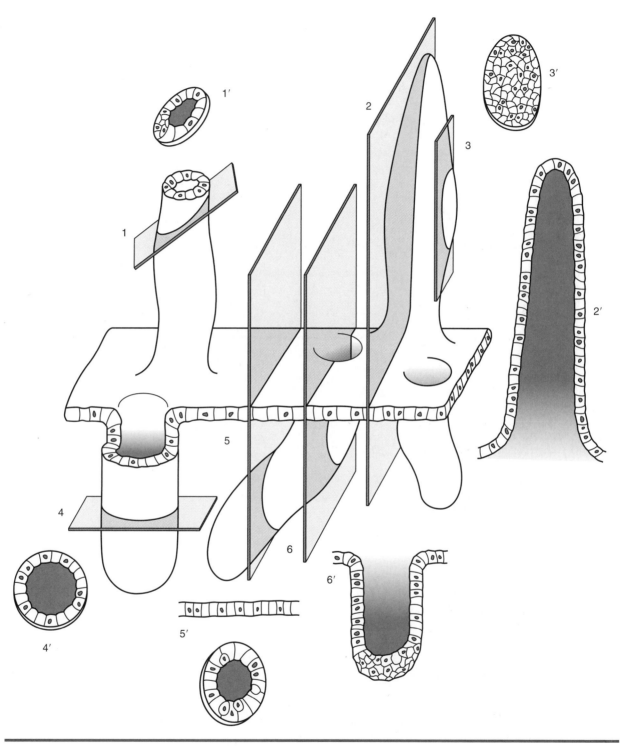

Figure 4.3: Appearances of cells and tissue when cut at various angles.

EXERCISE 4.1

DEMONSTRATION OF SECTIONING METHODS

Using an apple or lemon can help you visualize cuts made through tissues and organs. An apple is good as a model of a solid organ, whereas a lemon is good as a model of a hollow organ.

In step 1-4, an apple is cut in such a way as to demonstrate both the frontal and transverse sections of an organ. In steps 5 and 6, a lemon is used to illustrate a longitudinal cut.

Imagine the flesh of the apple as the exterior of an organ (the cortex) and the apple's core (the medulla) as the vascular center of the organ. Use the top of the apple, where the stem is located, as the front surface of the organ and the bottom of the apple as the rear surface. We will consider the front-to-rear of the apple to be the longitudinal axis of the organ. Mark the top, or superior, surface of the apple with a permanent marker or felt-tipped pen and do the same to indicate the inferior surface of the organ. As you perform the following cuts, you will notice how different slices through the apple look when the cut surface is viewed.

1. Mid-frontal cross section: Cut the apple in the middle, between the stem end and the bottom end, by dividing it equally, front and back. Both the cortex (outer pulp) and the medulla (core) are visible in our simulated organ (Figure 4.4).

2. Para-frontal cross section: Make a second slice in the rear half of the apple approximately 1 cm from the first cut and parallel to it. This section shows mostly cortex; only part of the medulla is visible when viewing the cut surfaces of the two pieces (see Figure 4.4).

3. Mid-transverse cross section: Using the front half of the apple, make a cut halfway between the top and bottom, dividing the apple equally. This cut should go through the stem of the apple. Both the cortex and medulla of our organ can be seen in this section.

4. Para-transverse cross section: Make another cut parallel to the previous cut, halfway toward the bottom of the apple. This section only engages the cortex.

5. Mid-sagittal longitudinal section: Take the lemon and cut it lengthwise, dividing it into left and right halves (starting at the stem end and cutting toward the bottom). The fruit of the lemon represents the cortex of an organ and the white fibrous middle area the medulla. This cut is down the lemon's longitudinal, or long, axis (Figure 4.5).

6. Para-sagittal longitudinal section: Take one half of the lemon and make a cut parallel to the first cut. This section shows mostly cortex and no medulla.

7. Take the remaining half of the lemon and remove the fruit, leaving only the skin. It now has the appearance of a hollow organ, such as the urinary bladder. You could even imagine that the pointed end of the lemon is the start of the urethra, and the skin represents the layers of cells lining the bladder's cavity. The yellow exterior of the lemon would be the *serosal surface*, that is, the part of the visceral peritoneum that gives the bladder surface its moist, shiny appearance inside the abdominal cavity of an animal (see Figure 4.5).

Figure 4.4: Cross-sectioning using an apple.

Figure 4.5: Longitudinal or sagittal sectioning using a lemon.

LONGITUDINAL SECTION OF BOWMAN'S CAPSULE USING A PLASTIC MODEL

1. Examine the plastic model of Bowman's capsule surrounding the glomerulus of the nephron of the kidney. The simulated cut is a longitudinal section. Note that you are able to see the shape and relative size of the bisected cells that form this capsule when the model is viewed sagittally (Figure 4.6). These cells are called *simple squamous epithelial cells.* Note that they are only a single layer thick.

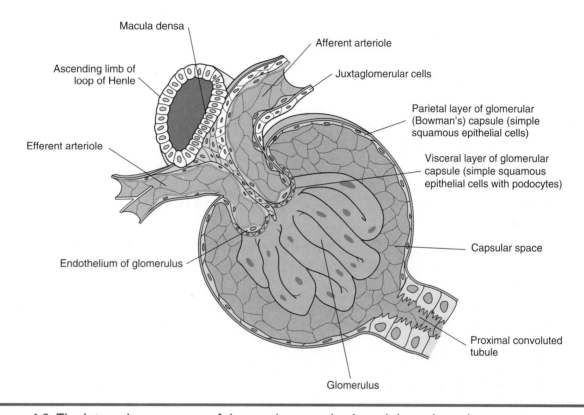

Figure 4.6: The internal appearance of the renal corpuscle viewed through a microscope.

2. If you were to look at Bowman's capsule from the exterior surface rather than from the cut surface (Figure 4.7), it would appear as a mosaic of cells in the shape of a ball, similar to a soccer ball. From the exterior, unless Bowman's capsule was transparent, we could not tell whether it was hollow or how many layers of cells were present.

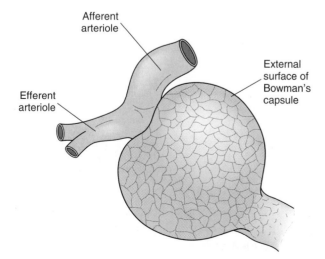

Figure 4.7: External view of the renal corpuscle.

Cross-Sectional Appearance of a Sheet of Cells:

Sometimes cells are arranged in sheets (epithelial cells are the best example of this) and appear as flat cells, as seen in Figure 4.8. As with the exterior of Bowman's capsule, you do not know how many layers are present based on this superficial view. Think of standing on a tile floor in a room. Looking at the tile, you cannot tell whether there is only one layer present or if the builders layered the tile on top of a previous floor. To determine whether you are observing one or many layers, a cross-section or longitudinal section is necessary. In Figure 4.8, the dark dots are the nuclei of the cells, the light purple areas are the cytoplasm, and the borders are the cell membranes. Note that each cell is equal in size.

Cross-section #1: This cross-section of a sheet of cells bisects each cell and nucleus. Each cell is of equal size and the nuclei can be observed to be centered in each cell.

Cross-section #2: This cross-section does not bisect the cells or nuclei equally; therefore, the cells no

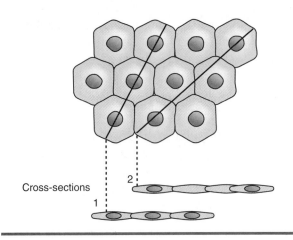

Figure 4.8: Flat surface of squamous cells and their cross-sectional appearance.

longer appear to be equal in size, and some cells do not appear to have nuclei. The situation in cross-section #1 rarely, if ever, occurs when sectioning and observing tissue under the microscope. Cross-section #2 is what an observer is likely to see in real-life cellular histology.

Clinical Significance

Often, veterinarians want to look at tissues of tumors or other sites of pathology before sending the specimens to the laboratory for histopathological analysis. One method of doing this is to make an impression smear *of the tissue prior to fixation. The mass is sectioned so a freshly cut surface is available. The section is then blotted dry on a clean paper towel and touched lightly several times in a row on a single glass slide. It is then ready for staining.*

An alternative method is to use a scalpel blade to scrape a freshly cut surface of the extracted tissue so an accumulation of cells collects on the blade surface. The tissue on the blade is then smeared along approximately 3/4 of the surface on a glass slide (leaving room for the slide to be labeled).

Another technique for viewing abnormal tissue of an animal's body is a fine needle aspirate. *A 25- or 22-gauge needle is introduced into the mass, and negative pressure is applied to a syringe attached to the needle. Several passes through different tangents of the mass should be conducted. Following withdrawal, the needle is removed from the syringe, and the syringe is filled with air and reattached. The cellular contents of the needle are then blown onto a clean glass slide. A squash-prep procedure is used to spread the contents on the slide. This is done by placing a second glass slide on top of the specimen on the first slide at a perpendicular angle. The slides are then drawn apart by sliding the top slide away from the bottom slide, leaving a smear of cells and tissue fluid on both slides.*

EXERCISE 4.3

SECTIONING METHODS FOR TUMORS

Veterinary technicians are sometimes required to perform a *post-mortem necropsy* on a deceased animal to help ascertain the cause of death. The veterinarian may be engaged in other diagnostic examinations and thus may only be present for critical parts of the necropsy. At the end of the procedure, the technician is expected to lay out the organs so the veterinarian can examine them for pathology. Tissue sections of selected organs are preserved in formalin so they can be processed for histopathological examination. If a tumor is found during the necropsy or is removed during surgery, it is often necessary to preserve the entire structure to view its borders and verify complete excision. However, sections greater than 1 cm in diameter will not undergo complete fixation because formalin cannot penetrate the structure. The following exercise will demonstrate the proper method of tissue preparation.

1. Using a crabapple, make serial slices that yield sections no greater than 1 cm thick. Do *not* cut all the way through the crabapple. Instead, leave a 1-cm thick area at the bottom of the cuts so the fruit remains in one piece (Figure 4.9). If this were a tumor, a pathologist would be able to view the tissue in its entirety in its proper structural alignment and observe all its borders, and the formalin would be able to penetrate the entire tumor.

2. When fixing a piece of tissue, the amount of fixative necessary is 10 times the volume of the tissue. A 1-cubic centimeter (cc) piece of tissue needs to be placed in 10 milliliters (ml) of fixative. You could measure the tumor to calculate its volume, or weigh it. Assuming 1 gram (g) of tissue displaces 1 cc of fluid, weigh the crabapple and place it in the appropriate amount of water using the following guide.

   a. weight of crabapple:_____ g = _____cc

   b. _____cc × 10ml = _____ml

   c. What is the minimum number of slices necessary for a tumor 8 cm in length?

      Answer:_____

   d. How many milliliters of fixative are necessary for a perfectly round tumor 4 cm in diameter? (V = r²l, V = volume, r = radius, l = length)

      Answer:_____

3. The tissue is fixed for 24 hours, and then it can be transferred to a container with a small amount of fixative for shipping. The answer to question *c* is seven slices and *d* is 250 ml.

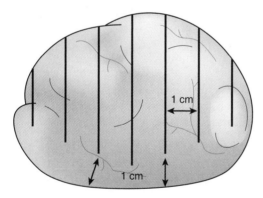

1 cm

1 cm

Figure 4.9: Method of sectioning of a tumor for fixation.

SUMMARY

A veterinarian in surgery cannot stop to prepare a specimen for viewing or shipment, then go back to finish the surgical procedure. This is one reason why educated veterinary technicians are so valuable in a veterinary hospital; they know how to do these procedural tasks. In this chapter you have learned what various sectioning techniques may look like when viewed with a microscope. You have also learned how to cut a tumor for fixation and how to determine the amount of fixative necessary to properly fix a specimen.

CHAPTER 5

MICROSCOPIC ANATOMY OF TISSUES

OBJECTIVES:

- define and describe the four types of tissues and their major subcategories
- describe the difference between simple, stratified, and pseudostratified layers of epithelial cells
- identify all forms of epithelial tissue, their component cells, and other structures, from prepared slides
- identify all forms of connective tissue (both connective tissue proper and special connective tissues), their component cells, and other structures
- identify the three forms of muscle tissue, their component cells, and other structures
- identify a nerve cell and its component parts
- locate the various tissue types in an animal's body

MATERIALS:

- compound light microscope
- immersion oil
- colored pencils
- prepared slides of the following epithelial cells: simple squamous (or use kidney section with glomeruli or "air sacs of the lungs"); simple cuboidal (or use kidney section with renal tubules); simple columnar (or use lining of small intestine); pseudostratified ciliated columnar (or use trachea); stratified squamous (nonkeratinized, such as tongue); stratified cuboidal; stratified columnar (or use urethra, male); and transitional (or use lining of the urinary bladder)
- three-dimensional model of a glomerulus and Bowman's capsule
- prepared slides of the following connective tissues: mesenchyme; loose irregular/areolar (use loose smear or subcutis of intestinal tract); dense irregular (or use dermis of skin); dense regular (or use tendon or ligament); elastic (use nuchal ligament); hyaline cartilage (or use trachea); elastic cartilage; fibrocartilage; reticular; blood smear; and compact bone
- prepared slides of skeletal, cardiac, and smooth muscle (longitudinal sections)
- differential cell counter

Introduction

As stated in the previous chapter, a tissue is a group of **cells** of similar structure united in the performance of a particular function. There are four primary tissue types: (1) **epithelium,** (2) **connective tissue,** (3) **nervous tissue,** and (4) **muscle.** Each of these tissue types has distinctive structures, patterns, and functions. Each type can be further divided into subcategories, which we will cover in this chapter.

To perform specific body functions, tissues are organized into **organs.** The intestinal tract absorbs nutrients through its lining cells, the heart muscle pumps **blood** throughout the body, and the liver cells detoxify the waste products of the body. All of these are examples of the special cellular functions that identify the nature of these organs. Because most organs contain numerous types of tissues, the arrangement of these tissues is critical to the structural shape of the organ and how it functions. Only by understanding the normal cellular arrangement of tissues are we able to recognize the abnormal when it presents itself. Also, to understand the physiology of a particular organ, you need a mental picture and an understanding of how that organ is structurally arranged. You can appreciate how the skin acts as a barrier to microorganisms entering the body if you know the epithelial cell histology and that the skin is thick and multilayered. In comparison, if you know the lining of the respiratory system, where the **alveoli** of the lungs are located, is only one cell-layer thick, you can understand that this is an easier portal for invasion.

Epithelial Tissue Histology

The **epithelia,** or **epithelial tissues,** are a special group of tissues designed to cover the external surface and line the internal surfaces (including the tubules and vessels) of an animal's body. All epithelial cells are supported by a **basement membrane** that separates them from the underlying **connective tissues.** Cells are the principle component of epithelial tissue. The intercellular substance is sparse and only consists of a very thin layer of material located between the cells that helps hold them together. This is known as *intercellular cement.* There is always a **free surface** to epithelial tissues, which faces the exterior of the body, the cavity of a hollow organ, the lumen of a tubular organ, or the lumen of a duct or vessel. All cells of epithelial tissues that are only one cell-layer thick line (touch) the free surface. In types of epithelial tissues that are multi-layered, most of the cells do not touch the free surface. Instead, only the *surface cells* line the free surface. The free-surface cells may possess **cilia, microvilli,** or **stereocilia.**

The function of the epithelium is to protect the body; absorb, filter, excrete, and secrete substances; and for sensory purposes. The cell's accessory structures also aid in their function. For example, the cilia attached to the epithelium lining the lungs help sweep away dust and other foreign particles. The following are characteristics that distinguish epithelia from other types of tissues.

1. Intercellular contacts: The cells fit closely together with one another, forming membranes, or sheets of cells, and are bound together by intercellular cement and specialized junctions.

2. Surface cells: The membranes always have one free surface, called the **apical (or free) surface.** (Note: Unfortunately, in anatomy, as well as other disciplines of medicine, it is not uncommon to find two or three words meaning the same thing.) Either term is correct, and it is not uncommon to find two authors using different terms to mean the same thing. For example, in two A&P (anatomy & physiology) books, both published in the year 2000, one author used the term *apical surface,* the other *free surface.*

3. Supported by connective tissue: Epithelial tissues are attached on their side opposite the free surface (the *tissue side,* also called the *basal surface*) to a basement membrane. The basement membrane is an amorphous material secreted by the basal epithelial cells *(basal lamina)* and connective tissue cells *(reticular lamina)* that lie adjacent to each other.

4. Avascularity: Generally speaking, epithelial cells have no blood supply; they are avascular. There are no capillaries that extend into the epithelium. Instead, epithelial cells are dependent on diffusion of nutrients from the underlying connective tissue. The exception is glandular epithelium, which is very vascular.

5. Regeneration: If epithelial cells are well nourished, they can easily regenerate themselves. This is an important characteristic considering where these cells are located and the trauma to which they are exposed.

One method of classifying epithelial tissues is based on the number of layers each possesses. **Simple epithelium** consists of a single layer of cells, whereas **stratified epithelium** contains two or more layers. The second major method of classifying epithelia is according to cell shape. **Squamous cells** are flat, thin, or scale-like. On cross-section, they are usually only as tall as their nucleus. **Cuboidal cells** are cube-like, equal in height and width. **Columnar cells** are column-shaped, much taller than they are wide. These two classifications are combined to designate and classify most of the various types of epithelial tissue: **simple squamous, simple cuboidal, simple columnar, stratified squamous, stratified cuboidal,** and **stratified columnar.** Note that *stratified epithelium* is named for the shape of the cells on its apical surface, rather than those on their basal surface.

Two other types of epithelial cells are less easily classified. **Pseudostratified** epithelium is actually a form of simple columnar epithelium. All its cells attach to the basement membrane, but because the cells vary in height, and their **nuclei** lie at different levels above the basement membrane, it gives a false appearance of being stratified. Pseudostratified epithelium is usually ciliated on its free surface. **Transitional** epithelium is a variation of stratified squamous epithelium and is only found in the urinary bladder. When the bladder is not distended with urine, the cells are rounded. However, when the bladder is filling, these cells have the ability to change shape and slide over one another, allowing the organ to stretch. The superficial cells are flattened (like true squamous cells) when the bladder is completely full. Therefore, transitional epithelium appears, in the distended form, as though the number of cell layers it contains has diminished.

The epithelial cells that form glands are highly specialized to remove compounds from the bloodstream and manufacture them into new substances, which they then secrete. **Endocrine glands** lose their ducts as they develop in the fetus and secrete their product into the blood or lymph directly. These glands produce hormones. **Exocrine glands** retain their ducts and secrete onto an epithelial surface. Exocrine glands can be found on the surface of the body (sweat and oil glands) or internally, such as those found in the pancreas, which secrete their products through the pancreatic ducts onto the epithelial surface of the duodenum (the first part of the small intestine).

EXERCISE 5.1
EPITHELIAL TISSUE HISTOLOGY

The procedure for this and subsequent exercises in this chapter is as follows.

1. When viewing the recommended types of tissues, first go to low power, focus, and orient yourself on the slide (i.e., find the free surface vs. the deeper structures of the tissue). Then, change to high power and view the necessary cellular detail to understand the morphology of the tissue. To see certain cells, such as blood cells, macrophages, mast cells, or plasma cells, it will be necessary to use the oil immersion lens.

2. For the types of tissues in these exercises, obtain the recommended slide(s) and locate all the things you are instructed to view and that are labeled in the diagrams and simulated photomicrographs featured as figures throughout the chapter. Then, draw and label what you see in the space provided at the end of each section. Items you will be instructed to find and observe will include a certain type of cell, its nucleus and cytoplasm, the basement membrane, connective tissue fibers, and ground substance. You should find all items labeled in the drawings. Structures to be identified are listed in colored bold print. Subsequent slides may also contain the same structures or other related structures, which may be listed in bold print or italics; they are of no less importance and therefore should also be identified. For instance, when viewing the epithelial slides, the basement membrane is only in colored bold under simple squamous epithelium. However, it is still important to try to find it in all the slides of epithelial tissue, even though it is difficult to see on certain slides.

Simple Squamous Epithelium

Slides recommended: slides labeled simple squamous epithelium (surface view), kidney, and/or air sacs of lungs.

Description: Simple squamous epithelium contains a single layer of flattened cells with rounded or oval central **nuclei,** and sparse **cytoplasm.** It is the thinnest of the epithelial tissues (Figure 5.1). Often it is hard to see the **basement membrane,** but it is always present on the opposite side of the free surface.

Location found: Simple squamous epithelial cells are found in the kidney, alveoli of the lungs, blood vessels, lining of the heart, lymphatic vessels, lining the thoracic and peritoneal cavities, and on the serosal surface of

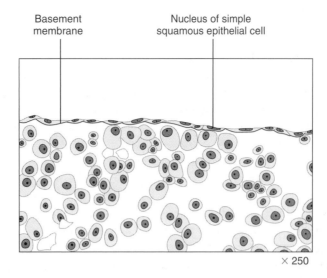

Basement membrane

Nucleus of simple squamous epithelial cell

× 250

Figure 5.1: Cross section of simple squamous epithelial cells that make up the serosal surface of the liver (i.e., the covering layer).

the abdominal and thoracic organs. This is also called the mesothelial surface, and the **mesothelial cells** are simple squamous in nature. The epithelial lining of the vascular system has a special name: **endothelium,** or **endothelial cells.**

Function: Because it is composed of a thin, single layer of cells, simple squamous epithelium permits passage of materials by diffusion and filtration in sites where protection is not necessary. For example, the mesothelial cells secrete a lubricating substance for the pleural and peritoneal cavities.

Slide: simple squamous epithelial cells (surface view)

This is a view of a sheet of epithelial cells looking both down on its surface and from the side. The cells appear to have asymmetrical shapes (Figure 5.2). All cells may or may not appear to have *nuclei*, depending on the level at which the cut was made. Making a cut so that each nucleus is visible within the cell is easier when the nuclei are in the same location in each cell or when the cells are short rather than tall.

Slide: kidney

There are several places to observe simple squamous epithelial cells easily. These include the kidney's **nephron** (specifically **Bowman's capsule**) and the capillaries (Figure 5.3). In the section of kidney you should easily find many large **renal tubules.** Look between these tubules and you will see a cross-sectional cut of numerous smaller tubules formed from very thin cells. These tubules are the **peritubular capillary network** of the kidneys. Each capillary is formed by two or three simple squamous epithelial cells linked in a circle.

In other parts of the kidney, note the rounded tufts of capillaries with a white space around them. On the slide they may appear as islands of tissue. Each is called a **glomerulus.** On the perimeter of the white space are the squamous epithelial cells that form *Bowman's capsule.* The cells may be so flat that they only bulge where their nuclei are located. The white space is the inside of Bowman's capsule, where the fluid filtered from the capillaries (called **glomerular filtrate**) enters the renal tubules. Find these cells and this space on the plastic model of the nephron.

Figure 5.2: Flattened epithelial cells.

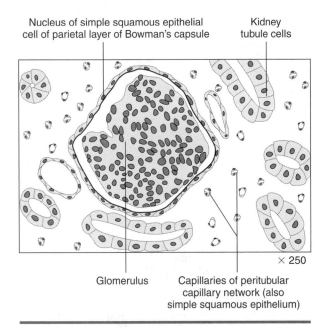

Nucleus of simple squamous epithelial cell of parietal layer of Bowman's capsule

Kidney tubule cells

× 250

Glomerulus

Capillaries of peritubular capillary network (also simple squamous epithelium)

Simple squamous epithelial cell

Air sacs

Nucleus of simple squamous epithelial cell

Figure 5.3: Kidney epithelial tissue (Bowman's capsule), showing simple squamous epithelium.

Figure 5.4: Lung epithelial tissue, showing simple squamous epithelium.

Slide: air sacs of the lungs

The cells that form the walls of the air sacs, or *alveoli,* of the lungs are made of simple squamous epithelial cells (Figure 5.4). On the slide you will note many large, white, irregularly shaped spaces. These are the alveoli, which are lined with a thin monolayer of simple squamous epithelial cells. Also look for small capillaries between the alveoli; these also are composed of simple squamous epithelium.

In the space below, using the slide that best represents simple squamous epithelial tissue, draw and label the parts of the tissue and cells.

Simple Squamous Epithelial Cells

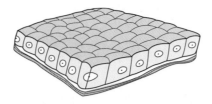

Figure 5.5: Simple cuboidal epithelial cells.

Simple Cuboidal Epithelium

Slides recommended: simple cuboidal epithelium or kidney tissue.

Description: Simple cuboidal epithelium is composed of a single layer of cube-shaped cells with large, spherical, centrally placed nuclei (Figure 5.5). Cytoplasm is abundant. The basement membrane, on the opposite side of the free surface, can sometimes be observed.

Location found: This type of tissue can be found in the kidney tubules, ducts and secretory portions of small glands, and peripheral cells of the ovaries just deep to the serosal surface.

Function: Its function is secretion and absorption of fluid and electrolytes.

Slide: kidney or simple cuboidal epithelium

The best example of simple cuboidal epithelial cells (Figure 5.6) are the cells that form the tubules of the nephron of the kidneys. In the section of kidney you should easily observe many large tubules, some oval and some more elongated. Because these tubules are convoluted in structure (look at the plastic model of the kidney's nephron), on a cut section some will be perfect cross-sections and others will be more oblique, giving the appearance of an elongated tubule. Because these cells are forming a tube, they are not perfectly square. Instead, they are usually a little narrower at the lumen of the tubule than at the basement membrane. Between the tubules are the *capillaries* mentioned previously as *connective tissue*.

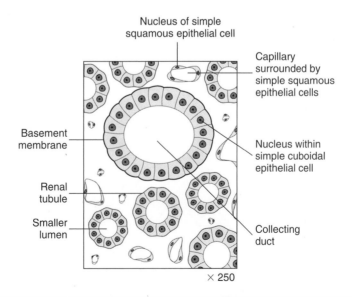

Figure 5.6: Simple cuboidal epithelial cells lining the tubules of the kidney.

In the space below, using the slide of simple cuboidal epithelial tissue, draw and label the parts of the tissue and cells.

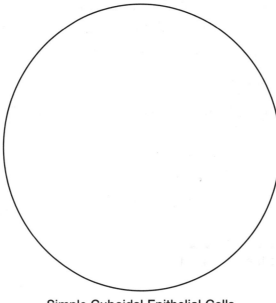

Simple Cuboidal Epithelial Cells

Simple Columnar Epithelium

Slides recommended: Slides labeled simple columnar epithelium or small intestine (duodenum, jejunum, or ileum).

Description: Simple columnar epithelium consists of a single layer of tall cells with oval-shaped *nuclei* (Figure 5.7). The *nuclei* are located toward the base of the cell, rather than in the center or toward the free surface. Cytoplasm is abundant. The *basement membrane,* on the opposite side of the free surface, can sometimes be observed.

Location found: Non-ciliated simple columnar epithelium lines most of the digestive tract (stomach to anus), gallbladder, and the excretory ducts of some glands. Ciliated types line parts of the small tertiary bronchi of the lungs, the oviducts, and the uterus.

Function: This tissue's function is the absorption of nutrients and fluids and the secretion of mucus, enzymes, and other substances. The ciliated types propel mucus (or reproductive cells) by ciliary action.

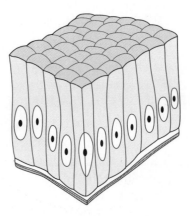

Figure 5.7: Simple columnar epithelial cells.

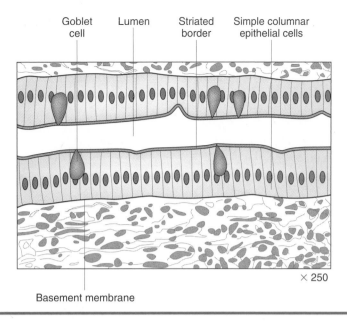

Figure 5.8: Simple columnar epithelial tissue from the small intestine.

Slide: small intestine (jejunum) or simple columnar epithelium

The best example of simple columnar epithelial cells are the cells that line the lumen of the intestinal tract (Figure 5.8). Search the slide until you find the place where the tissue section perfectly bisects the intestinal lining and the cells and their *nuclei* form a single row, side by side. The border of the free surface should appear as a thickened membrane; however, under oil immersion, if you look closely, you will note that it actually appears striated because it has a brush-like surface. This is due to the presence of **microvilli** and is called a **striated border.** *Microvilli* increase the surface area and absorptive capacity of the cells. As you look at the rows of columnar cells you will see oval areas between them. These are **goblet cells,** which are mucus secreting, unicellular glands. On some slides they may appear hollow, and on others they may contain a granular substance, which is mucus after processing.

In the space below, using the slide of simple columnar epithelial tissue, draw and label the parts of the tissue and cells.

Simple Columnar Epithelial Cells

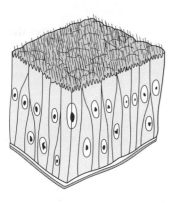

Figure 5.9: Pseudostratified ciliated simple columnar epithelial cells.

Pseudostratified Columnar Epithelium

Slides recommended: Those labeled pseudostratified columnar epithelium or trachea.

Description: Pseudostratified columnar epithelium consists of a single layer of cells of varying heights (Figure 5.9). All cells are attached to the basement membrane, which is sometimes visible. Not all cells, however, reach the free surface. The nuclei may be seen at different levels, and some even appear to be stacked on top of one another. This gives the appearance that this tissue is stratified, but it is not, hence the name *pseudostratified*. Most of the cytoplasm is seen just below the free surface and above the nuclei. *Goblet cells* also are seen interspersed between these cells. **Cilia** are present on the free surface; therefore, this tissue is also called **pseudostratified ciliated columnar epithelium.**

Location found: The ciliated type of pseudostratified columnar epithelium lines the trachea and continues down the respiratory tract to the first part of the tertiary bronchi. The non-ciliated type is found in the male's sperm-carrying ducts and the ducts of large glands.

Function: This tissue is active in secretion, particularly of mucus, and in propulsion of debris out of the respiratory system via ciliary action.

Slide: trachea or pseudostratified columnar epithelium

Note that some prepared slides labeled ciliated columnar epithelium are, in fact, pseudostratified ciliated columnar epithelium. Find the free surface of the section of trachea (Figure 5.10). This is best done by finding the cells with the small, hair-like *cilia* on the surface. The epithelial cells lining the respiratory (such as these from the trachea), digestive, and reproductive tracts are also known as **mucosal cells.** Mucosal cells line mucous membranes. Note the numerous *goblet cells* present. On some slides you may note that the goblet cells are releasing some of the mucus from inside the cell onto the free surface.

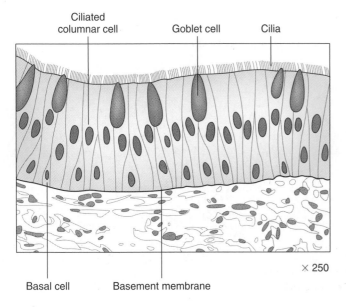

Ciliated columnar cell   Goblet cell   Cilia

Basal cell   Basement membrane

× 250

Figure 5.10: Pseudostratified ciliated columnar epithelial tissue from the trachea.

In the space below, using the slide of pseudostratified ciliated columnar epithelial tissue, draw and label the parts of the tissue and cells.

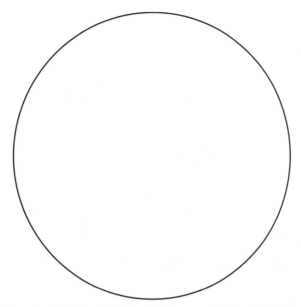

Pseudostratified Ciliated Columnar Epithelial Cells

Stratified Squamous Epithelium

Slides recommended: Those labeled stratified squamous, esophagus, or tongue.

Description: Stratified Squamous Epithelium is a thick epithelial surface composed of multiple cell layers (Figure 5.11). It is named for the appearance of the cells on the surface, which are squamous. Only the *basal cells* are attached to the *basement membrane*. The basal cells are cuboidal, or even columnar, and they are metabolically active. Mitosis occurs in the first few layers above the basement membrane and produces the cells to replace the cells of the more superficial layers as they die and slough off. The nuclei are easily seen in the lower layers of cells, but they tend to become smaller and disappear as they move toward the free surface where the cells are squamous.

Location found: The **non-keratinized** type of stratified squamous epithelium forms the moist lining of the mouth, esophagus, part of the stomach, the rumen, reticulum, omasum, and vagina. The **keratinized** type forms the epidermis of the skin.

Function: This tissue protects underlying tissues in areas subjected to abrasion.

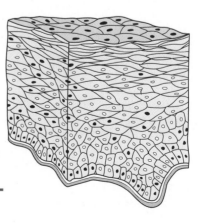

Figure 5.11: Stratified squamous epithelial cells.

Lumen of
esophagus

Stratified squamous
epithelium

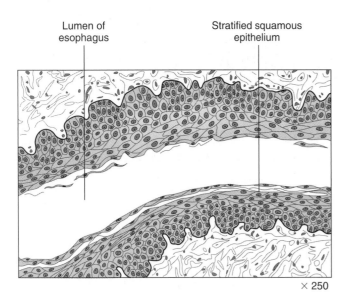

× 250

Figure 5.12: Stratified squamous epithelial tissue from the lumen of the esophagus.

Slide: tongue, esophagus, or stratified squamous epithelium

Locate the free surface of the tissue, where the cells appear more flattened or squamous in appearance (Figure 5.12). This is an example of *non-keratinized* stratified squamous tissue. Note that the cell layers go all the way to the free surface, where there is moisture present. If it were *keratinized* there would be two acellular layers of dry, **cornified** tissue, as is found in the epidermis of the skin. Considering the foods placed on the surface of these tissues, it is easy to understand why a thick, protective, mucosal surface is needed. The surface of the body also requires a thick layer of cells that are capable of regeneration to protect the body from dehydration, cellular desiccation, and from the ubiquitous external microorganisms of the world. This is why the epidermis of the skin has extra layers of cornified tissue for protection. We will study the histology of keratinized stratified squamous epithelium in Chapter 6.

In the space below, using the slide of stratified squamous epithelial tissue, draw and label the parts of the tissue and cells.

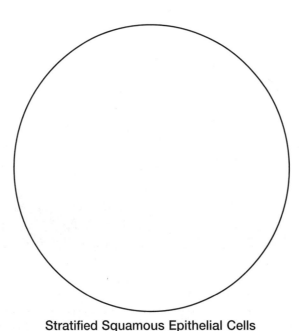

Stratified Squamous Epithelial Cells

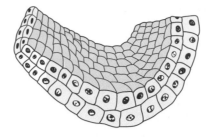

Figure 5.13: Stratified cuboidal epithelial cells.

Stratified Cuboidal Epithelium

Slides recommended: Those labeled stratified cuboidal, salivary glands, or esophagus.

Description: Generally there are two layers of cuboidal cells that form **stratified cuboidal epithelium** (Figure 5.13).

Location found: This tissue is found in the largest ducts of sweat glands, mammary glands, and salivary glands, and in the glands that are found in the esophagus.

Function: Their purpose is protection and the transportation of fluid.

Slide: salivary glands, esophagus, or stratified cuboidal epithelium

Search for ducts containing two layers of cuboidal cells (Figure 5.14). They will have a relatively large elliptical lumen surrounded by the bilayered, cube-shaped cells.

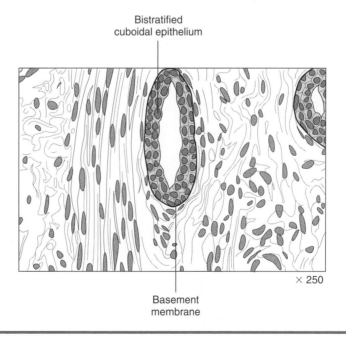

Bistratified
cuboidal epithelium

× 250

Basement
membrane

Figure 5.14: Stratified cuboidal epithelial tissue from the ducts of the glands of the esophagus.

In the space below, using the slide of stratified cuboidal epithelial tissue, draw and label the parts of the tissue and cell.

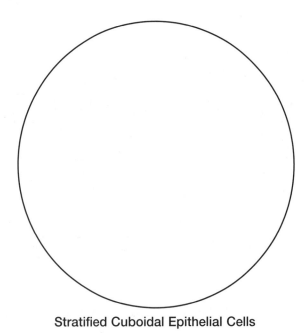

Stratified Cuboidal Epithelial Cells

Stratified Columnar Epithelium

Slides recommended: Those labeled stratified columnar or urethra.

Description: Stratified columnar epithelium appears to resemble both stratified columnar and stratified cuboidal cells. The surface cells are columnar, but the basal cells and intermediate cells are more cuboidal (Figure 5.15).

Location found: This tissue is found in areas of the urethra of male animals and in large ducts of some glands.

Function: It functions for protection, secretion, and transporting of fluid.

Slide: urethra or stratified columnar epithelium
On the free surface, look for a multilayered mucosal epithelium of tall columnar cells with numerous layers of cuboidal-shaped cells beneath it, deep to the *basement membrane* (Figure 5.16). The *nuclei* of the superficial columnar cells are at the same level in each cell, and they are centrally located. The deeper nuclei and their cells appear more random in their distribution.

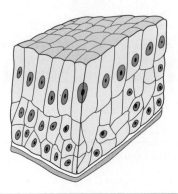

Figure 5.15: Stratified columnar epithelial cells.

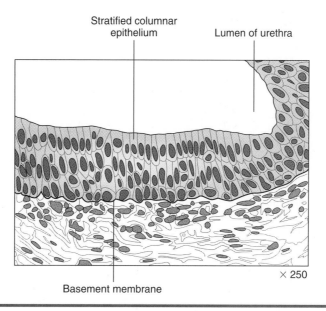

Figure 5.16: Stratified columnar epithelial tissue from the urethra of a goat.

In the space below, using the slide of stratified columnar epithelial tissue, draw and label the parts of the tissue and cells.

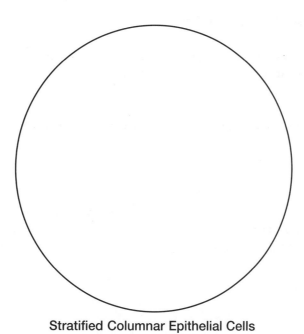

Stratified Columnar Epithelial Cells

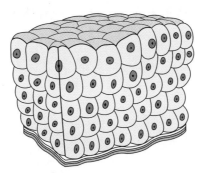

Figure 5.17: Transitional epithelial cells.

Transitional Epithelium

Slides recommended: Those labeled transitional epithelium or urinary bladder.

Description: Transitional epithelium resembles both stratified cuboidal tissue (when unstretched) and stratified squamous tissue (when stretched) (Figure 5.17). The basal cells are cuboidal or slightly rounded. Remember, as stated previously, these cells have the ability to slide across each other and flatten out to become squamous-like when the bladder fills with urine and the wall stretches.

Location found: Transitional epithelium is found in the urinary bladder, and it lines the ureters and part of the urethra.

Function: It participates in the storage of urine by readily permitting distention of the urinary bladder, and it helps with transporting of urine to the bladder through the ureters and out of the body via the urethra.

Slide: urinary bladder or transitional epithelium
Find the surface cells that are dome-like, with centrally placed *nuclei* (in the unstretched urinary bladder, which is probably what your slide shows) (Figure 5.18). In contrast, find the smaller *basal cells* and the *basement membrane* to which they are attached.

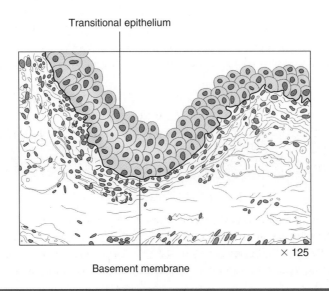

Transitional epithelium

× 125

Basement membrane

Figure 5.18: Transitional epithelial tissue, unstretched, from the urinary bladder of a cat.

In the space below, using the slide of transitional epithelial tissue, draw it and label the parts of the tissue and cells.

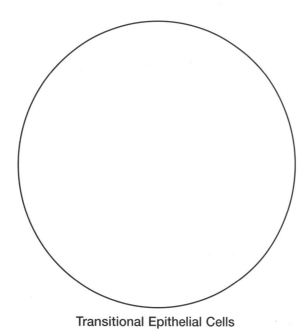

Transitional Epithelial Cells

CONNECTIVE TISSUE HISTOLOGY

Connective tissue is found everywhere in the animal's body. It binds tissues together, supports other types of tissues, and transports nutrients and oxygen (i.e., as the blood). All connective tissues have the three common characteristics: (1) they contain **cells,** which are their living component, (2) they contain **fibers,** which are produced by the cells, and (3) they contain a **ground substance,** which is composed largely of glycoproteins and glycosaminoglycans. In connective tissue, the latter two components compose the extracellular **matrix,** which predominates over the cellular elements. The ground substance forms a well-hydrated gel that fills the spaces between cells, fibers, and the vascular components of connective tissue. It varies in composition depending on the type of connective tissue, but generally it acts as a reservoir for interstitial fluid, providing a medium through which nutrients, oxygen, and metabolic by-products diffuse to and from cells between the connective tissue and the vascular system.

Connective tissue can be divided into three major types: **embryonal, proper,** and **special connective tissues. Mesenchyme** and mucous connective tissue are classified as **embryonal connective tissue. Connective tissue proper** (also known as *proper connective tissue*) includes the general types of connective tissue: *loose* and *dense; regular* and *irregular;* and *reticular, elastic,* and *adipose.* **Special connective tissue** includes tissues such as *blood, cartilage,* and *bone.*

There are three types of fibers found in connective tissue proper: (1) *collagenous,* (2) *reticular,* and (3) *elastic.* **Collagenous fibers** are composed of the protein **collagen** (which is actually the protein **tropocollagen,** which polymerizes to become collagen). Collagen is composed of strong, flexible fibers, which may be fine or coarse. They are characteristically unbranched, somewhat wavy, and they resist stretching. They stain pink with hematoxylin and eosin (H&E) stain. **Reticular fibers** are also composed of collagen; they are delicate, branching fibers possessing a coat of glycoproteins and proteoglycans. Silver-containing stains are used to differentiate these fibers from other fibers in connective tissue. **Elastic fibers** are formed from the protein **elastin.** They are typically fine fibers in **areolar** connective tissue, but in other places they can be coarse. They sometimes stain darker pink with H&E stains, or other special staining techniques also can be used to distinguish these fibers.

Fibroblasts are the most numerous type of cells found in connective tissue proper. They are responsible for producing the fibers and the ground substance.

Embryonal Connective Tissue, or Mesenchyme

Slides recommended: Those labeled either **mesenchyme** or **embryonal connective tissue.**

Description: This tissue is destined to become either connective tissue proper or special connective tissue (Figure 5.19). It has a soft, gel-like ground substance containing fibers and **stellate** (star-shaped) **mesenchymal cells.**

Location found: These cells are found primarily in the embryo.

Function: Serves as a precursor to all other connective tissue types.

Slide: mesenchyme or embryonal connective tissue

Note the many *stellate mesenchymal cells* with small *nuclei* located within (Figure 5.20). The clear background between the cells and fibers is the fluid-like ground substance. Also notice how fine and sparse the fibers are in this slide.

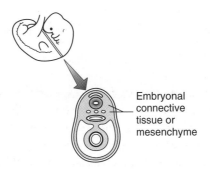

Embryonal connective tissue or mesenchyme

Figure 5.19: An embryo, showing where embryonal connective tissue is located.

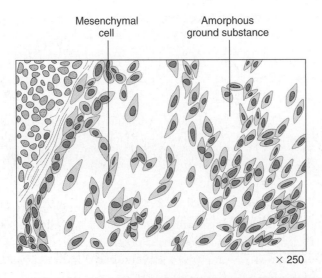

Mesenchymal cell

Amorphous ground substance

× 250

Figure 5.20: Embryonal connective tissue from a 72-hour-old chicken embryo.

In the space below, using the slide of embryonal connective tissue, draw and label the cells, fibers, and tissue parts.

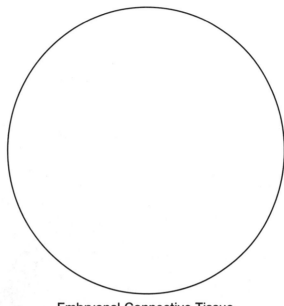

Embryonal Connective Tissue

Loose Irregular, or Areolar, Connective Tissue

Slides recommended: Those labeled areolar connective tissue.

Description: In areolar (loose irregular) connective tissue, a gel-like ground substance predominates. All three types of **fibers (collagenous, elastic,** and **reticular)** are present (Figure 5.21). There are many types of cells present, scattered about within the **matrix.** The majority of cells are **fibroblasts,** although **macrophages** (also known as **histiocytes,** or tissue *monocytes*—meaning white blood cells [monocytes] that have moved out of the bloodstream and into tissue), **plasma cells** (a type of tissue lymphocyte), and **mast cells** (cells that develop in the bone marrow in separate cell lines from white blood cells) can also be found. Other white blood cells, which include tissue *eosinophils, neutrophils, basophils,* and *lymphocytes,* can occasionally be seen, but they are in an unchanged state from what can be observed in blood.

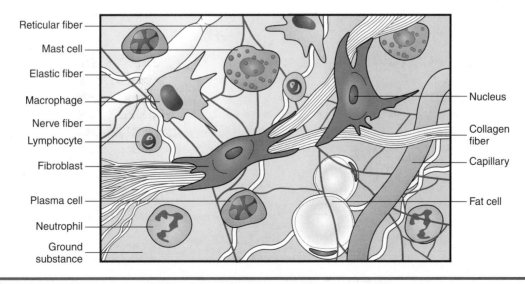

Figure 5.21: Areolar, or loose irregular, connective tissue.

Location found: Areolar connective tissue is found under the mucosal epithelium in most places of the body. It forms the lamina propria of mucous membranes, is found in the papillary layer of the dermis of the skin, is the connective tissue that attaches the serosa to organs, and surrounds capillaries.

Function: This tissue wraps and cushions organs, supports the mucosal epithelial cells, holds and conveys tissue fluids, and its immune cells help fight off infection by phagocytizing bacteria and other cells.

Slide: areolar connective tissue

Scan the slide and look for an area that contains the fibrous and cellular features found in this tissue: a **fibroblast,** with visible cytoplasm and is *fusiform* in shape *(spindle-shaped)*; thick **Collagen fibers;** thin **elastic fibers;** hair-like **reticular fibers;** and, if you can find them, a **mast cell;** a **macrophage;** and a **plasma cell** (Figure 5.22). Under H&E stain, the mast cell will appear as a cell with many red-granules, is irregularly shaped cell, and has a dark blue, round nucleus. Orcein or toluidine blue stains the mast cell's granules purple and the nucleus blue. The macrophage will be pinkish and elliptical to fusiform with a dark blue nucleus. The plasma cell is slightly larger than the others, is elliptical, and has an eccentrically placed nucleus in which the stained chromatin appears almost spoke-like.

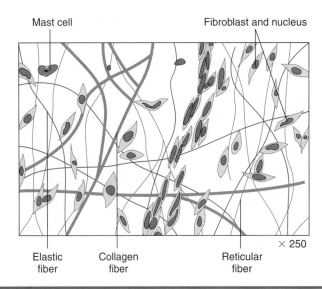

Figure 5.22: Areolar, or loose irregular, connective tissue from the mesentery of a cat.

In the space below, using the slide of areolar connective tissue, draw and label the cells, fibers, and tissue parts.

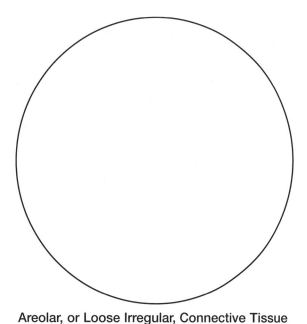

Areolar, or Loose Irregular, Connective Tissue

Dense Irregular Connective Tissue

Slides recommended: Those labeled skin.

Description: Fibers predominate in **dense irregular connective tissue.** There are primarily irregularly arranged cords of *collagen fibers,* with some *elastic fibers* interspersed (Figure 5.23). The major cell type is the *fibroblast.*

Location found: This tissue makes up the dermis of skin, the submucosa of the digestive tract, the capsule of organs and joints, the perichondrium around cartilage, and the periosteum.

Function: It supports the epidermis and provides the bulk of the strength for the skin, supports the joints, and provides strength and protection for bones and joints.

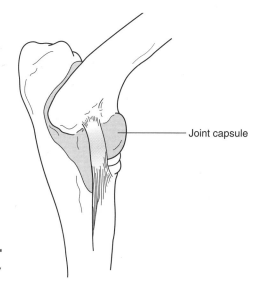

Joint capsule

Figure 5.23: Fibrous joint capsule (from the elbow of a dog), where dense irregular connective tissue is located.

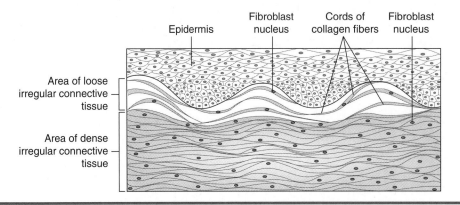

Figure 5.24: Dense irregular connective tissue from the dermis of a cow.

Slide: skin

Find the blue-stained epidermal cells (which are stratified squamous epithelium); just deep to these epithelial cells is where the dense irregular connective tissue is found (Figure 5.24). Note the pattern of abundant bands of thick, red *collagen fibers*. Note the irregular wavy pattern in which these bundles of collagen are weaved together. Also look for *elastic fibers*, which will appear thin. Locate the spindle-shaped *fibroblasts*.

In the space below, using the slide of dense irregular connective tissue, draw and label the cells, fibers, and tissue parts.

Dense Irregular Connective Tissue

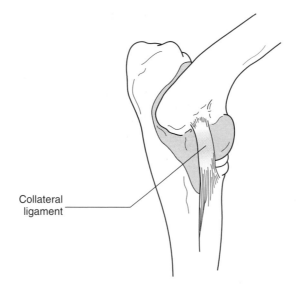

Figure 5.25: Ligament, where dense regular connective tissue is located.

Dense Regular Connective Tissue

Slides recommended: Those labeled tendon or ligament.

Description: Dense regular connective tissue is characterized by parallel *collagen fibers*. Few *elastic fibers* are present, and the major cell type is the *fibroblast* (Figure 5.25).

Location found: This tissue is found in ligaments, tendons, and aponeuroses.

Function: Tendons attach muscle to bone; ligaments attach bone to bone to support joints; and aponeuroses are sheets of tissue that give strength to the tissue to which they are attached. Dense regular connective tissue can withstand great tensile stress when force is applied in one direction.

Slide: tendon or ligament

Notice that the predominant feature of this slide is layer on layer of parallel *collagen fibers,* which stain intensely pink (Figure 5.26). The *nuclei* of the *fibroblasts* are thin and elongated and found between the fibers. Very little *ground substance* can be seen in this slide.

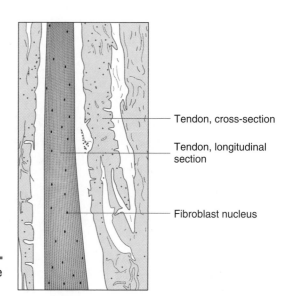

Tendon, cross-section

Tendon, longitudinal section

Fibroblast nucleus

Figure 5.26: Dense regular connective tissue from the tendon of a pig.

In the space below, using the slide of dense regular connective tissue, draw and label the cells, fibers, and tissue parts.

Dense Regular Connective Tissue

Elastic Connective Tissue

Slides recommended: Those labeled ligamentum nuchae.

Description: Elastic connective tissue is characterized by numerous regularly and irregularly arranged *elastic fibers* (Figure 5.27). The major cell type is the *fibroblast*.

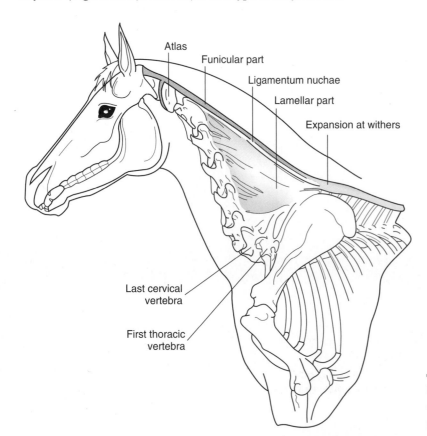

Atlas

Funicular part

Ligamentum nuchae

Lamellar part

Expansion at withers

Last cervical vertebra

First thoracic vertebra

Figure 5.27: Ligamentum nuchae (from a horse), where elastic connective tissue is located.

Nucleus of a fibroblast

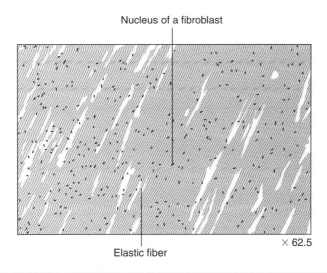

× 62.5

Elastic fiber

Figure 5.28: Elastic connective tissue from the ligamentum nuchae of a sheep.

Location found: This tissue is found in the ligamentum nuchae, the large ligament that runs from the **nuchal crest** of the skull to the dorsal spinous processes of the thoracic vertebrae in grazing animals. It is also found in the vocal ligaments.

Function: Elastic connective tissue supports the neck and head, and is responsible for phonation.

Slide: ligamentum nuchae

This slide shows the intensely pink-stained *elastic fibers.* Note that they lie parallel to each other but are not as perfectly layered as those in the ligament or tendon. Many *fibroblasts,* with their noticeable *nuclei,* lie between the elastic fibers (Figure 5.28).

In the space below, using the slide of elastic connective tissue, draw and label the cells, fibers, and tissue parts.

Elastic Connective Tissue

Adipose Tissue

Slides recommended: Those labeled **adipose tissue.**

Description: Adipocytes, better known as fat cells, are predominant in **adipose tissue** (Figure 5.29). The nucleus of each cell is pushed to the side by the large intracellular **vacuole** containing the fat. **White fat** cells contain one vacuole (called *monolocular*), whereas **brown fat** contains multiple vacuoles (*plurilocular*). The matrix is similar to that of areolar connective tissue but is extremely sparse.

Location found: This tissue is found under the skin (the subcutaneous fat), around the kidneys, within the omentum, in and around the mesentery of the digestive tract, around the kidneys (retroperitoneal fat), and in the mammary tissue. It also forms the *marbling* of muscle tissue.

Function: Adipose tissue is used to store fuel for the body; it also insulates against heat loss and supports and protects organs.

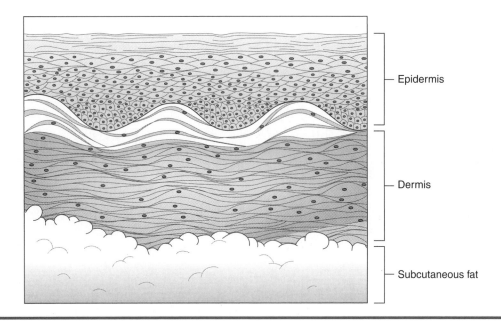

Figure 5.29: Subcutaneous fat, where adipose tissue is located.

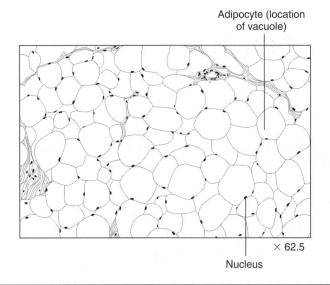

Figure 5.30: Adipose tissue from a cow.

Slide: adipose tissue

When viewing this slide, it may be necessary to dim your light or close down the condenser some to see the fat cells (Figure 5.30). Focus on an area of the slide where you can see clear definition of the cell walls and where the nuclei are visible. The large empty space between the cell walls is the vacuole that contains the fat. The nuclei are small and compressed by the large fat vacuoles of the cells.

In the space below, using the slide of adipose tissue, draw and label the cells and tissue parts.

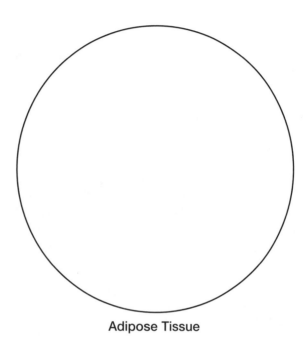

Adipose Tissue

Reticular Connective Tissue

Slides recommended: Those labeled reticular connective tissue. This slide should be a special stained slide to demonstrate these cells and fibers.

Description: The fibers of **reticular connective tissue** form a network of fine **reticular fibers,** which are made of *collagen* and possess a coat of glycoproteins and proteoglycans (Figure 5.31). The fibers lie in a loose *ground substance,* and the **reticular cells** lie on the network.

Location found: This tissue is found in the liver, kidney, and lymphoid organs such as lymph nodes, spleen, and bone marrow.

Function: The fibers of reticular connective tissue form the internal support structure, called the **stroma,** which supports the cells of these organs.

Slide: reticular connective tissue

Note the intensely dark-stained *reticular fibers* coursing between the bundles of cells. The *reticular cells* are smaller than the other cells, which also stain very dark, and they are adjacent to the fibers (Figure 5.32). The *nuclei* of the cells predominate and there is little visible cytoplasm.

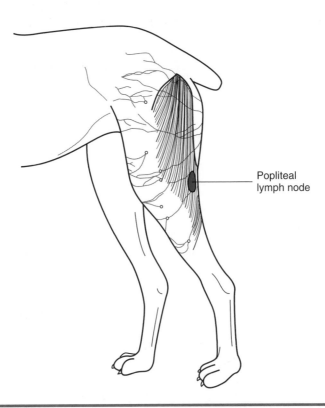

Figure 5.31: Popliteal lymph node of a dog, where reticular connective tissue is located.

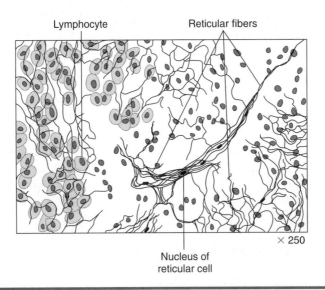

Figure 5.32: Reticular connective tissue from the lymph node of a cow (silver stain used).

In the space below, using the slide of reticular connective tissue, draw and label the cells, fibers, and tissue parts.

Reticular Connective Tissue

EXERCISE 5.3

SPECIAL CONNECTIVE TISSUE

SPECIAL CONNECTIVE TISSUE: CARTILAGE

There are three basic types of cartilage: (1) **hyaline,** (2) **elastic,** and (3) **fibrous,** or **fibrocartilage.** Each type meets the same requirements as previously discussed types of connective tissue. However, the cells found in cartilage are called **chondrocytes.** The matrix is composed of an **amorphous ground substance,** which is rich in *sulfated glycosaminoglycans,* **chondroitin sulfate,** and **hyaluronic acid.** This is combined with an adhesion protein called *chondronectin* forms large molecules called *proteoglycans,* which are bound electrostatically to individual fibrils of collagen and elastin. The matrix is firm yet flexible, so it can bend when necessary yet support weight and tensile stress.

Hyaline Cartilage

Slides recommended: Those labeled hyaline cartilage or trachea.

Description: The ground substance makes up the bulk of **hyaline cartilage** tissue, and it can be separated into pale and darkly stained areas called the **interterritorial** and **territorial matrix,** respectively (Figure 5.33). The latter, darker-staining areas contain a higher concentration of sulfated glycosaminoglycans. The **chondrocytes** (cartilage cells) are confined to small pockets called **lacunae** within the matrix. In a cluster of chondrocytes, often found are two newly divided cells within one lacuna called **isogenous groups.** The cartilaginous matrix is surrounded by a **perichondrium.** The outer portion of this is dense irregular connective tissue, and the inner layer is **chondrogenic,** containing cells with the capacity to become **chondroblasts.**

Location found: This tissue is found in the cartilages of the trachea, nose, and larynx. In addition, it covers the ends of the long bones in synovial joints, forms the costal cartilage of the ribs, and forms most of the embryonic skeleton.

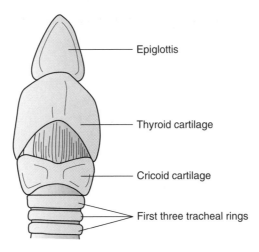

Figure 5.33: Ventral view of the larynx and trachea of a dog, where hyaline cartilage is found.

Function: Hyaline cartilage supports the body of the growing fetus; forms a smooth surface for articular joints; forms the structure of the costal ribs, nose, and larynx; and supports the structure of the trachea.

Slide: hyaline cartilage or trachea

Find the area of the slide where the cartilage is located. If you are viewing the trachea, it is the cross-sectional area of the tracheal rings (Figure 5.34). You should see a large, round mass of solid, pinkish-violet structure (the *matrix*) with numerous white holes (the *lacunae*) containing a cell within (a *chondrocyte*). The *nucleus* of the cell is dark and the cytoplasm is very light. The collagen and elastic fibers are not visible because they blend imperceptibly within the matrix.

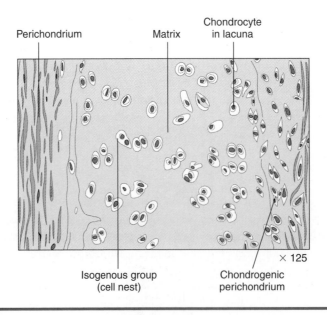

Figure 5.34: Hyaline cartilage from the trachea of a cow.

In the space below, using the slide of hyaline cartilage or trachea, draw and label the cells and tissue parts.

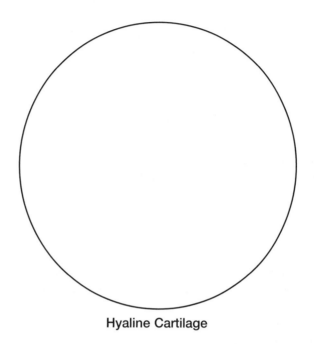

Hyaline Cartilage

Elastic Cartilage

Slides recommended: Those labeled elastic cartilage.

Description: The matrix forms the bulk of **elastic cartilage** tissue, and it contains many of the same substances as hyaline cartilage except that it has a much larger amount of *elastic fibers,* for which it is named (Figure 5.35). The *chondrocytes* are inside *lacunae* located within the *matrix*.

Location found: This tissue is found in the cartilages of the epiglottis, parts of the larynx, and the pinna of the ear.

Function: It is able to support and maintain the shape of the structure, while allowing great flexibility.

Pinna

Figure 5.35: Pinna of a cat's ear, where elastic cartilage is found.

Elastic fibers   Chondrocyte   Isogenous
                  in lacuna   group

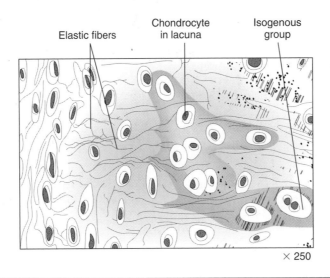

× 250

Figure 5.36: Elastic cartilage from the epiglottis of a dog.

Slide: elastic cartilage

Find the area of the slide where the cartilage is located. The *matrix* is most recognizable because of the large, dark-stained purple area within its center (Figure 5.36). Upon close examination, small red *elastic fibers* may be seen at the periphery of the purple area and should appear to course transversely across the cartilage. The *chondrocytes' nuclei* and *cytoplasm* are viewable within the *lacunae*.

In the space below, using the slide of elastic cartilage, draw and label the cells, fibers, and tissue parts.

Elastic Cartilage

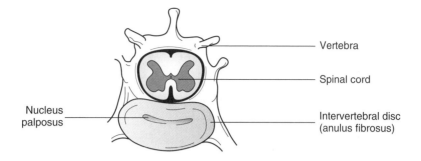

Figure 5.37: Intervertebral disc (found between vertebrae), where fibrous cartilage is located.

Fibrous Cartilage (Fibrocartilage)

Slides recommended: Those labeled fibrocartilage.

Description: Fibrous cartilage is distinctively different from the others. It contains large amounts of fibrous connective tissue in the form of *collagen*, which is embedded in a *ground substance* not unlike that of the other cartilages (Figure 5.37). The fibers are distributed linearly throughout the ground substance (i.e., they run in the same direction). The *chondrocytes* lie in the *lacunae*, which are also distributed linearly throughout the *matrix*.

Location found: Fibrocartilage is found in intervertebral discs, pubic symphysis, discs of the knee joint, and within some tendons close to where they attach to bone.

Function: It has high tensile strength for support and to absorb compressive shock.

Slide: fibrocartilage
Note the large amount of *collagen fibers* located within this cartilage. Fibrous cartilage might be easily confused with dense irregular connective tissue if not for the large number of *lacunae* present with their resident *chondrocytes* (Figure 5.38).

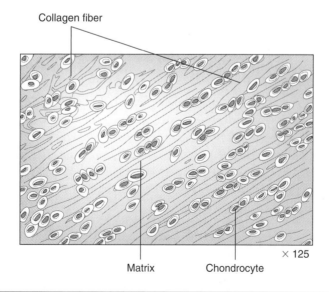

Figure 5.38: Fibrous cartilage from the intervertebral disc of a horse.

In the space below, using the slide of fibrous cartilage or trachea, draw and label the cells, fibers, and tissue parts.

Fibrous Cartilage

SPECIAL CONNECTIVE TISSUE: BLOOD

This section will introduce you to the types of blood cells found in peripheral blood. Courses in clinical laboratory procedures, including hematology, will go into this subject matter in more depth.

Blood is classified as a connective tissue that is composed of cells in a fluid *matrix.* **Plasma** is the fluid portion, which is called **serum** when depleted of **fibrinogen.** The formed elements include **erythrocytes (red blood cells), leukocytes (white blood cells),** and **platelets** (also known as **thrombocytes**). Blood cells and platelets usually are studied using stained blood smears. After a drop of blood is spread thinly on a glass slide and dried, it is stained with a Romanovsky-type stain such as **Giemsa** or **Wright's stain.** On one end of the slide is a thin monolayer of cells that are more flattened and less crowded. This is the area in which cell morphology should be studied. Blood smears should be scanned using the high-power objective lens; then oil immersion can be used for studying specific cells in detail or for performing a **differential cell count** of the leukocytes. The *differential cell count* is done by counting the various cell types among the first 100 white blood cells encountered. They are expressed as in the following example: segmented neutrophils - 66, band neutrophils - 3, lymphocytes - 18, monocytes - 9, eosinophils - 3, basophils - 1. Because these total 100 cells it also means they can be expressed as a percentage (e.g., of the total white blood cells, 66% were segmented neutrophils).

Description: Mammalian *erythrocytes* are the largest in the dog (7.0 μm) and smallest in the goat (4.1 μm). Mature red blood cells are *anucleated.* As the newly forming erythrocyte develops, it gradually loses its nucleus. Visible nucleated red blood cells (RBCs) in the peripheral blood are an indication that the bone marrow is working hard to produce new RBCs. Erythrocytes of birds, reptiles, and amphibians are nucleated. *Leukocytes* are larger, ranging in size from 6.0 μm in the smallest lymphocytes to 20 μm in the larger monocytes. They are divided into **granulocytes (neutrophils, eosinophils,** and **basophils)** and **agranulocytes (lymphocytes** and **monocytes)** (Figure 5.39). Leukocytes tend to accumulate along the periphery of blood smears, so they are more readily visible in these areas. Neutrophils are the predominant leukocyte of the dog, cat, and horse,

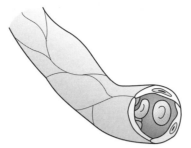

Figure 5.39: Capillary with red blood cell passing through it.

whereas lymphocytes are the predominant type of ruminants and pigs. Birds, reptiles, and amphibians have a granulocyte known as a **heterophil** rather than a neutrophil. The rest of the leukocytes are the same types as found in mammals.

Slide: human or dog blood smear

You will notice numerous small, red non-nucleated **erythrocytes.** These cells are round with central, biconcaved surfaces on the top and bottom. This makes them appear lighter in the center, which is known as having a **central pallor.** This is most observable in dog and human blood; in other species it in not as predominant. Scan the slide and find the various types of white blood cells. You should be able to find all of them, except perhaps the basophil, which is hard to find because they only stay in peripheral blood for a short time after leaving the bone marrow.

The first granulocytes to find are the **segmented neutrophils** (also known as **segs**) (Figure 5.40). They have dark-purple-stained nuclei in which the lobes are separated by slight indentations or thin strands of nucleoplasm. The stage of development prior to segmentation of the nucleus is called a *band neutrophil,* which differ from the segs only in the shape of the nucleus. The **band neutrophil** nucleus is U-shaped and is the same width throughout its entire length. As with seeing nucleated RBCs, seeing many band neutrophils in peripheral blood indicates that the bone marrow is producing numerous cells. The cytoplasm of the neutrophil (both bands and segs) is slightly granulated.

The **eosinophil** has a similar nucleus to that of the neutrophil, although it tends to be slightly less dense and have fewer lobes. The most noticeable feature of the eosinophil is its predominant red, reddish-purple, or lavender granules (depending on the stain and the species of animal's blood used). The final granulocyte to locate is a **basophil.** The nucleus may be irregular, bilobed, or highly segmented. The granules vary in size, number, and intensity of staining depending on the type of stain and species of the animal's blood. They are often large, round to oval, and they stain reddish-purple to dark purple, except for in the cat in which they are dull gray to lavender. Because the nuclei of the granulocytes have many forms, these cells are also called *polymorphonucleated leukocytes* (known as *polymorphs* or *PMNs*). However, these terms have come to be used synonymously for the neutrophil.

The **monocytes** are the largest of the white blood cells (WBCs). The shape of their nucleus is highly variable; it can be oval, irregular, kidney-shaped, or horseshoe-shaped. The nuclear chromatin is diffuse and lacy or patchy in appearance. The cytoplasm is generally a pale gray-blue and may contain small dust-like **azurophilic granules.** A feature that helps distinguish this cell from the large lymphocyte is that it often contains *vacuoles* that give it a foamy appearance. **Small lymphocytes** have a large, dense, often eccentrically placed nucleus that is generally round, although it is oval in the pig, sometimes slightly indented in the cat, and may be binucleated in the ruminant. Most of the lymphocytes of carnivores, horses, and pigs are small, whereas the large type predominates in the ruminants. In the small lymphocyte, only a thin rim of blue cytoplasm with a light perinuclear halo may be seen. The **large lymphocyte** has a less dense nucleus with a pale blue, more abundant cytoplasm. The nucleus may be round, oval, or kidney-shaped. In both types of lymphocytes, nonspecific azurophilic granules may be seen in certain medical conditions.

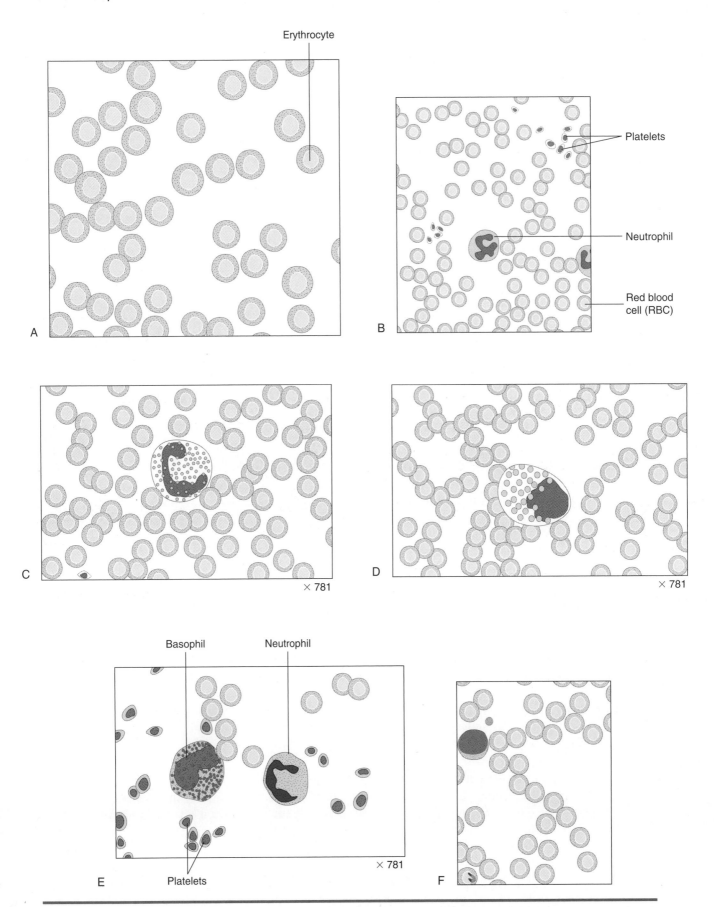

Figure 5.40: *A.* Erythrocyte from cat blood. *B.* Neutrophil (segmented) and platelets from cat blood. *C.* Eosinophil from cow blood. *D.* Eosinophil from horse blood. *E.* Basophil from horse blood. *F.* Small lymphocyte from cat blood.

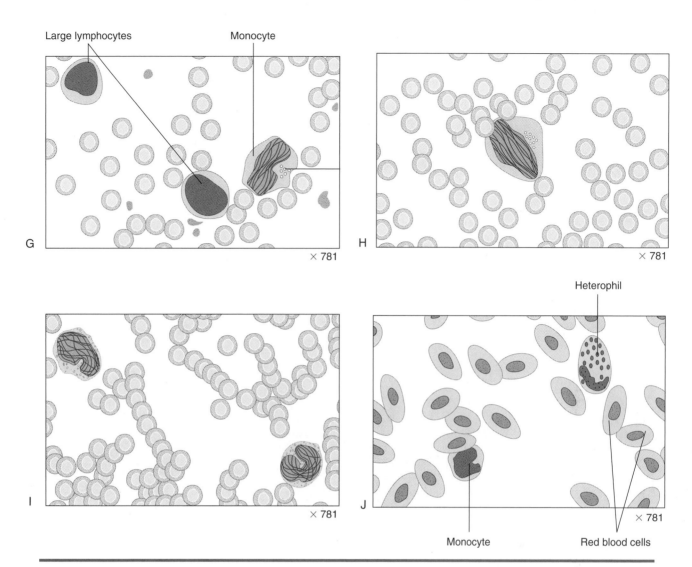

Figure 5.40, *continued*: *G.* Large lymphocytes and monocyte, pig. *H.* Monocyte from horse blood. *I.* Monocytes from cat blood. Red blood cells are stacked in rouleaux. *J.* Heterophil, monocyte, and red blood cells (nucleated) from chicken blood.

Platelets are very small and occur singly or in clusters between the RBCs. They are pale blue and contain purple central granules. They play an important role in *hemostasis*, the clotting of blood.

Draw all the types of cells seen in peripheral blood and described previously in the spaces below.

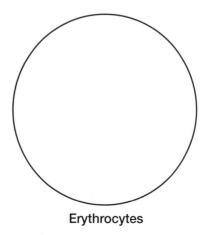

Erythrocytes

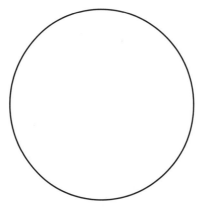

Segmented and Band Neutrophils

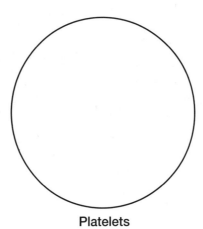

Platelets

Eosinophil

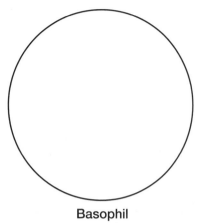

Basophil

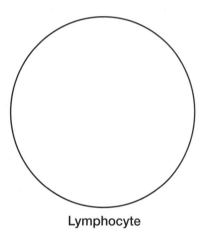

Lymphocyte

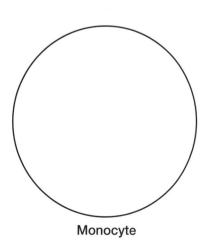

Monocyte

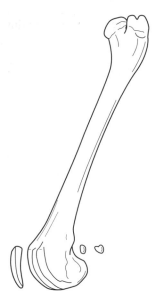

Figure 5.41: Bone tissue from the femur of a cat.

Special Connective Tissue: Bone

The skeletal system will be discussed in an upcoming chapter. A brief overview of the histology of compact bone will be covered here.

Description: Bone has a hard, calcified matrix containing many collagen fibers and **osteocytes** that lie within lacunae (Figure 5.41). Bones are well vascularized.

Location found: The bones of the skeletal system.

Function: Its functions are as follows: (1) provides form and structure, (2) provides protection, (3) provides mineral storage, (4) acts as site of blood formation (for oxygen transport and immune function), (5) enables the body's leverage and mobility.

Slide: compact bone

Observe the multiple circles of bone throughout the tissue; these are called **osteons,** and all of these together make up the **Haversian system** of bone. At the center of each osteon is a **central canal,** which contain a blood vessel (Figure 5.42). The dark, linear dots of the osteon, which encircle the central canal, are the **lacunae,** which

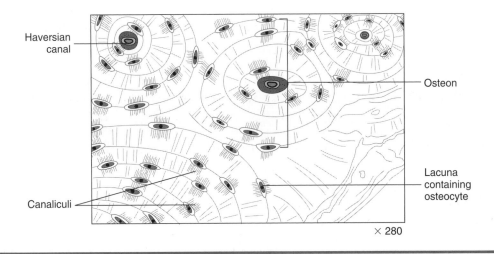

Figure 5.42: Cross section of compact bone from a cat.

contain the **osteocytes.** Note the thread-like lines that radiate toward each lacuna; these are **canaliculi,** or tiny canals that link the lacunae together and provide nutrition for the osteocytes.

In the space provided, draw a section of compact bone and label the cells, canals, and other parts discussed previously.

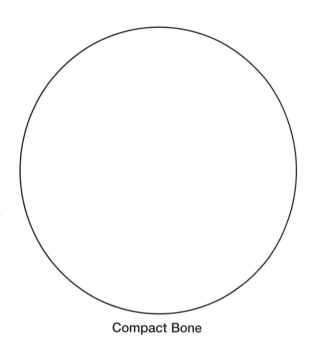

Compact Bone

EXERCISE 5.4
MUSCLE TISSUE HISTOLOGY

There are three types of muscle: *skeletal, smooth,* and *cardiac.* **Muscle** is a highly specialized tissue designed to contract, and it produces most types of body movement.

Skeletal Muscle

Skeletal muscle is attached to the skeleton. It is under voluntary control, and its contractions move the limbs and other external body parts.

Description: Skeletal muscle cells are long, cylindrical, and have multiple *nuclei.* The most prominent feature of this tissue is the transverse striations across each cell. Therefore, skeletal muscle is classified as **striated muscle** (Figure 5.43).

Location found: Specifically, skeletal muscle can be found attached to bone and tendon, and in some areas, to skin.

Function: It allows for voluntary movement and locomotion, facial expression, and skin movement.

Slide: skeletal muscle
Note the long, red *muscle cells* with *cross striations.* On the outside of these cells, notice the numerous *nuclei* (Figure 5.44).

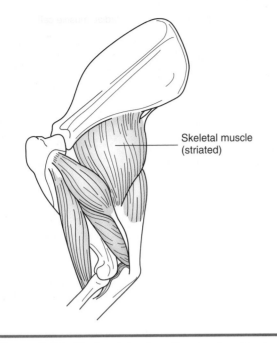

Figure 5.43: Skeletal (striated) muscle attached to bone.

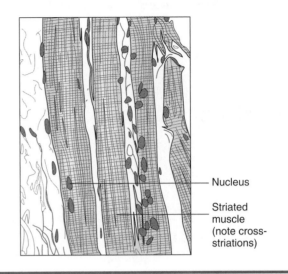

Figure 5.44: Cross section of skeletal muscle from the tongue of a horse. Note the cross striations.

In the space provided, draw a section of skeletal muscle and label the parts of the cells.

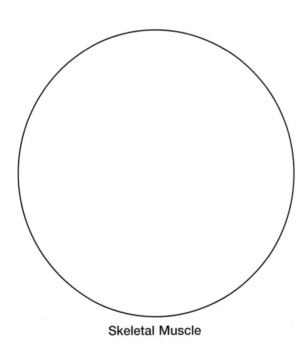

Skeletal Muscle

Cardiac Muscle

Cardiac muscle is *involuntary* and makes up the heart's **myocardium** (or heart muscle).

Description: Cardiac muscle cells are also *striated*, are generally *uninucleated*, and may be *branched*. They are separated from one another by specialized junctions called **intercalated discs** (Figure 5.45).

Craniodorsal aspect

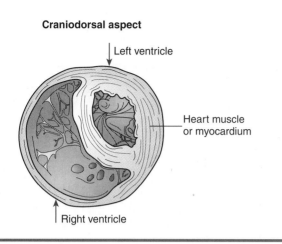

↓ Left ventricle

Heart muscle or myocardium

↑ Right ventricle

Intercalated disc   Cardiac muscle cell

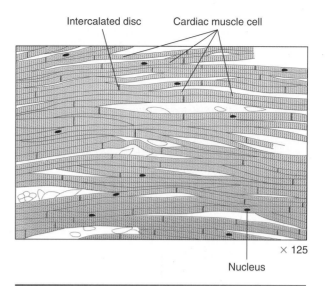

× 125

Nucleus

Figure 5.45: Cross section of myocardium of the heart.

Figure 5.46: Cross section of feline cardiac muscle.

Location found: Located exclusively in the heart.

Function: Cardiac muscle contracts the atria and ventricles, moving blood into the heart, through it, and out again.

Slide: cardiac muscle

The striations on cardiac muscle cells are not as noticeable as they are on skeletal muscle (Figure 5.46). Instead they appear as multiple, fine-lined, **cross striations** in each cell. Note that the cells are separated on each end by dark, heavy lines, which are the *intercalated discs.* Notice also that some of the cells **branch.** The **nuclei** are large in relation to the cells and cannot be found in every cell because of the way the slide sample was sectioned.

In the space provided, draw a section of cardiac muscle and label the parts of the cells.

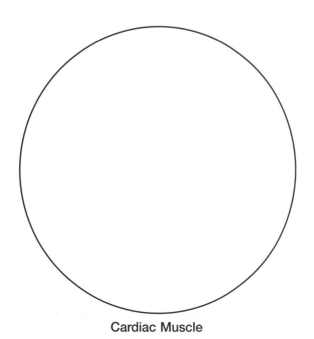

Cardiac Muscle

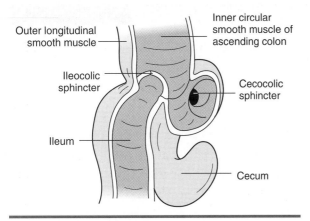

Outer longitudinal smooth muscle

Inner circular smooth muscle of ascending colon

Ileocolic sphincter

Cecocolic sphincter

Ileum

Cecum

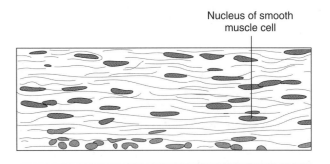

Nucleus of smooth muscle cell

Figure 5.47: Longitudinal view of an intestinal tract, showing the layers of smooth muscle.

Figure 5.48: Cross section of smooth muscle from the small intestine of a sheep.

Smooth Muscle

Smooth muscle is *involuntary* and is found in organs that need to contract to move fluids. It can also act as a **sphincter muscle** to control the diameter of the opening of vessels, airways, or between tubular organs.

Description: *Smooth muscle cells* are *not striated,* have *spindle-shaped* cells with a central *nucleus,* and are arranged closely to form sheets of tissue (Figure 5.47).

Location found: Smooth muscle is found in the walls of tubular organs, around bronchi, in the walls of arteries and veins, in the uterus, and it forms the precapillary sphincter muscles.

Function: The functions of smooth muscle are to propel fluids, substances (such as foodstuffs or urine), and the neonatal animal along internal passageways. This muscle also controls the amount of fluid or air entering or leaving certain areas of the body.

Slide: smooth muscle or intestinal tract

Smooth muscle cells are long and thin and tapered at both ends. They course in the same directions but are not perfectly parallel in configuration (Figure 5.48). The *nuclei* are located within the cells and sometimes will appear rippled if the cell is contracting.

In the space provided, draw a section of smooth muscle and label the parts of the cells.

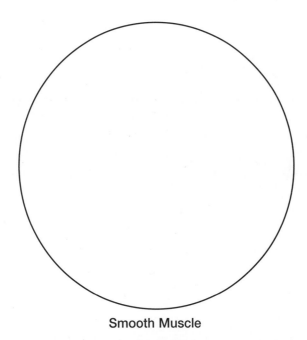

Smooth Muscle

EXERCISE 5.5

NERVOUS TISSUE HISTOLOGY

Nervous Tissue

The brain and spinal cord possess billions of **neurons,** or **nerve cells** (Figure 5.49). You have already looked at a nerve cell in Chapter 3, Exercise 3.2.

Description: Each neuron has a **nerve cell body** complete with a *nucleus,* and many short, thin processes called **dendrites** come off of it. The dendrites collect electrochemical signals from other nerves and route them through the cell body and down a long fiber, called an **axon,** to the next nerve cell, organ, or muscle.

Location found: In addition to being found in the brain and spinal cord, neurons are found in nerves throughout the body.

Function: The nerve cells' function is to transmit information by electrochemical signals from sensory receptors to the brain and from the brain to the effectors (muscles, organs, and glands) that control their activity.

Slide: nerve cells (neurons), brain

Find the following structures (these were described previously in Chapter 3, Exercise 3.2): *nerve cell body, nucleus, nucleoli, axon, axon hillock, dendrite, Nissl bodies, cytoskeletal elements* (Figure 5.50).

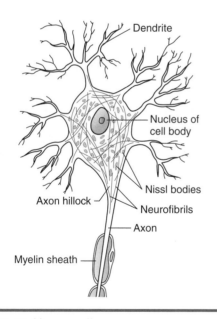

Figure 5.49: Nerve cells.

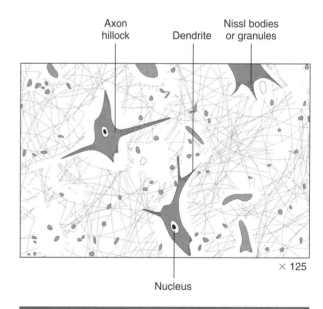

Figure 5.50: Nerve cells from the spinal cord of a sheep.

In the space provided, draw a section of nervous tissue and label the parts of the cells.

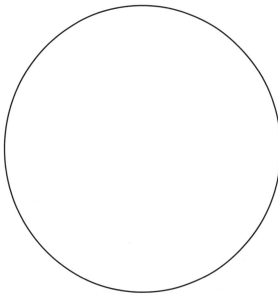

Neuron Cell Body & Processes

Clinical Significance

The clinically significant part of this chapter is an exercise in performing a differential white blood cell count. Because certain disease processes can be characterized by their pattern of leuko- cyte increase or decrease in peripheral blood, it is *necessary to obtain this information as quickly as possible. Performing an in-house total white blood cell count and differential can be very helpful in obtaining a rapid diagnosis.*

EXERCISE 5.6
DIFFFERENTIAL WHITE BLOOD CELL COUNT

Take out your blood smear slide again and count the first 100 recognizable leukocytes using a five-punch differential cell counter to keep track. A bell should ring on the counter when 100 total cells have been counted. (Keep track on paper if you do not have a cell counter.) Fill in the results below:

segmented neutrophils:	_____	monocytes:	_____
band neutrophils:	_____	eosinophils:	_____
lymphocytes:	_____	basophils:	_____

This count is the **relative cell count.** Differential leukocyte counts should always be expressed as total cell counts rather than percentage counts (which is what the relative cell count is called). The relative cell count should be just a means to obtain a total cell count; it has little interpretive value.

Assume the total white blood cell (leukocyte) count is 12,000 cells/μl (this count may be obtained by per- forming manually a total white blood cell count using a hemocytometer or by an electronic blood cell counter).

Now multiply each of the previous values by the total leukocyte count. This will yield the total cell count for each type of leukocyte (also called the **absolute cell count**).

segmented neutrophils: _____ × 12,000 = _____

band neutrophils: _____ × 12,000 = _____

lymphocytes: _____ × 12,000 = _____

monocytes: _____ × 12,000 = _____

eosinophils: _____ × 12,000 = _____

basophils: _____ × 12,000 = _____

Assuming this is a dog (even if your slide is a human blood smear), the normal range of neutrophils (segs and bands combined) is 3,000 to 12,000 cells/μl, lymphocytes is 1,000 to 5,000 cells/μl, monocytes is 150 to 1,350 cells/μl, and eosinophils is 100 to 1,250 cells/μl. The normal range of basophils is not listed because they are rare in peripheral blood. Which, if any, of these levels are out of the normal range?

SUMMARY

This chapter has been a general study of microscopic anatomy and should form the basis for the study of the organ systems we will encounter in later chapters. You learned the basic structure of epithelial tissues, their functions, and where they are located. We covered connective tissue, and you found that the fibers and matrix predominates over the cellular components in this tissue. You learned the difference between embryonal, proper, and special connective tissue. We also reviewed the structure and function of nerve cells. The final exercise was a practical application of the knowledge gained from this study and a technique most technicians should put to great use in the future.

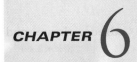

THE INTEGUMENTARY SYSTEM

OBJECTIVES:

- name and histologically identify the layers of the epidermis and dermis from prepared slides
- identify the cutaneous glands and hair from prepared slides
- understand the factors determining skin color
- identify the structure and parts of the equine foot; note which parts are epidermal derivatives and which are dermal derivatives
- understand the structure of a horn as compared to an antler; identify the parts and note which are epidermal derivatives and which are dermal derivatives
- understand the structure of the claws, pads, and nose of a dog
- know the anatomical structure of a cow's mammary gland, including the names of all the ducts, lobes, lobules, and alveoli

MATERIALS:

- three-dimensional model of the skin (if available)
- three-dimensional model of the equine foot (if available)
- model of the cow's mammary gland
- longitudinal and cross-sectional displays of horn and antler
- prepared slides of bovine skin, hairy mammal skin, sweat gland, and human heavily pigmented skin
- compound light microscope
- immersion oil
- magnifying glass
- colored pencils
- prepared specimen of an equine hoof
- live dog(s), some with unpigmented nails, and others with pigmented nails
- live cat
- live horse

Introduction

The **integument** includes the skin and its derivatives, which are the accessory organs of the skin. The skin consists of the **epidermis** and **dermis.** The dermis is attached to the underlying **hypodermis,** or **subcutis (subcutaneous tissues),** which is composed mostly of **adipose tissue,** and this is attached to muscle and bone underneath. The accessory organs of the skin include the sweat, sebaceous, and mammary glands, as well as the hair or feather follicles, claws, hooves, beak, and scales.

EXERCISE 6.1

HISTOLOGY OF THE INTEGUMENTARY SYSTEM

As recommended in the previous chapter, when viewing the recommended slides of tissues, first go to low power, focus, and orient yourself on the slide (i.e., find the free surface and the deeper structures). Then change to high power to see the cellular detail needed to understand the morphology of the tissue.

For the following types of tissue, obtain the recommended slide(s), locate all the structures on the slide you are instructed to find and that are labeled in the diagrams and photomicrographs shown in the figures throughout the chapter. Then, draw and label what you see in the spaces provided. The things you will be instructed to find will include the cell, its nucleus and cytoplasm, the basement membrane, connective tissue fibers, and ground substance. The most important structures to be identified are listed in colored bold print. Subsequent slides may also contain the same structures or other related structures, which may be listed in bold print or italics; they are of no less importance and therefore should also be identified.

Epidermis

Slides recommended: Those labeled bovine skin or human skin, scalp.

Description: You may want to begin by reviewing the section of the previous chapter covering stratified squamous epithelium. The epidermis viewed in the previous chapter on page 45 was non-keratinized. However, the epidermis of thick skin (as seen here) is keratinized to assist in its protective function. Hence, the most abundant type of cells in this tissue are the **keratinocytes** (Figure 6.1), which are a special type of squamous cell with the ability to produce **keratin.** This is a tough, fibrous, waterproof protein, which imparts to the skin its strength and resiliency.

Slide: skin

The layers of the epidermis, from deep to superficial, are the *stratum basale, stratum spinosum, stratum granulosum, stratum lucidum, and stratum corneum* (Figure 6.2).

Locate the bottom layer of the epidermis, or the basal cell layer, next to the red, fibrous dermis. This is the **stratum basale;** it consists of only one layer of keratinocytes, which are constantly undergoing mitosis and producing millions of new cells daily. This tissue's alternative name is the *stratum germinativum.* This stratum is attached to the *basement membrane.* Up to 25% of the cells in this layer are **melanocytes,** depending on the amount of color the skin possesses (the darker the skin, the more melanocytes). Other cells located at this level are the **Merkel cells,** which are associated with sensory nerve endings extending from the dermis. The Merkel cell and nerve ending combine to form a **Merkel disc,** or a touch receptor.

The **stratum spinosum** is a relatively thick layer of cells just superficial to the stratum basale. These cells appear stippled, an appearance that is a result of processing for histological examination: The cells shrink, but their **desmosomes** hold tight. Contained within these cells are thick, web-like bundles of intracellular tonofilaments of prekeratin protein. These cells are also undergoing mitosis but at a slower rate than the stratum basale. A cell called a **Langerhans' cell** is found within this layer, and it functions as a macrophage. Only these two layers of the epidermis receive adequate nourishment via diffusion from the dermis. As the cells multiply, the daughter cells are pushed upward and away from their source of nutrition, and they gradually die.

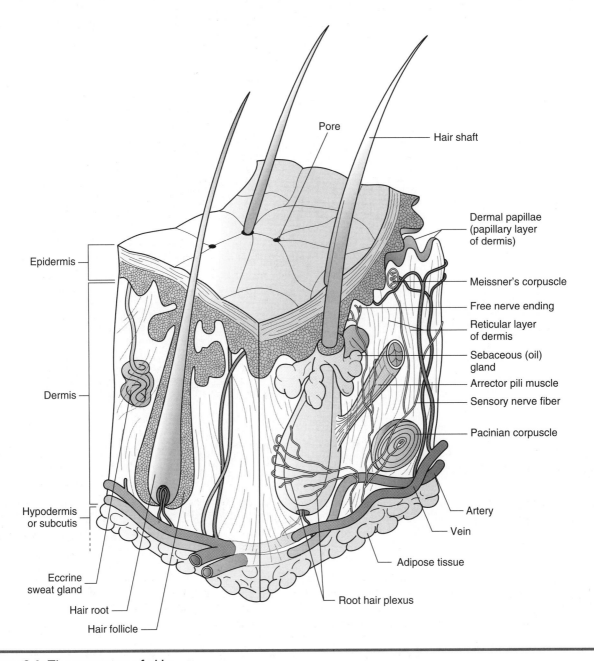

Figure 6.1: The structure of skin.

The next layer of cells, just above the stratum spinosum, is the **stratum granulosum,** named for the abundant number of granules its keratinocytes contain. These cells are more flattened in appearance than the previous layers. The granules within them are of two types: (1) *lamellated granules,* which contain a waterproofing glycolipid that is excreted into the extracellular space; and (2) *keratohyaline granules,* which combine with the tonofilaments in the more superficial layers to form keratin fibrils within the cells. At the upper border of this level, the cells begin to die.

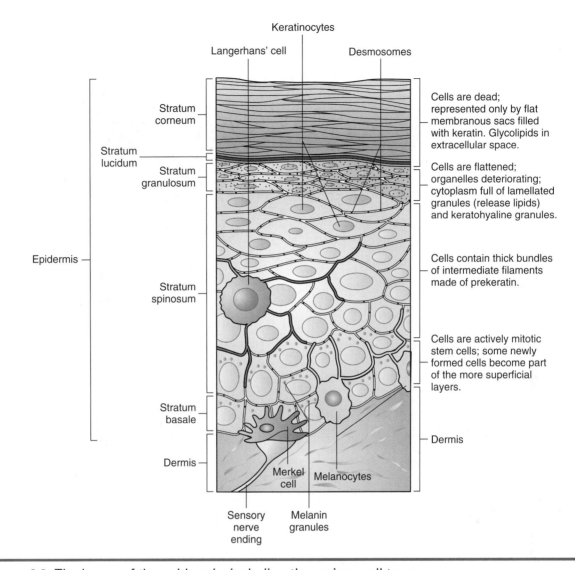

Keratinocytes

Langerhans' cell

Desmosomes

Stratum corneum

Stratum lucidum

Stratum granulosum

Epidermis

Stratum spinosum

Stratum basale

Dermis

Dermis

Merkel cell

Melanocytes

Sensory nerve ending

Melanin granules

Cells are dead; represented only by flat membranous sacs filled with keratin. Glycolipids in extracellular space.

Cells are flattened; organelles deteriorating; cytoplasm full of lamellated granules (release lipids) and keratohyaline granules.

Cells contain thick bundles of intermediate filaments made of prekeratin.

Cells are actively mitotic stem cells; some newly formed cells become part of the more superficial layers.

Figure 6.2: The layers of the epidermis, including the various cell types.

The thinnest layer is the next layer, called the **stratum lucidum.** It is a clear or translucent band of dead keratinocytes with indistinct boundaries and keratin fibrils located within. It is present in areas of thick skin, but not thin skin.

The outermost layer of the epidermis is called the **stratum corneum,** which consists of 20 to 30 cell layers and accounts for much of the epidermal thickness. These cells are dead and their flattened, scale-like remnants are fully keratinized. These cells are constantly shed and replaced by division of the deeper cells (Figure 6.3).

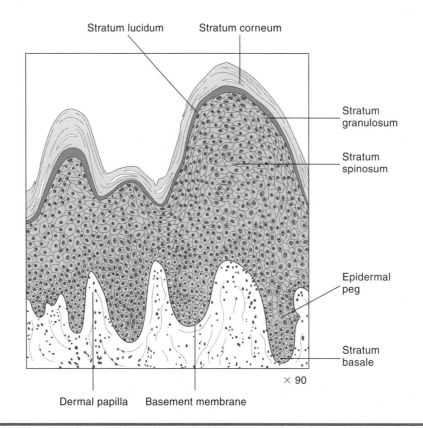

Figure 6.3: The layers of the of epidermis in the nose of a cat.

In the space below, using the slide of the epidermis, draw and label the layers and cells observed.

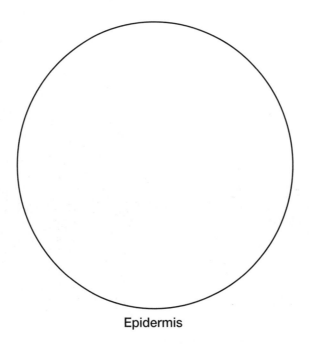

Epidermis

Dermis

Slides recommended: Those used in the previous activity, labeled bovine skin or human skin, scalp.

Description: Below the epidermis is the **dermis.** It is composed of two layers: (1) the **papillary layer,** and (2) the **reticular layer.** The *papillary layer* is made of areolar (also known as loose irregular) connective tissue and the *reticular layer* of dense irregular connective tissue. The reticular layer accounts for 80% of the total thickness of the dermis. Like the epidermis, the dermis varies in thickness depending on its location within the body. For example, a dog's foot pads are much thicker than an eyelid.

Slide: skin

The *papillary layer* is the most superficial of the two dermal layers. Its name is derived from the word *papilla*, which means small, nipple-shaped projection. This layer contains multiple papillary projections, called **dermal papillae,** which rise into the epidermis and form a wave-like pattern between the epidermis and dermis (Figure 6.3). The shape of this interface helps ensure the connection between these two structures. Within the papilla are looping blood vessels that provide the nourishment to the cells of the epidermis, as mentioned previously. These vessels also help remove waste products and assist with temperature control of the body. Also, within this layer are nerve endings for pain and touch, called **Meissner's corpuscles** (Figure 6.1), and receptors sensitive to temperature changes.

The boundary between the two layers of the dermis is indistinct because the *collagen fibers* from the *papillary layer* blend with the bundles of collagen fibers of the *reticular layer.* (You may want to review the material on deep irregular connective tissue in the previous chapter on page 56.) Note that in the reticular layer these layers of heavy, wavy bundles (or cords) of collagen fibers tend to run parallel to each other (see Figure 6.4). The orientation of these fibers is in the direction of the stress placed on them. It also consists of many small arteries and veins, *sweat glands* (in species containing them), and *sebaceous glands.* These vessels allow the skin to play a major role in temperature regulation of the body. When the body's temperature is high, the arterioles dilate, and the capillary network becomes engorged with the heated blood to allow the heat to radiate from the skin's surface. If the environment is cool and the body's heat must be conserved, the arterioles constrict so that blood bypasses the dermal capillary networks.

Both layers of the dermis, papillary and reticular, are heavily invested with *collagenic* and *elastic fibers,* which give the skin its exceptional elasticity in youth. As the body ages, the number of elastic fibers decreases and the subcutaneous layer loses fat; this leads to wrinkling and inelasticity of the skin and can be observed most easily in primates.

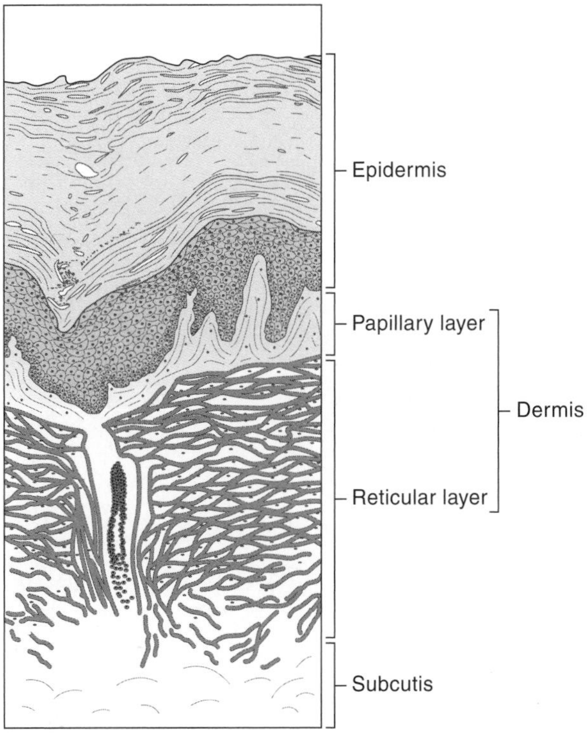

Figure 6.4: Dermis in the digital pad of a cat.

In the space below, using the slide of the dermis, draw and label the layers, cells, and other structures observed.

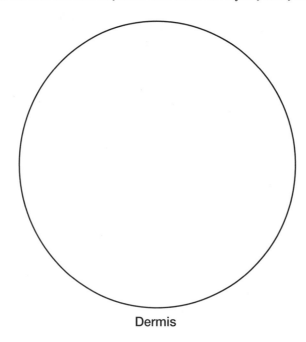

Dermis

Hypodermis, or Subcutis

Slides recommended: Those used previously, labeled bovine skin or human skin, scalp.

Description: The **hypodermis,** or **subcutis,** is the thick layer of *subcutaneous areolar connective tissue* that is rich in adipose tissue and lies deep to the dermis (Figure 6.5). You may want to review the section of the previous chapter on page 60 on *adipose tissue.* Blood and lymphatic vessels and nerves are also found in this region. At the dermis-subcutis border is a special type of touch receptor, called a **Pacinian corpuscle,** which is sensitive to heavy pressure. The boundary between the dermis and subcutis is blurred because the fibers of both merge with one another. This layer is important because it permits the skin to move freely over the underlying muscle and bone without putting tension on the skin that might result in tearing. The hypodermis also acts to help insulate the body from environmental temperature variations.

Slide: skin
Look beneath the dermis to view the subcutaneous fat and the structures within.

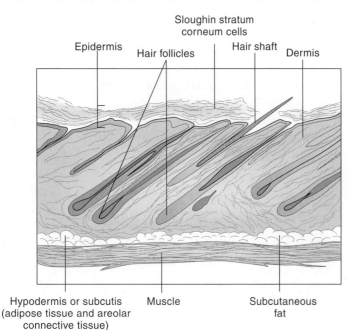

Figure 6.5: Hypodermis from the skin of the back of a cat.

In the space below, using the slide of the hypodermis, draw and label the layers, cells, and other structures observed.

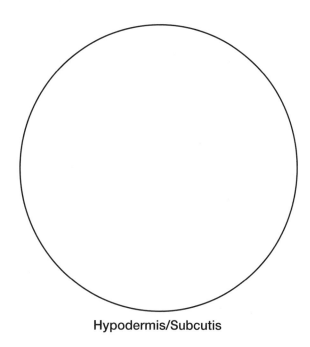

Hypodermis/Subcutis

Accessory Structures of the Integument: Hair

Slides recommended: Those labeled hairy mammal skin.

Description: Most skin covered with hair or fur consists of three epidermal layers: the *stratum basale, stratum spinosum,* and *stratum corneum.* However, there are regions in furry mammals where all five layers exist. These are places where the keratinization process has slowed and the skin is very thick. The surface of the skin of a hairy mammal is covered in scale-like folds, and the hair shaft emerges from beneath the scales and is directed away from the opening. Dogs usually have a cluster of three hair follicles per scale. A **tylotrich hair** (also known as a **tactile hair**) is a special type of hair that emerges from a knob-like elevation called a *tactile elevation* (or epidermal papilla). These hairs are important to animals because they assist with their perception of touch and space.

Slide: hairy mammal skin

Observe the shaft of the hair first; it is made of hard *keratin* and has a central **medulla,** surrounded by a **cortex,** which is enveloped by the **cuticle** layer (Figures 6.6 and 6.7). The amount of **melanin** in the cortex determines whether the hair is brown, red, yellow, or black. As pigment is lost from the cortex, which occurs in old age, the hair becomes progressively more gray. If it loses pigment entirely, and the medulla fills with air, the hair becomes white.

Next, observe the hair follicles and try to find one cut directly through the middle longitudinally. The portion of the hair enclosed within the follicle is the **root.** The **hair bulb** is a collection of well-nourished, germinal epithelial cells at the basal end of the follicle. As the daughter cells are pushed farther away from this growing region, they die and become keratinized. Thus, the bulk of the hair (i.e., the shaft) is composed of hard keratin and is no longer living material. The follicle contains both epidermal and dermal cells.

Look closely at the hair bulb and find the hair's **connective tissue papilla,** which is a continuation of the **connective tissue sheath** that surrounds and anchors the hair in the dermis. Proceeding from the exterior inward, the next layer is a thin, clear layer called the **glassy membrane.** The first layer of cells in the follicle proper is

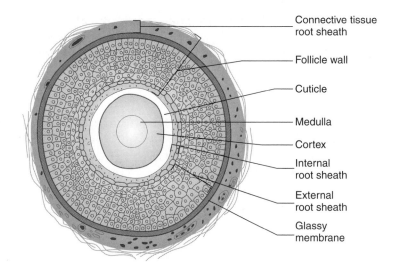

Figure 6.6: Cross section of a hair shaft.

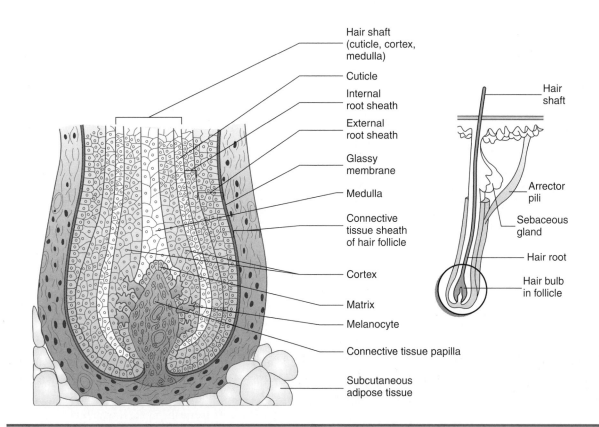

Figure 6.7: Longitudinal section of a hair follicle.

the **external root sheath.** Beneath this is the **internal root sheath.** Everything inside of this becomes part of the hair shaft as it grows out, but these cell layers are not distinct in the follicular area. A thin area just inside the internal root sheath is the **cuticle,** the area directly in the center is the **medulla,** and the cells in between, which are the most abundant part of the hair root and shaft, compose the **cortex.** The cells of the bulb, located around the connective tissue papilla, are called the **matrix.** This is also where *melanocytes* are found. These add the pigment *melanin* to the cortex.

The part of the root where cells begin to lose their cellular appearance is called the *keratogenous zone.* Above this area, just peripheral to the internal root sheath, you should find the pale-stained, glandular epithelial cells.

These are **sebaceous gland cells** (modified squamous epithelial cells) into which the whole cell is shed. It ruptures and emits a white, semi-liquid mixture called **sebum,** the oil that lubricates the hair shaft. In the sheep, this product ultimately becomes **lanolin.** This type of gland is known as a **holocrine gland** (an exocrine-type gland) and is named for its secretory process. *Holocrine* refers to a gland in which the whole cell is liberated during secretion.

Other types of exocrine glands named for their secretory process are *merocrine* and *apocrine glands.* In **merocrine glands** the secretory product passes through the cell walls without any loss of cytoplasm; examples of this are sweat glands. **Apocrine glands** lose a slight amount of cytoplasm and/or cell membrane during secretion, although the whole cell is not lost in the process; examples of this include the prostate gland.

In the space below, using the hairy mammal slide, draw the follicle and hair shaft and label the layers, cells, and other structures observed.

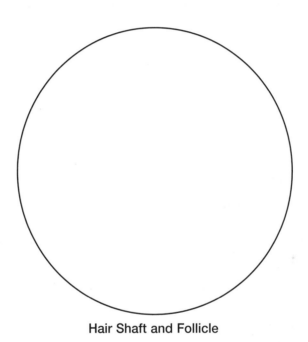

Hair Shaft and Follicle

EXERCISE 6.2
THE INTEGUMENTARY SYSTEM MODEL

Locate all the items found on the slides in the previous exercise on the plastic model of the integument. The model also should have an example of the **arrector pili muscle.** This is composed of small bands of smooth muscle cells that connect each hair follicle to the papillary layer of the dermis. When these muscle cells contract, which is usually during cold, fright, or bluffing behavior (e.g., the animal is making itself appear larger), the slanted hair follicle is pulled upright and dimples the skin surface. In areas of small, fine hairs this dimpling appears as *goose bumps,* but in heavily furred areas it will not be seen. Contraction of the arrector pili muscles also exerts pressure on the sebaceous glands surrounding the follicles, causing a small amount of sebum to be released.

Note some of the other glands present on the model. There are the **sweat glands,** also known as **sudoriferous glands.** These are found over the entire body of most domestic species, including the dog, pig, horse, cow, and sheep. However, only the horse produces a profuse amount of sweat. There are two types of sweat glands, *apocrine* and **eccrine.** The *eccrine* sweat gland consists of a simple, coiled tube located in the dermis or subcutis and connected to the surface via a long duct. In dogs, this type of sweat gland is found only in the deep layers of fat and connective tissue of the foot pads. *Apocrine* sweat glands have a coiled, excretory portion found deep in the dermis or subcutis, and they empty into the hair follicles rather than to the skin's surface.

EXERCISE 6.3
PAW PADS, CLAWS, AND NOSE OF THE DOG AND CAT

Pigmented Skin

There are numerous regions in the dog, including the skin, mucous membranes, and claws, that are darkly pigmented. Pigmentation is determined by the relative amount of two pigments, **melanin** and **carotene,** in the skin and accessory structures. Another factor that may affect the color of the skin is the degree of oxygenation of the blood, which can give a reddish tint to areas of white skin. This may be especially noticeable in areas of inflammation.

Melanin is produced by *melanocytes* and becomes incorporated into the cells of the *stratum basale* and the deeper layers of the *stratum spinosum.* Because these cell layers also form the germinal layers of claws (and hooves in large animals), the amount of melanin deposited determines the color of these structures also. The release of the melanin granules is controlled by the release of melanocyte-stimulating hormone, which is controlled by the pituitary gland. *Carotene* is a yellow-orange pigment present primarily in the *stratum corneum* and in the *adipose tissue* of the *hypodermis.*

Obtain a live dog and observe the tongue and other areas inside the mouth for black pigmentation. Which breeds are likely to have this pigmentation? Then, look at the top of the nose and the claws to see if they are pigmented.

The Nose of a Dog

Using a magnifying glass, observe the surface of the nose and pads of the dog. The top of the nose in dogs, cats, pigs, and sheep is called the **planum nasale.** In the cow and horse, this area is called the **muzzle** and technically is named the **planum nasolabiale.** Note that it is composed of many **polygonal plates** packed together, and that it may be further divided by grooves into **polygonal plaques** (Figure 6.8). The nose is usually pigmented black and dark brown and is composed of three layers of epidermis: the *stratum basale, stratum spinosum,*

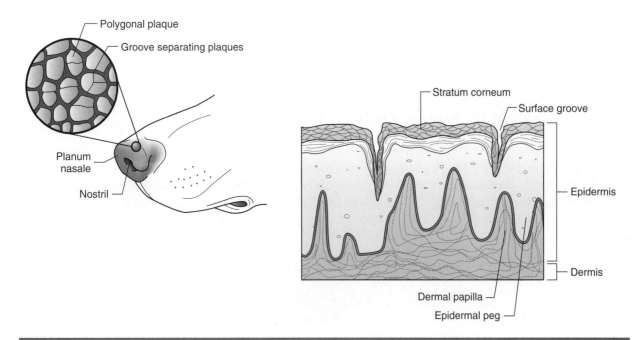

Figure 6.8: The nose of a dog and cross section of the epidermis and dermis.

and *stratum corneum.* Surprisingly, considering the heavy use of the nose, the stratum corneum is composed of only four to eight layers of cells. The nasal epidermis and dermis *interdigitate* in a manner similar to other areas of the skin; they form prominent dermal papillae deep into the epidermal cells. In the dog, the *planum nasale* contains no glands. However, in the sheep, pig, and cow, tubular glands are found.

Foot Pads of a Dog or Cat

Now observe the pads on a dog's or cat's paws. The pads adjacent to the claws are called the **digital pads;** the large central pad is called the **metacarpal pad** (or **metatarsal pad** in the rear feet), and the small pad on the caudal surface of the leg (located a short distance proximal from the metacarpal or metatarsal pad) is called the **carpal pad** or **tarsal pad,** depending on its location, front or hind feet (Figure 6.9). The digital, metacarpal, and metatarsal pads are the weight-bearing pads in the dog and cat. Using a magnifying glass, observe the many minute conical papillae covering the pads. Sometimes the central areas of the pads are worn smooth from walking on hard, rough surfaces. When this happens, the central papillae are rounded or flattened. All five layers of the epidermis are present, and the stratum corneum is the thickest. There are multiple dermal papillae at the epidermal–dermal interface also. Deep to the dermis is a thick layer of *subcutaneous adipose tissue.* This insulating fat, plus the thick, tough keratinized stratum corneum, forms an effective protective barrier against abrasion and thermal changes. This enables the animal to walk on rough, hot, or cold surfaces. The interior of the pads is composed of exocrine sweat glands and lamellar corpuscles. The ducts pass through the dermis to the stratum basale. The secretions of these glands are then expelled onto the surface of the pad.

The Claw of a Dog and Cat

Now look at the claws and compare a non-pigmented claw of a dog to a pigmented one. Then, look at the claw of a cat and compare it to the dog and to the diagram in Figure 6.10. In both animals, the claw is attached to the **toe** at the **coronary band** (or bed). This composes the germinal, epidermal cell matrix that gives rise to the *horny wall* (or **unguis**). Beneath the wall is the sensitive **corium,** which is modified dermis, otherwise known as the *quick* of the claw. This is the area that bleeds if you accidentally cut the claw off too short during a nail trim. The corium surrounds a bony process attached to the **distal phalanx** called the **unguicular process.**

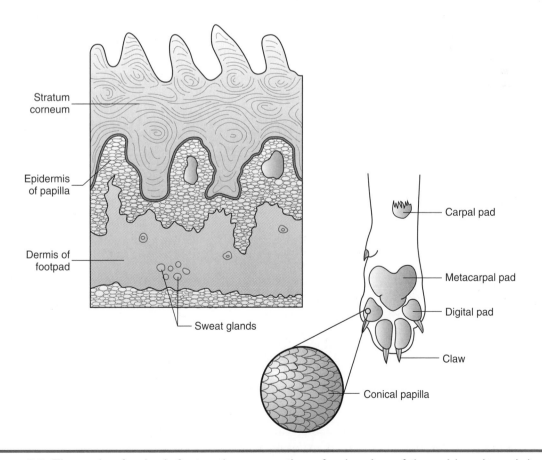

Figure 6.9: The pads of a dog's feet and cross section of a drawing of the epidermis and dermis.

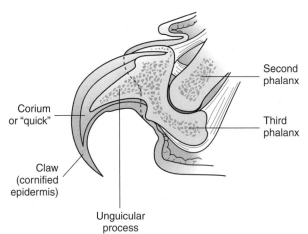

Second phalanx

Corium or "quick"

Third phalanx

Claw (cornified epidermis)

Unguicular process

Figure 6.10: The claw of a cat.

ASSOCIATED STRUCTURES OF THE INTEGUMENT IN LARGE ANIMALS

Chestnuts and Ergots

The chestnut and ergot are dark, horny structures found on the legs of members of the equine family. **Chestnuts** are brown to tan or grayish, depending on the hair color of the animal. They are found on the medial side of the legs at the level of the carpus or knee (note that this is not the true knee, which is the stifle joint of the hind leg) and the tarsus or hock (Figure 6.11). The **ergot** is the horny tissue found buried in the long hairs below the fetlock joint.

Chestnuts are thought to be vestiges of the carpal (front leg) and tarsal (hind leg) pads of the first digit; and similarly, the ergots are vestiges of these pads of the second and fourth digit. Remnants of the fifth digit do not exist. As the horse progressed through its evolutionary development, the multi-toed species progressively lost their digits as their need for speed increased. Now the equine species supports all of its weight on the third digit. Using the live horse, find the chestnuts and ergots.

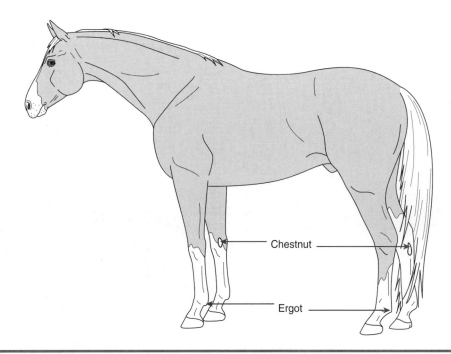

Chestnut

Ergot

Figure 6.11: Location of the chestnuts and ergots of a horse.

Horns and Antlers

The **horns** of cattle, sheep, goats, and similar species are formed over the **horn process,** or **os cornua,** which is an outgrowth of bony tissue from the frontal bone. Horns are covered by a thin layer of modified dermis called the **corium** of the horn. The corium at the base of the horn where it joins the skin is thickest; this area is the germinal area. The horn itself is made of highly keratinized *stratum corneum* of the epidermis and is constructed of many tubules of horn tissue that extend from the base of the horn to the tip. The medullary cavity of the os cornua is continuous with the frontal sinus within the frontal bone. A soft type of horn, called the **epikeras,** covers the surface of the horn at the base and extends for a variable distance toward the horn's apex.

Seasonal variations in the animal's nutritional state are reflected as variations of the horn's growth, which appear like the rings of a tree. The age of the animal may be estimated by counting the rings in the horn. Horns occur in both males and females and are not sex-linked characteristics. Observe the cross section and longitudinal section of the display of horn, and compare it to the diagram in Figure 6.12.

Antlers are bone developed by a process called *endochondral ossification* (bone growth on a cartilaginous matrix). Nutrition for this developmental process comes via the antler's *velvet.* The velvet is modified skin and contains both epidermis and dermis. The antler's progression to full size coincides with the aging and death of the velvet tissue, which the animal rubs off to expose the bony antler. Antlers undergo cyclic growth, maturation, and shedding annually and are associated with the breeding habits of the species. Antlers in the deer and elk are sex-linked; only the males of the species have them. Observe the cross section and longitudinal section of the display of antler.

The Foot of the Horse

The *foot* of the horse is defined as the hoof and all the structures contained within it. The horse walks on its third digit, which is equivalent to the middle finger in a human. The main bones inside of the hoof are the **distal phalanx (coffin bone),** the distal tip of the **middle phalanx (short pastern bone),** and the **distal sesamoid bone (navicular bone).**

Observe the plastic model and the prepared specimen of the horse's foot and identify the parts listed as follows. Use the diagrams to aid in this identification. After doing this, find these structures on the foot of a live horse.

1. **hoof wall:** The *wall* is the portion of the hoof visible when the horse is standing up. On the outside it is divided arbitrarily into a **toe** region in front, medial and lateral **quarters** on the sides, and the medial

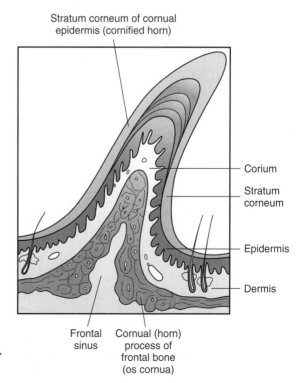

Figure 6.12: A horn.

and lateral **heels.** Internally, it is divided into three layers: the **stratum externum** (outer layer—the *periople* and a thin layer of horny scales that gives the outer surface of the wall below the periople its smooth, glossy appearance), the **stratum medium** (middle layer—forms the bulk of the wall), and **stratum internum** (inner layer—consists of the horny lamina). The wall is constructed of parallel tubular horn cemented together by intertubular horn, all of which is highly keratinized material, making it hard and durable.

Why would the hoof be constructed on thousands of small tubules of horny material? Take a piece of paper and try to get it to stand without support. Now take that same piece of paper and roll it into a coiled cylinder; it should stand on its own. The tubular structure is stronger. The *stratum internum* connects to the corium deep by interdigitating with the **primary** and **secondary lamina** of the **laminar corium.** This forms a solid connection between these two layers (Figure 6.13).

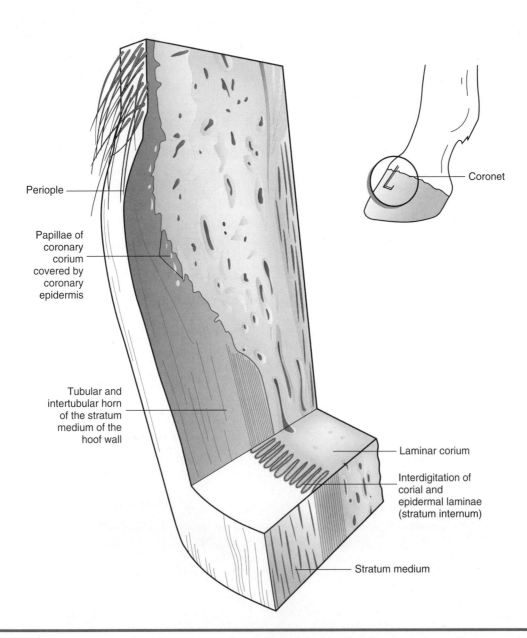

Figure 6.13: Longitudinal and cross section of the hoof wall, showing the primary and secondary lamina.

Note the rippled laminar surface of the *stratum internum* on the model and specimen of the inside of the hoof wall. This layer is also called the **insensitive lamina,** or **horny lamina.** The **coronary band** *(coronet)* produces the wall in an area just inside where the hair meets the wall. This is actually the **coronary corium,** which contains multiple papillae with the germinal cells that produce the tubular wall. This is why the hoof wall grows downward and needs to be trimmed if it does not wear off. The **coronary groove** is the convex surface on the top edge of the wall that connects to the coronary corium. See Figures 6.13 through 6.16 for examples of these structures.

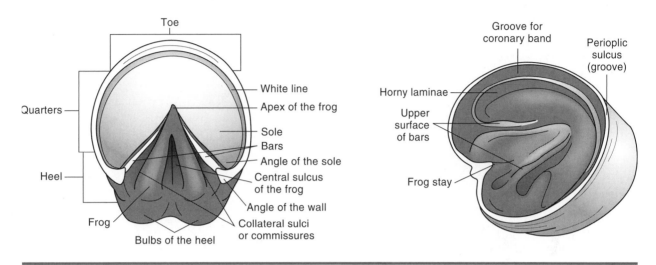

Figure 6.14: The bottom and inside of the hoof.

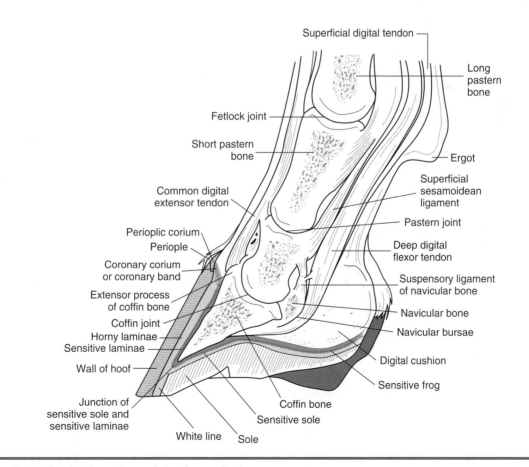

Figure 6.15: Sagittal section of the foot of a horse.

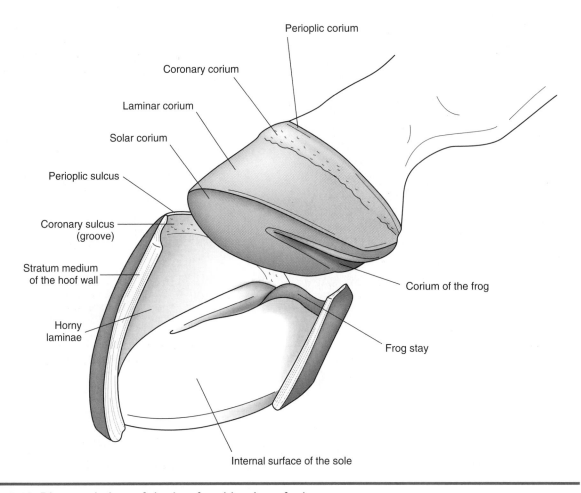

Figure 6.16: Dissected view of the hoof and lamina of a horse.

2. **periople:** The *periople* is a thin layer of tubular horn that covers the top of the wall for a short distance distally from where the hair meets the wall. It turns a milky white color when the hoof is soaked in water. The periopole is produced by the **periopic band,** a small area of **periopic corium,** which is just above and concentric with the *coronary band.*

Note, structures 3 through 9 can be viewed from the bottom of the hoof.

3. **angle of the wall:** This area is where the walls turn inward at the heel, forming the bars.

4. **bars:** Bars are the extensions of the wall toward the apex of the frog.

5. **sole:** This is the horny bottom of the hoof between the wall, the bars and the frog.

6. **frog:** The frog is a cornified tissue that forms a point in the middle of the bottom of the foot. It is split into two halves at the heel by the **central sulcus,** and is more elastic than the sole. The frog, with its underlying **digital cushion,** aids in the absorption of concussion as the foot impacts the surface of the ground. As the frog strikes the ground, both the digital cushion and the frog are compressed between the distal phalanx and the ground. As they become thinner and wider, pressure is placed on the *bars,* the **collateral cartilages,** and the *wall.* The blood within the corium has an hydraulically shock-absorbing effect on the hoof wall.

The compression on the frog, digital cushion, and blood within the hoof cause the blood to be forced out of the vascular corium. This effectively pumps it into the veins of the legs and upward against gravity. Because the leg veins of horses do not have valves, movement of the horse's legs and resulting compression within the foot keeps the blood flowing up the legs and prevents *edema,* or the fluid portion of blood filling the extracellular spaces.

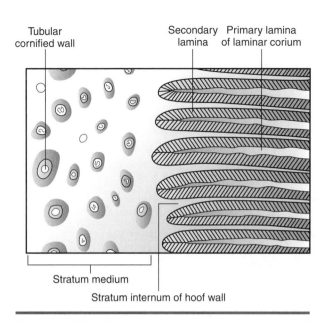

Figure 6.17: Microscopic appearance of hoof wall showing the primary and secondary lamina of the horse's foot.

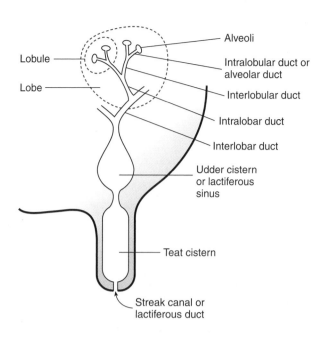

Figure 6.18: Anatomy of the bovine mammary gland.

7. **commissures:** There are two commissures, one on each side of the frog, which separate the frog from the bars. They are also know as the **collateral sulci.**

8. **bulbs of the heel:** These are paired, rounded projections—one medial, one lateral—and are located proximocaudally from the caudal aspect of the hoof proper.

9. **white line:** This is the continuation of the stratum internum to the bottom of the hoof, where the sole and the hoof wall meet.

10. **corium:** The corium, like other coria previously mentioned, is modified dermis. Like the dermis of the skin, which contains the blood vessels and nerves and provides nutrition to the epidermis, the corium also is well vascularized and innervated and supplies nutrition to the wall. The corium is not considered part of the hoof proper, but it is part of the foot. The corium can be subdivided into different regions as follows.

The **laminar corium** interweaves with the stratum internum and lies between the wall and the third phalanx. It consists of primary and secondary lamina (Figure 6.17) and supplies nutrition to the stratum internum.

The **perioplic corium** lies in the perioplic groove (located above the coronary groove), contains the germinal cells, and supplies nutrition to the periople.

The **coronary corium** lies in the coronary groove, contains the germinal cells for the wall, and supplies nutrition to the stratum externum and medium.

The **sole corium** lies superior to the sole and supplies it with nutrition.

The **frog corium** lies superior to the frog and supplies it with nutrition.

Mammary Gland of a Cow

Compare the structures listed below to the model of a cow's **mammary gland.** If a model is not available, use only the diagram in Figure 6.18.

The mammary gland of a cow represents the ideal structure to use as an example of a compound gland because many other glands use the same terminology to name their ducts, lobes, and lobules. A mammary gland is a compound, alveolar, exocrine gland. It is multicellular, and its secretions are merocrine and serous in nature. It is actually classified as a modified sweat gland that produces milk. See Figure 6.18 for the anatomy of the cow's mammary gland.

Comparative Anatomy of Mammary Glands

1. Cow: As pictured in Figure 6.18, there are technically four mammary glands in the cow. However, it is conventional to describe the gland as having four **quarters,** each with one **teat.**

2. Mare: The female horse has a smaller **lactiferous sinus** at the base of each teat, which opens to the exterior via two or three **lactiferous ducts.** There are two mammary glands, one on each side of the median line in the prepubic region.

3. Ewe and doe: The mammary glands of the female sheep and goat are similar to the cow's in structure, but they have only two mammary glands.

4. Sow: A female pig's mammary glands contain a small lactiferous sinus at the base of each teat, and each teat has two lactiferous ducts. There are usually 10 to 12 glands or 5 to 6 pairs.

5. Bitch and queen: The mammary glands of the dog and cat contain a small lactiferous sinus at the base of each teat. The teats are short and contain 6 to 12 lactiferous ducts. There are usually 10 mammary glands or 5 pairs.

Milk Production

The hormones *estrogen* and *progesterone* cause the mammary glands to enlarge during pregnancy. Shortly before birth these hormone levels decrease, and *prolactin* is released, stimulating *lactogenesis.* Maintenance of milk production is dependent of the secretion of prolactin, growth hormone, adrenocorticotropic hormone, and thyroid-stimulating hormone. Prolactin release is stimulated by the suckling of the offspring.

Milk Let-Down

Suckling also causes milk let-down; it stimulates nerves that send a message to the posterior pituitary gland to release *oxytocin* into the blood stream. This in turn causes a contraction of the myoepithelial cells surrounding the alveoli and smaller ducts to cause milk ejection (let-down).

CLINICAL SIGNIFICANCE

The skin, because it is the largest organ of the animal's body and because of its location (i.e., covering the exterior), is subjected to many forms of injurious agents. Dermatology *is the study of the skin and its associated diseases. The skin must protect the animal from becoming overhydrated during immersion in water, from becoming dehydrated during exposure to heat or sun, and from a variety of bacterial, viral, fungal, and parasitic agents.*

When an animal becomes infested with the mite Demodex canis *this condition is known as demodectic mange. These mites have the ability to burrow into the skin along the hair shafts and reside around the follicle. This parasite is present in the hair follicles of almost all dogs and in some cats. Their presence indicates a form of immunity called* **premunition,** *which is seen in some parasitic conditions and depends on the continued presence of the parasite in or on the host. In other words, having a few mites on the dog stimulates the immune system and prevents an overwhelming infestation. We occasionally see dogs that have generalized demodectic mange all over the body, and often there is an associated pyoderma with it. Many of these dogs have some form of immune deficiency. Treatment is aimed at getting rid of the mites and the secondary bacterial complications. Often these are young dogs, and it is hoped that as the animal matures, the immune system will also mature and prevent future infestations and secondary infections.*

Veterinary Vignettes

The profession of veterinary medicine is a job, and like all jobs, there are things about the job we like and things we do not. Personally, I enjoy surgery. The saying "a chance to cut, a chance to cure" really does define the psyche of the surgeon, and I think I have this affliction. If I had the opportunity to do it all over again, I would become a board-certified veterinary surgeon—either that or a dermatologist. Dermatologists never lose their patients; they never die and they never become completely cured.

The day Mel Parker burst into my hospital carrying a profusely bleeding Irish Setter was the day I had the opportunity to be both. And after looking at this dog, I wanted to be neither.

At first I thought it was Mel's own Setter, which struck me with fear.

"Take him to the treatment room," I directed, bypassing the exam room.

"What happened, Mel? Where did you get this dog?" I asked as I immediately started to assess the problems presented and stem the flow of blood.

"I don't know where he came from. All of a sudden he was just there, right in front of my truck. I was on top of him before I could hit the brakes. You know what my truck looks like, raised up. I went right over him and stopped I don't know how far down the road—40 feet or so, I guess. I looked back in my rear view mirror, thinking I would see the dog lying there, but it was gone. I jumped out to look and heard this crying underneath my truck. Oh God! It was horrible. The dog's collar had gotten caught on something, and I had drug it the entire distance. I got bit twice trying to get the dog loose." Mel showed me the wounds on his hand.

The dog was a male, and he was not being aggressive right now. Nevertheless, we put a muzzle on him to be safe. It is hard to treat an injured animal if the veterinarian or technician is bleeding as badly as the dog. It was the dog's collar getting caught that had saved his head from injury, but his right front leg and hip suffered the rest of the damage. The bone was exposed from his elbow down to his carpus; the pavement had ground it right down to the marrow cavity. The outer muscles were in shreds and bleeding profusely.

Mel was a contractor and an old friend; in fact, we had worked together for awhile. When my hospital was being built and I was between jobs as a relief veterinarian, I pounded nails, carried lumber, and helped him build a house or two. It was just like being back on the farm. I knew Mel, and I knew he would do anything he could to help this dog recover.

After administering some hefty doses of pain medication, which had the beneficial effect of also tranquilizing the dog, we were able to stop the bleeding and clean up his wounds. His rear legs had many abrasions and a few cuts that needed suturing, but nothing of the magnitude of the damage to the front leg. We administered fluids and other anti-shock treatment to stabilize him, and within a few hours he was doing quite well. The problem, of course, was that there was not enough skin left to cover the open wound, and it was so large that sliding skin grafts would be of no help.

Fortunately, the dog's owner was located within a couple of hours. After some discussion between Mel and the owner, I got permission to do surgery. The owner was not exactly rich; and in fact, you would be stretching it to call him financially stable. So I think Mel, out of the goodness of his heart, paid for part of the bill, even though technically he did not have to. The dog was running loose in a city with a leash law.

I decided a full-thickness skin graft was our best hope. I took a strip of skin from the dog's back, removed all the subcutaneous fat down to the bottom layer of the dermis, and sewed it over the open wound.

For skin grafts to survive, like all tissues, they must receive blood to nourish the cells and supply oxygen for metabolism. This is the critical factor that determines whether the graft will take. A *sliding skin graft,* in which a section of skin is slid from one area to another, retains its vascular supply. Sometimes surgeons will use a *tube graft,* which involves making a tubular stalk of skin containing the blood vessels, spreading the end out, and sewing it in place. When that section of skin develops a new blood supply in its new location, the tubular connection is severed and removed. This type of graft can used in areas of the body that are not connected but are close to each other, such as from the animal's side to an area around the knee.

If these two grafts are not possible, either a full-thickness or split-thickness graft can be used. For a split-thickness graft, a dermatome knife is used to cut a strip of skin down to the level of the dermis to prepare it for transplant. Because I did not have a dermatome and there was so much tissue missing, a full-thickness graft was my only choice in this situation. Removing the subcutis so the underlying vessels could attach directly to the dermis was the critical step in the surgical procedure.

I had marked the skin with a surgical marking pen so I knew which direction the hair grew. This way I could orient the skin strip so the hair grew down the leg, not up. My technician and I put a thick, cotton bandage, called a Robert-Jones bandage, on the leg temporarily to stabilize it and apply compression without cutting off circulation. After a few days, we switched to another type of splint to support both the elbow and carpal joints. Fortunately, the damage was to the side of the leg and not to the cranial surface, which might have damaged the radial nerve.

As it turned out, about 60% of the graft took, and there were no large areas that did not survive, just smaller patchy areas. Those formed scab-like crusts, and new skin from the periphery filled them in. The dog walked with a limp for about a year, and then one day the limp mysteriously went away.

All in all, we were quite lucky: Mel for not getting hurt any worse than he did, the owner because Mel helped him pay for his dog's surgery, the dog because he could have just as easily been killed, and finally me because it was the first and only time I have ever done this surgery. I'm one for one!

SUMMARY

The integumentary system is the animal's first line of defense when it encounters the various parasites, bacteria, viruses, fungi, toxins, and other potentially injurious agents that come in contact with the skin. This system also includes the structures that support weight, enable traction during limb progression, and enable functions such as digging and personal defense.

Because so many disease conditions occur to the structures of this system, an understanding of both the structure and function of all the components is necessary. In this chapter we covered the structure and function of skin and its accessory organs. This included the horse's foot; the dog's and cat's pad, claws, and the external surface of the nose; horns; and antlers.

CHAPTER 7

THE SKELETAL SYSTEM

OBJECTIVES:

- know the five functions of bone
- name and histologically identify the parts of mature compact bone from a model, diagram, or prepared slide
- identify the all the bones of the cat's skeleton, including the skull bones
- identify the major processes, depressions, and foramina in the cat's skeleton as specified in this chapter
- identify the comparative nature of the bones of the skulls of the dog, cat, horse, ox, and pig
- identify the comparative nature of the bones of the foreleg and the hind leg of the dog, cat, horse, ox, sheep, and pig
- identify the bones of the fowl and recognize their analogous parts in the mammal and the major differences
- identify the names (anatomical and common) of the specified joints in this chapter
- locate specified ligaments, tendons, and cartilage
- understand that ionized calcium is the critical form for metabolism and how it is associated with serum protein levels

MATERIALS:

- three-dimensional model of mature, compact bone (if available)
- prepared slides of compact and cancellous bone
- compound light microscope
- immersion oil
- colored pencils
- three-dimensional model of the equine foot (if available)
- three-dimensional model of the knee joint (human or animal)
- whole and bisected skulls of a cat
- whole skulls of a dog, horse, and cow (if available)
- articulated and disarticulated skeletons of a cat
- articulated skeleton of a dog
- articulated skeleton of a chicken

- articulated skeleton of an equine foreleg, from the carpus distally, and equine hindleg, from the hock distally

- articulated foreleg and hindleg of a sheep (The legs of sheep are suggested because they are small, store easily, are relatively inexpensive, and yet are good representations of large animal legs.)

- blood chemistry machine

- assay cards for calcium, albumin, and total serum protein

Introduction

There are two forms of bone formation: **intramembranous ossification** and **endochondral ossification.** During intramembranous ossification the **osteoblasts** originate from **embryonal mesenchymal cells,** whereas in endochondral ossification, an embryonal bone is constructed initially of cartilage, which eventually is replaced by true bone. The cartilaginous bone serves as a temporary support structure and is replaced gradually so no part is left unsupported at any time. In both types of ossification, the first type of bone formed is **woven bone;** it resembles spongy bone in its appearance. This bone then goes through a process of **erosion** and **remodeling** to become either mature **cancellous (spongy) bone** or **compact bone.**

Osteoblasts are cells actively producing bone. They are destined to become **osteocytes** and are so named when found in a **lacuna** in mature bone. **Osteoclasts** are cells that tear down bone so it can be rebuilt. The combined actions of the osteoclasts and osteoblasts have three important effects: (1) Bone is modeled, then remodeled in fetal life; (2) old bone is removed and replaced with new bone in adult life; and (3) bones heal by laying down a large callus of spongy bone, which is remodeled into compact bone as needed to bear weight. In addition to helping remodel bone, osteoclasts are active in the formation of **bone marrow cavities and spaces,** which are formed to provide sites of **hematopoiesis** (blood cell formation).

The skeleton has two divisions: the **axial skeleton** and the **appendicular skeleton.** The axial skeleton is composed of the bones that lie around the body's center of gravity; this includes the bones of the skull, vertebrae, hyoid apparatus, ribs, and sternum. The appendicular skeleton is composed of the bones of the limbs.

There are five functions of bone, which are as follows.

1. form (shape): bones help define the shape and appearance of animals

2. protection: certain bones have a critical role in protecting internal organs, such as the skull protecting the brain and the ribs protecting the thoracic organs (heart and lungs)

3. mineral storage: bone is a primary storage site of calcium and phosphorus

4. blood formation (oxygen-carrying cells and immune function): the marrow cavities of bones are sites of blood cell formation, producing both red and white blood cells

5. leverage (mobility): the muscles and bones work together to enable movement

Types of Bones

Bones are classified according to their shape and structure as follows.

1. **Long bones** are proportionally longer than they are wide. Each has a central marrow cavity and a proximal and distal epiphysis. Examples of long bones include the femur and metacarpals.

2. **Short bones,** relatively speaking, are about as short as they are wide, and each has only one growth center. Examples of short bones are carpals and tarsals.

3. **Flat bones** have two plates of compact bone with spongy bone in between. This forms the trabeculae crossing from one side of the bone to the other. These bones have no marrow cavity but have small, irregular marrow spaces. Examples of flat bones include pelvic bones, skull, and ribs.

4. **Irregular bones** are all the irregularly shaped bones, such as the vertebrae and some skull bones. Sesamoid bones are a type of irregular bone and are interposed in tendon. An example of a sesamoid bone is the patella (kneecap).

5. **Pneumatic bones** are bones with air spaces in them. These include bird bones.

Projections, Processes, Depressions, and Foramens

To understand the comparative nature of bones, you should know the terminology used to describe the skeletal system. The following list of definitions is provided to aid in the study of bones and their projections, depressions, and foramens.

Articular Projections

head: The proximal end of the bone; it may have a **neck** attached.

condyle: A rounded projection on the distal aspect of a bone.

trochlea: A grooved, sliding surface that acts like the surface of a pulley to increase the mechanical advantage in the movement of a joint.

Non-Articular Projections

process: A prominence or projection on a bone.

tuberosity: A large prominence on a bone; for example, the *ischiatic tuberosity* on the ishium of the pelvis.

tubercle: Large and small prominences on the humerus, known as the greater and lesser tubercles.

trochanter: Large and small prominences on the femur, known as the *greater* and *lesser trochanters.*

epicondyle: Roughened area on the sides of a bone, just proximal to the condyles.

spine: A pointed projection off a bone, as in the spine of the scapula.

crest: A ridge on a bone, as in the crest of the ilium.

Depressions

cavities: Generally articular spaces, where the head of a bone fits; for example, the glenoid cavity of the scapula.

notch: If non-articular, this is a groove or concavity, as in the alar notch of the atlas. If articular, it is inside the joint where the ligaments attach; for example, the trochlear notch of the ulna.

facet: An almost flat surface, as found on the carpals.

fovea: A small depression on a bone, such as the fovea capitis.

fossa: A large shallow depression, as in the supraspinous fossa of the scapula.

foramen: A hole in a bone.

canal: A small, tubular hole through a bone that is longer than a foramen.

Parts of the *Long Bone* (Figure 7.1)

diaphysis: The *shaft* of the bone. This consists of compact bone, a central medullary canal, and spongy bone at the proximal and distal ends of the shaft area.

epiphyses (singular is epiphysis): The proximal and distal ends of the bone. They have a thin outer layer of compact bone with centers composed of mostly cancellous (spongy) bone.

metaphyses (singular is metaphysis): The regions in mature bone where the diaphysis joins the epiphyses. This is where the bone widens at the proximal and distal ends of a long bone prior to the **epiphyseal plates,** or **growth plates.** The metaphyseal region includes the *epiphyseal plate,* the area in which the cartilage laid down by chondroblasts is replaced by bone via ossification in a growing animal.

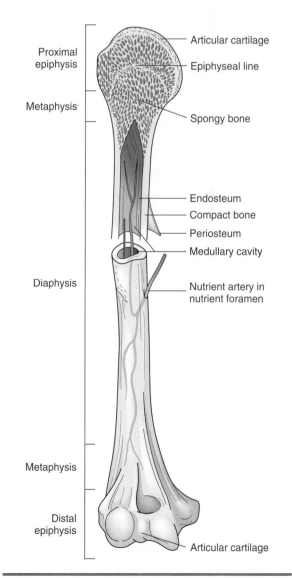

Figure 7.1: Parts of the long bone.

When viewing the following recommended bone tissues, first go to low power, focus, and orient yourself on the slide. Then change to high power to view the cellular detail necessary to understand the morphology of the tissue. The high-power, dry, objective lens should be sufficient for viewing these slides.

Obtain the recommended slide(s) and locate all the items on the slide you are instructed to find, including those labeled in the diagrams and photomicrographs contained in the figures throughout the chapter. Draw and label what you see on each slide in the spaces provided.

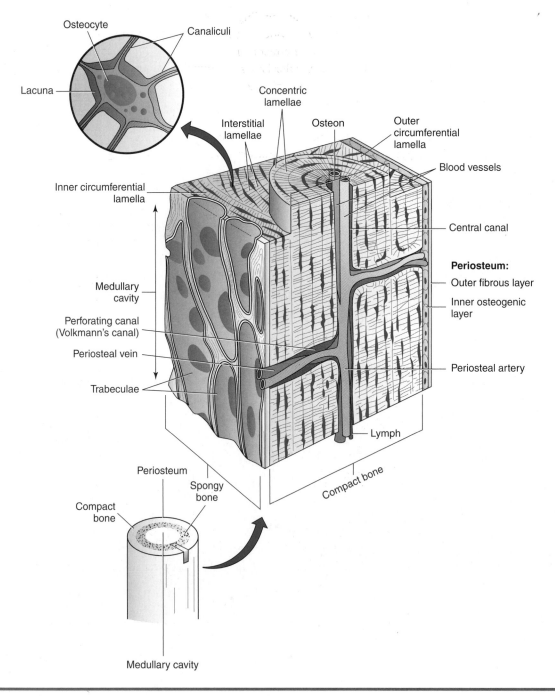

Figure 7.2: Diagram of cross section of compact bone.

Compact Bone

Obtain a slide labeled compact bone, and use the plastic model and diagram (Figure 7.2) as a reference to locate the parts of the bone described in the following section.

Slides recommended: Those labeled compact bone.

Description: **Compact bone** is a cylinder of bone containing a hard, calcified **matrix** and many collagen fibers. The cells, or **osteocytes,** lie in **lacunae** similar to those in cartilage. Bones are well vascularized by vessels from the bone marrow canal and the **periosteum.** Approximately the outer third of the blood supply comes from the periosteum, and the inner two-thirds comes from the bone marrow.

Slide: compact bone

First, observe the dense irregular connective tissue surrounding the bone; this is called the **periosteum.** Because it is an important source of blood to the bone and provides nutrition to the outer third as mentioned previously, veterinarians attempt to preserve as many periosteal attachments to the bone as possible during fracture-repair surgery. Just inside the periosteum are multiple circular layers of bone that go around the entire circumference of the bone. These are called the **outer circumferential lamellae.** Similarly, just inside the **endosteum** are the **inner circumferential lamellae** (Figure 7.3).

As mentioned in Chapter 5, there are many round cylinders of bone throughout the bony tissue, called **osteons,** and these combine to make up the **Haversian system** of bone (Figure 7.3). The layers of bone between the osteons are the **interstitial lamellae.** At the center of each osteon is a **central** (or **Haversian**) **canal** that contains a blood vessel. The dark, elliptical dot-like structures are the lacunae, each containing an *osteocyte*. Note the thread-like lines that radiate toward each lacuna; these are **canaliculi,** or tiny canals that link the lacunae together and provide nutrition for the osteocytes. The central canals of the osteons are connected by transverse canals, called **Volkmann's canals,** or **perforating canals.** These canals also contain blood vessels that interconnect to the vessels in the central canals. These canals, together with the canaliculi, create a complex system for providing nutrition throughout the bone.

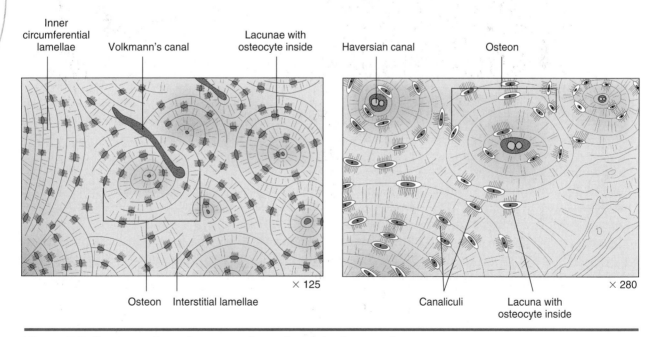

Figure 7.3: Cross section of compact bone from the femur of a cat.

In the space provided, draw a section of compact bone and label the cells, canals, and other parts discussed previously.

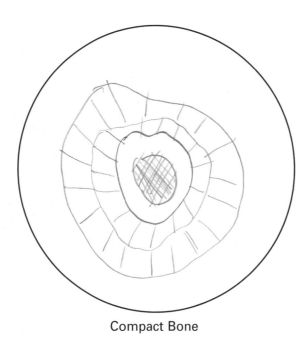

Compact Bone

Cancellous, or Spongy, Bone

Whereas compact bone contains more bone (compact bony tissue) than interosseus spaces, **cancellous bone** contains more interosseus spaces than bone.

Slides recommended: This type of bone is found in the epiphysis or metaphysis (just inside the epiphyseal plate), so slides of these regions can be used. Recommended slides are labeled compact and spongy bone, epiphyseal plate, or the epiphysis.

Description: Cancellous (spongy) bone is characterized by a sponge-like interconnection of bony spicules; hence the name *spongy bone*.

Slide: cancellous bone

On the slide, note the areas of bony matrix surrounded by white areas containing either undifferentiated cells of bone marrow or *osteoblasts*. If you are looking at the area beneath the growth plate, you will see large, dark osteoblasts lining the irregularly shaped, purple-stained areas of bone. Within the purple-stained areas are the elliptical lacunae containing *osteocytes*. Immediately below the area of *hypertrophied chondroblasts* of the growth plate is the zone of ossification, which involves calcification (Figure 7.4). Above the area of hypertrophied chondroblasts are areas of spongy bone with small, elliptical lacunae and osteocytes within.

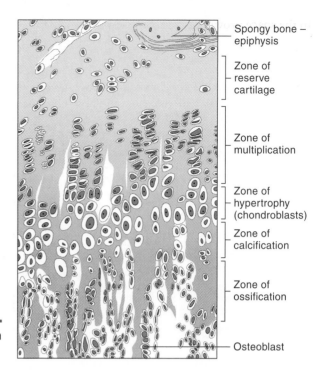

Figure 7.4: Cross section of epiphyseal plate from the humerus of a cat.

In the space provided, draw a section of cancellous (spongy) bone and label the cells, canals, and other parts discussed previously.

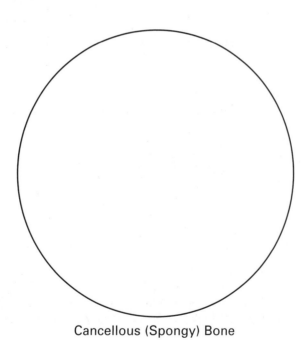

Cancellous (Spongy) Bone

EXERCISE 7.2

THE SKELETON OF THE CAT

Locate and identify the following bones and bone parts (such as notches, foramina, protuberances, and condyles) for a cat. The names of the bones and bone parts listed here are in bold print because these are the items are important to identify and know.

1. Identify, name, and number each of the vertebrae in the following groups of vertebrae (Figure 7.5):
   a. **cervical:** 7 vertebrae; includes the **atlas** (first vertebra) and **axis** (second vertebra)
   b. **thoracic:** 13 vertebrae
   c. **lumbar:** 7 vertebrae
   d. **sacral (sacrum):** 3 vertebrae (These three vertebra are fused into one bone, that is why they may be also called the *sacrum*.)
   e. **coccygeal:** variable number of vertebrae depending on breed of cat (You may see this abbreviated Cy++.)

2. **Parts of a Vertebra:** Using the cat skeleton, identify the parts of the vertebra listed as follows for each of the types of vertebrae listed in the previous activity (Figure 7.6).
   a. **vertebral arch**
      1. **lamina**
      2. **pedicle**
   b. **vertebral canal**
   c. **vertebral body**
   d. **spinous processes**
   e. **cranial and caudal articular processes**
   f. **lateral vertebral foramen**
   g. **transverse process**
   h. **transverse foramen** (cervical vertebrae only)
   i. **costal fovea and costal facet** (thoracic vertebrae only)
   j. **accessory process** (lumbar vertebrae only)
   k. **atlas (C1)**
      1. **dorsal and ventral arches**
      2. **ventral tubercle**
      3. **transverse processes** *(wings)*
      4. **alar notch**
      5. **lateral vertebral foramina**
      6. **vertebral canal**

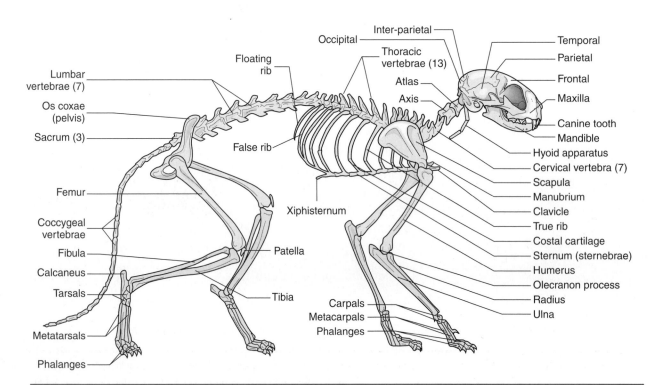

Figure 7.5: Side view of the skeleton of the cat.

7. cranial articular fovea
8. caudal articular fovea
l. **axis (C2)**
 1. **dens or odontoid process**
 2. **cranial articular surface**
 3. **spinous process**
 4. **transverse foramen**

m. **sacrum**
 1. **median sacral crest**
  (the 3 spinous processes)
 2. **intermediate sacral crest**
 3. **dorsal and ventral sacral foramina**
 4. **promontory**
 5. **lateral sacral crest**
 6. **auricular face**

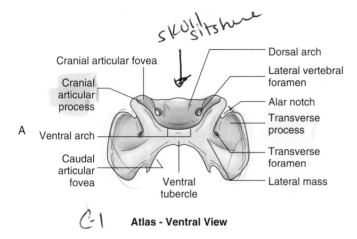

A — Cranial articular fovea / Cranial articular process / Ventral arch / Caudal articular fovea / Dorsal arch / Lateral vertebral foramen / Alar notch / Transverse process / Transverse foramen / Lateral mass / Ventral tubercle

Atlas - Ventral View

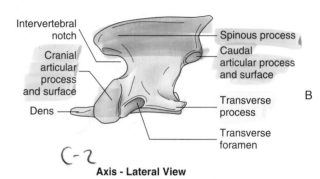

B — Intervertebral notch / Cranial articular process and surface / Dens / Spinous process / Caudal articular process and surface / Transverse process / Transverse foramen

Axis - Lateral View

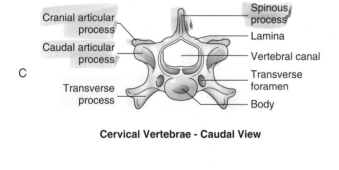

C — Cranial articular process / Caudal articular process / Transverse process / Spinous process / Lamina / Vertebral canal / Transverse foramen / Body

Cervical Vertebrae - Caudal View

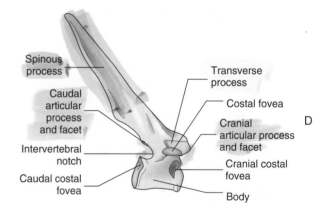

D — Spinous process / Caudal articular process and facet / Intervertebral notch / Caudal costal fovea / Transverse process / Costal fovea / Cranial articular process and facet / Cranial costal fovea / Body

Thoracic Vertebra - Lateral View

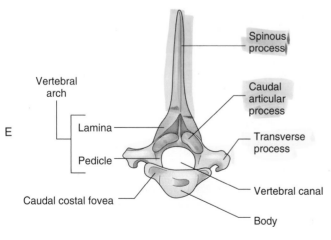

E — Vertebral arch / Lamina / Pedicle / Caudal costal fovea / Spinous process / Caudal articular process / Transverse process / Vertebral canal / Body

Thoracic Vertebrae - Caudal View

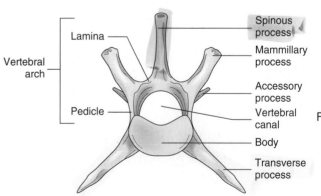

F — Lamina / Vertebral arch / Pedicle / Spinous process / Mammillary process / Accessory process / Vertebral canal / Body / Transverse process

Typical Lumbar Vertebra - Cranial View

Figure 7.6: Vertebrae of the cat. *A.* Atlas, ventral view. *B.* Axis, lateral view. *C.* Cervical vertebrae, cranial view. *D.* Thoracic vertebrae, lateral view. *E.* Thoracic vertebrae, cranial view. *F.* Lumbar vertebrae, cranial view. *Continued*

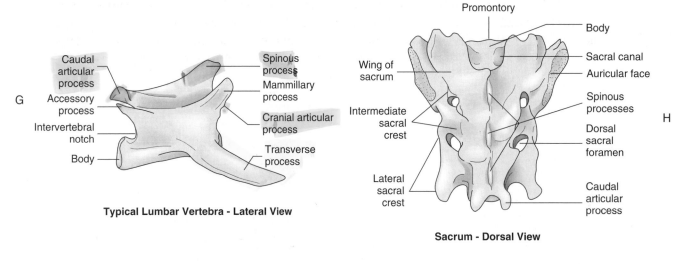

Typical Lumbar Vertebra - Lateral View

Sacrum - Dorsal View

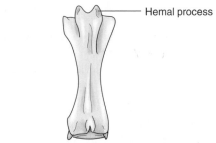

Coccygeal Vertebrae - Dorsal View

Figure 7.6, cont'd: *G.* Lumbar vertebrae, lateral view. *H.* Sacrum, dorsal view. *I.* Coccygeal vertebra, dorsal view.

3. **Ribs, Sternum, & Associated Structures:** Using the cat skeleton, identify the parts of the ribs, sternum, and associated structures listed as follows (see Figures 7.5 and 7.7). Also, determine the number of each rib and sternebrae (i.e., the fourth rib is attached to the fourth thoracic vertebra).

a. **head**
b. **tubercle**
c. **neck**
d. **shaft or body**
e. **costal cartilage**
f. **costochondral junction**
g. **true (1-9), false (10-12), and floating ribs (13)**

h. **sternum**
   1. **sternebrae**
   2. **manubrium sterni**
i. **xiphoid process**
j. **xiphoid cartilage**
k. **intersternal cartilage**

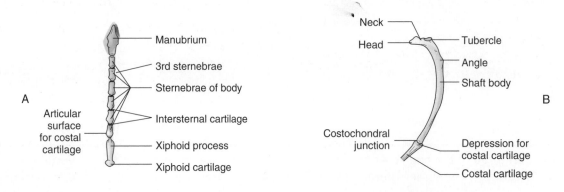

Figure 7.7: Sternum and ribs of the cat. *A.* Sternum, ventral view. *B.* Ribs, lateral view.

4. **Skull:** Using the whole and bisected cat skulls, identify the bones of the *skull,* their processes, and the foramens listed as follows (Figure 7.8). Remember, the bones of the skull are irregular bones. These bones are three-dimensional, whereas the drawings provided are two-dimensional; and therefore, the same bone may be seen on two views of the skull in apparently different places. Remember that bones bend, curve, and occupy space on both the outside and inside of the skull.

a. **bones of the skull**
  1. **incisive**
  2. **nasal**
  3. **lacrimal**
  4. **maxilla**
  5. **palatine**
  6. **frontal**
  7. **parietal**
  8. **temporal**
  9. **occipital**
  10. **tympanic bulla**
  11. **zygomatic bone**
  12. **interparietal**
  13. **presphenoid**
  14. **basisphenoid**
  15. **pterygoid and the hamulus of the pterygoid**
  16. **nasal concha** (also known as nasal turbinate bones; bones visualized through external nares)
  17. **mandible**
  18. **alisphenoid**
  19. **ethmoid**

b. **processes, crests, sutures, and other protuberances**
  1. **frontal suture**
  2. **sagittal suture**
  3. **coronal suture**
  4. **nuchal crest**
  5. **sagittal crest**
  6. **external occipital protuberance**
  7. **occipital condyle**
  8. **zygomatic process of the frontal bone**
  9. **zygomatic arch**
  10. **frontal process of the zygomatic bone**
  11. **zygomatic bone**
  12. **zygomatic processes of the temporal bone and maxilla**
  13. **bregma**
  14. **paracondylar process**
  15. **retroarticular process**
  16. **mastoid process**

c. **parts and processes of the mandible**
  1. **body**
  2. **ramus**
  3. **mandibular symphysis**
  4. **angular process**
  5. **condyloid process**
  6. **coronoid process** (the top of the ramus)

d. **foramina and concavities of the skull**
  1. **infraorbital foramen**
  2. **maxillary foramen** leading to the **lacrimal canal**
  3. **frontal sinus**
  4. **maxillary sinus**
  5. **oval foramen** (or *foramen ovale*)
  6. **round foramen** (or *foramen rotundum*)
  7. **orbital fissure**
  8. **optic foramen**
  9. **external acoustic meatus**
  10. **jugular foramen** (best seen from the inside)
  11. **tympano-occipital fissure leading to jugular foramen**
  12. **foramen magnum**
  13. **external nares**
  14. **internal nares**
  15. **orbit**
  16. **mandibular fossa**

e. **foramina and depressions of the mandible**
  1. **massenteric fossa** (on lateral side)
  2. **mandibular foramen** (on medial side caudally)
  3. **mental foramina** (on lateral side cranially)

Note: Teeth will be covered in the chapter on the digestive system.

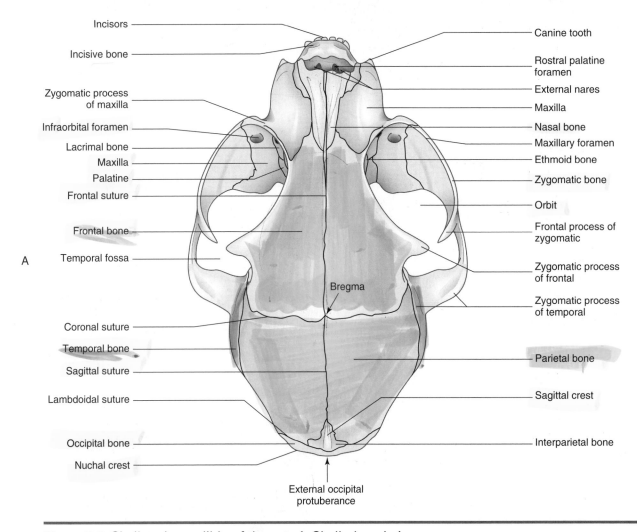

Incisors

Incisive bone

Zygomatic process
of maxilla

Infraorbital foramen

Lacrimal bone

Maxilla

Palatine

Frontal suture

Frontal bone

A   Temporal fossa

Coronal suture

Temporal bone

Sagittal suture

Lambdoidal suture

Occipital bone

Nuchal crest

Canine tooth

Rostral palatine
foramen

External nares

Maxilla

Nasal bone

Maxillary foramen

Ethmoid bone

Zygomatic bone

Orbit

Frontal process of
zygomatic

Zygomatic process
of frontal

Zygomatic process
of temporal

Parietal bone

Sagittal crest

Interparietal bone

Bregma

External occipital
protuberance

Figure 7.8: Skull and mandible of the cat. *A.* Skull, dorsal view.

5. **Scapula:** Using the cat skeleton, identify the parts of the *scapula* (or shoulder blade) listed as follows
   (Figure 7.9).
   a. **coracoid process**
   b. **glenoid cavity**
   c. **clavicle**
   d. **spine**
   e. **acromion**
      1. **hamate process** (of the acromion
         process)
      2. **suprahamate process** (of the
         acromion process)

   f. **infraspinous fossa**
   g. **supraspinous fossa**

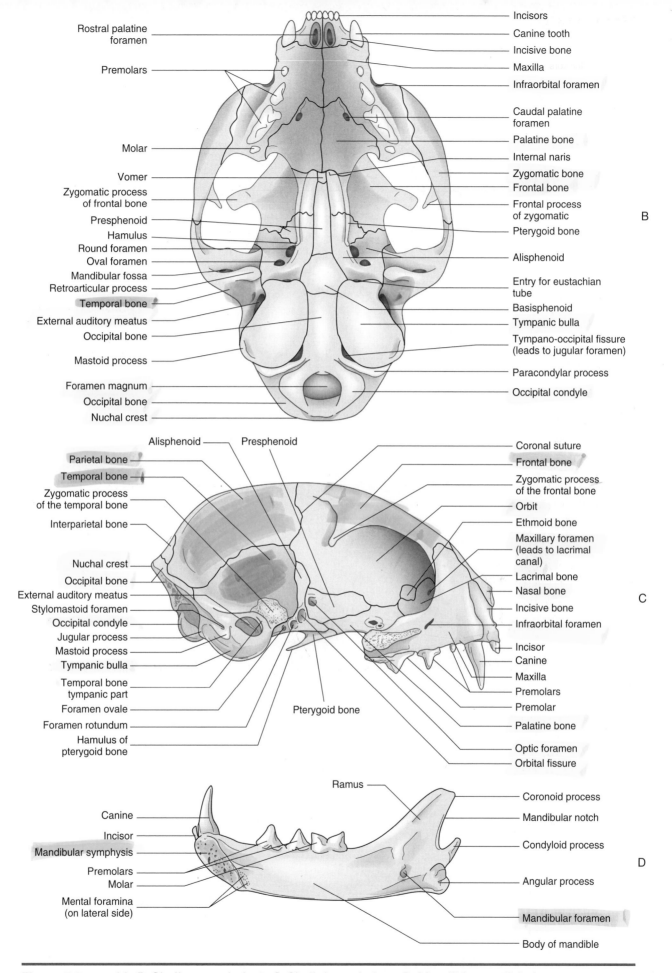

Figure 7.8, cont'd: *B.* Skull, ventral view. *C.* Skull, lateral view. *D.* Mandible, medial view.

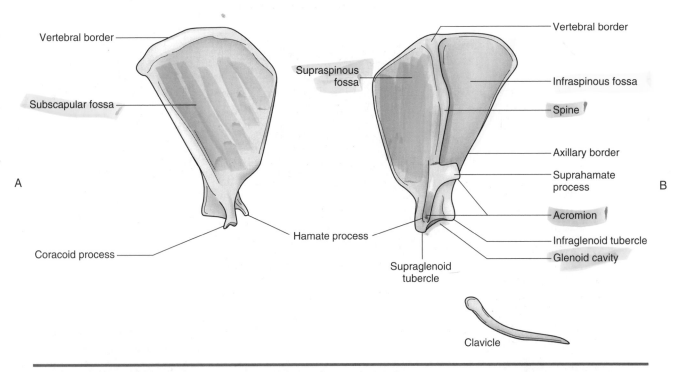

Figure 7.9: Scapula of the cat. *A.* Left medial view. *B.* Left lateral view.

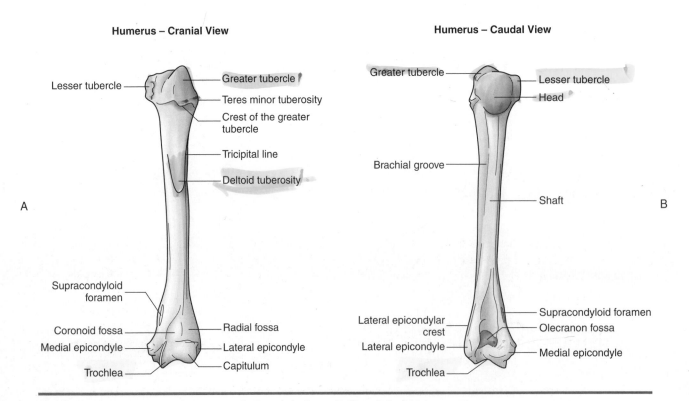

Figure 7.10: Humerus of the cat. *A.* Cranial view. *B.* Caudal view.

6. **Humerus:** Using the cat skeleton, identify the parts of the *humerus* listed as follows (Figure 7.10).
   a. **head**
   b. **greater and lesser tubercles**
   c. **tricipital line**
   d. **deltoid tuberosity**
   e. **brachial groove**
   f. **shaft**
   g. **supracondyloid foramen**
   h. **medial and lateral epicondyles**
   i. **coronoid fossa**
   j. **radial fossa**
   k. **trochlea**
   l. **capitulum**

7. **Radius:** Using the cat skeleton, identify the parts of the radius listed as follows (Figure 7.11).
   a. **head**
   b. **radial tuberosity**
   c. **shaft**
   d. **styloid process** (medial)

8. **Ulna:** Using the cat skeleton, identify the parts of the *ulna* listed as follows (see Figure 7.11).
   a. **trochlear notch**
   b. **olecranon process** *(Tuber olecranon)*
   c. **anconeal process**
   d. **medial and lateral coronoid processes**
   e. **lateral styloid process**
   f. **body of the ulna**

9. **Carpus, metacarpal bones, and phalanges of the digits:** Using the cat skeleton, identify the parts of the *carpus, metacarpal bones,* and *phalanges of the digits* listed as follows (Figure 7.12).
   a. **intermedioradial carpal bone**
   b. **ulnar carpal bone**
   c. **accessory carpal bone** (on lateral aspect of leg in all species)
   d. **carpal bones I to IV** (medial to lateral)
   e. **metacarpal bones I to V**
   f. **head** (proximal end)
   g. **body** (shaft)
   h. **base** (distal end)
   i. **proximal sesamoid bones**
   j. **proximal phalanges of digits I to V**
   k. **middle phalanges of digits II to V** (the dew-claw, digit I, has no middle phalanx)
   l. **distal phalanges of digits I to V**
   m. **unguicular process of the distal phalanges**

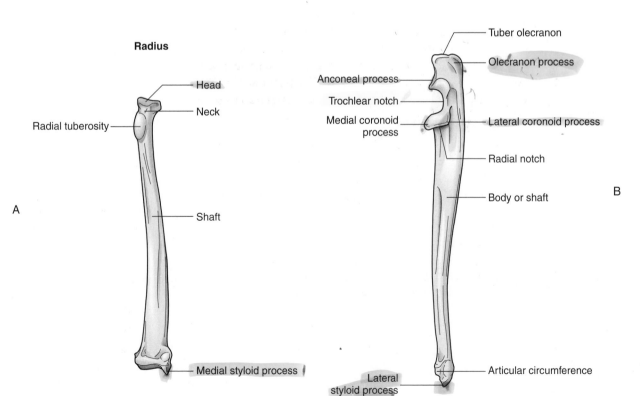

Figure 7.11: *A.* Radius, and *B.* Ulna of the cat.

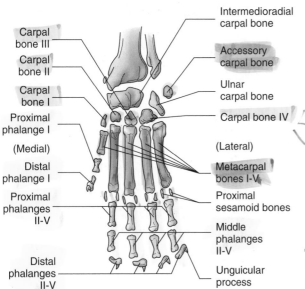

Carpal bone III
Carpal bone II
Carpal bone I
Proximal phalange I
(Medial)
Distal phalange I
Proximal phalanges II-V
Distal phalanges II-V

Intermedioradial carpal bone
Accessory carpal bone
Ulnar carpal bone
Carpal bone IV
(Lateral)
Metacarpal bones I-V
Proximal sesamoid bones
Middle phalanges II-V
Unguicular process

Figure 7.12: Carpal bones, metacarpal bones, and phalanges of the cat.

10. **Pelvis (os coxae):** Using the cat skeleton, identify the parts of the *pelvis (os coxae)* listed as follows (Figure 7.13). There are four bones of the pelvis: the ilium, ishium, pubis, and acetabular bones. The acetabular bone is in the center of the acetabulum, approximately where the acetabular notch is located. There is no clear distinction as to where one bone starts and another leaves off.

a. **acetabulum**
b. **acetabular fossa** (the cavity of the acetabulum)
c. **acetabular notch** (the center area)
d. **obturator foramen**
e. **ilium**
f. **iliac crest**
g. **gluteal surface**
h. **tuber sacrale** (very prominent and important in large animals)

i. **tuber coxae** (very prominent and important in large animals)
j. **greater ischiatic notch**
k. **ischium**
l. **ischiatic spine**
m. **lesser ischiatic notch**
n. **ischiatic tuberosity** (tuber ischii)
o. **pubis**
p. **symphysis pubis** (or pubic symphysis)

11. **Femur:** Using the cat skeleton, identify the parts of the *femur* listed as follows (Figure 7.14).

a. **head**
b. **neck**
c. **fovea capitis**
d. **greater trochanter**
e. **lesser trochanter**
f. **trochanteric fossa**
g. **intertrochanteric crest**

h. **shaft**
i. **trochlea** (*medial* and *lateral trochlear ridge* and trochlear groove)
j. **medial and lateral condyles**
k. **medial and lateral epicondyles**
l. **patella** (or *kneecap*)
m. **sesamoid bones**

12. **Tibia:** Using the cat skeleton, identify the parts of the tibia listed as follows (Figure 7.15).

a. **medial and lateral condyles**
b. **tibial tuberosity**

c. **shaft**
d. **medial malleolus**

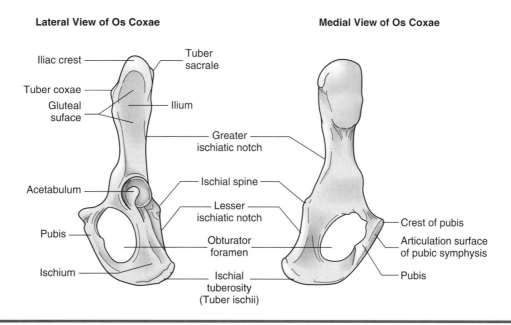

Figure 7.13: Pelvis (or os coxae) of the cat.

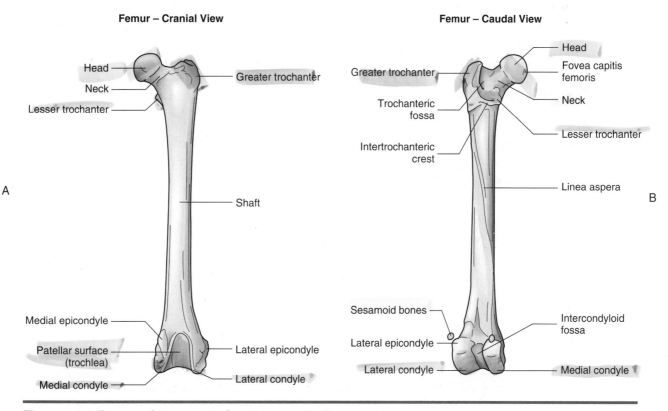

A

B

Figure 7.14: Femur of the cat. *A.* Cranial view. *B.* Caudal view.

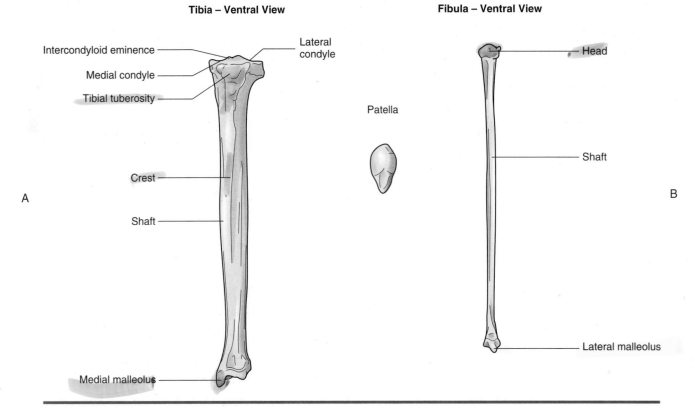

Figure 7.15: *A.* Tibia, ventral view; and *B.* Fibula, ventral view of the cat.

13. **Fibula:** Using the cat skeleton, identify the parts of the *fibula* listed as follows (see Figure 7.15).
    a. **head**
    b. **shaft**
    c. **lateral malleolus**

14. **Tarsus:** Using the cat skeleton, identify the parts of the *tarsus* (the tarsal bones), also known as the **hock,** listed as follows (Figure 7.16).
    a. **talus (tibial tarsal bone)**   e. **sustentaculum tali**
    b. **trochlea**   f. **central tarsal bone**
    c. **calcaneus (fibular tarsal bone)**   g. **tarsal bones I to IV**
    d. **tuber calcanei**

15. **Metatarsals and phalanges:** Using the cat skeleton, identify the *metatarsals* and *phalanges,* listed as follows (see Figure 7.16).
    a. **metatarsal bones I to V**   e. **middle phalanges of digits II to V**
    b. **body**   f. **distal phalanges of digits II to V**
    c. **proximal sesamoid bones**
    d. **proximal phalanges of digits II to V**
       (medial to lateral; digit I usually is miss-
       ing in the cat)

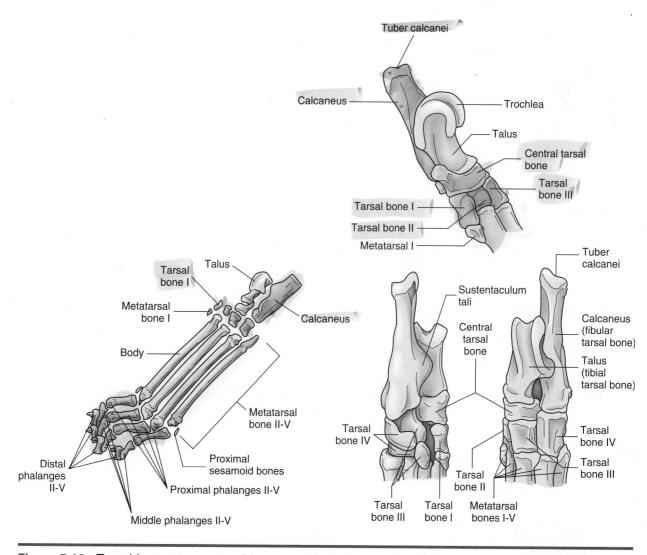

Figure 7.16: Tarsal bones, metatarsal bones, and phalanges of the cat.

EXERCISE 7.3
COMPARATIVE OSTEOLOGY

1. Using skulls from the dog, horse, and ox, identify the following bones, and compare them to each other and to the skull of the cat (Figure 7.17).

   a. **maxilla**
   b. **frontal and frontal sinus**
   c. **parietal**
   d. **occipital**
   e. **temporal**

   f. **nasal**
   g. **lacrimal**
   h. **zygomatic arch**
   i. **mandible**

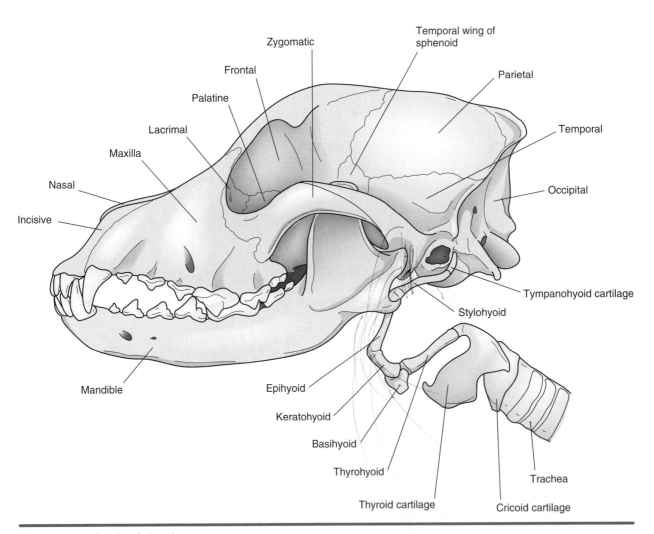

Figure 7.17: Skull of the dog.

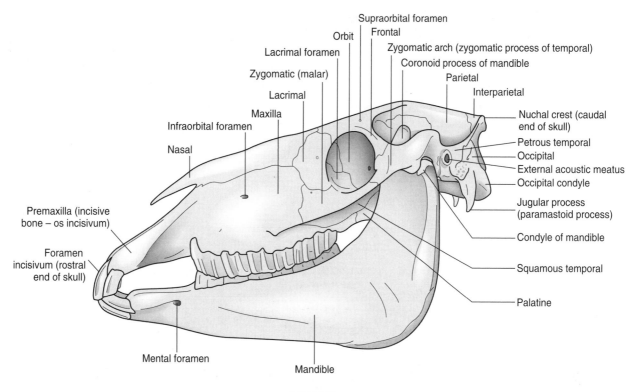

Lateral View

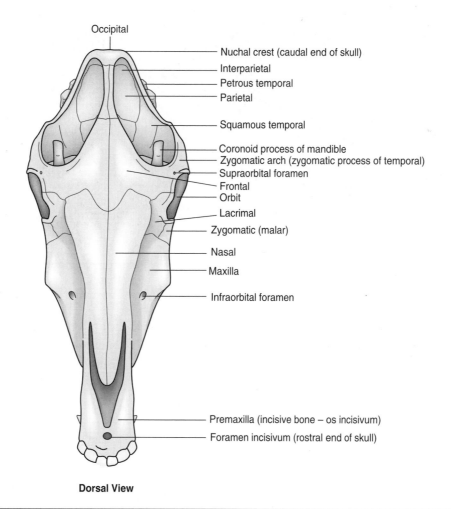

Dorsal View

Figure 7.17, cont'd: Skull of the horse.

Continued

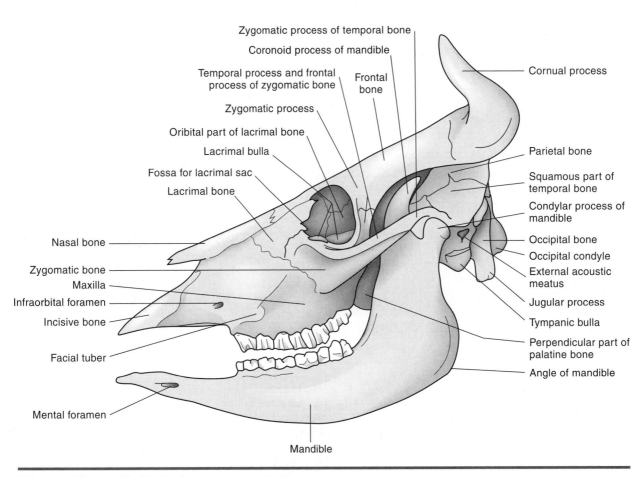

Figure 7.17, cont'd: Skull of the ox, lateral view.

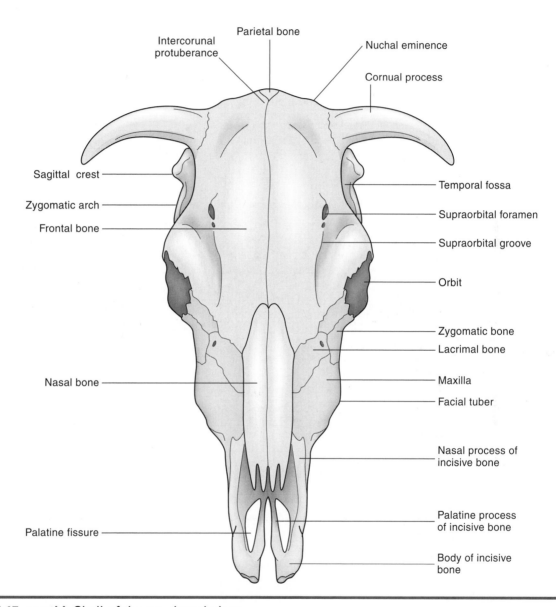

Figure 7.17, cont'd: Skull of the ox, dorsal view.

2. Identify the following bones using the following items: (1) the front leg and hind leg of the sheep, (2) the model of the leg of a horse, and (3) the articulated skeleton of a horse and cow from the carpus and tarsus distally (Figure 7.18).

a. **scapula**

b. **humerus**

c. **fused radius and ulna**

d. **carpal bones** (The equine **carpus** contains the **accessory, radial intermediate,** and **ulnar carpal bones,** as well as **carpal bones I to IV;** see Figure 7.18*B*)

e. **cannon bone (Metacarpal [Mc] III or metatarsal [Mt] III;** Mc III and IV are fused in the ruminant.)

f. **splint bones (Mc or Mt II and IV in the horse)**

g. **proximal sesamoid bones**

h. **long pastern bone** (This is the **proximal phalanx, digit III** in the horse and **digits III** and **IV** in the ruminant.)

i. **short pastern bone** (This is the **middle phalanx, digit III** in the horse and **digits III** and **IV** in the ruminant.)

j. **coffin bone** (This is the **distal phalanx, digit III** in the horse and **digits III** and **IV** in the ruminant.)

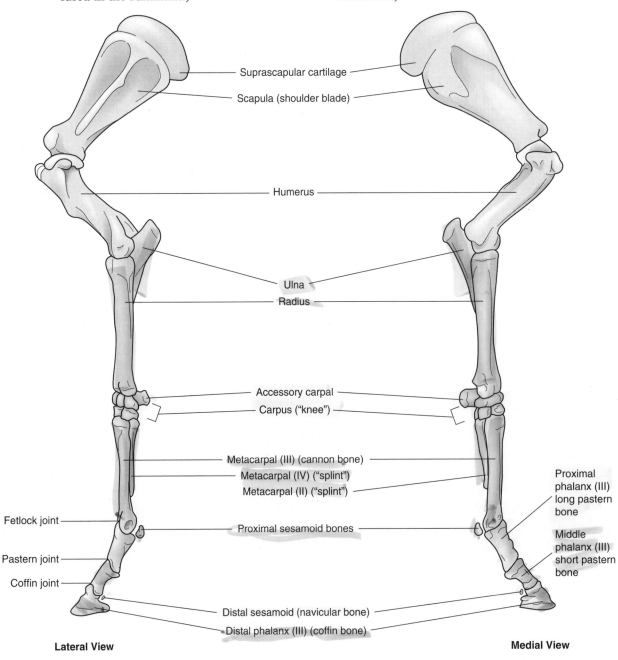

A

Lateral View

Medial View

Figure 7.18: *A.* Foreleg of the horse, lateral and medial views.

k. **navicular bone or distal sesamoid bone(s)** (There is one in the horse and two in the ruminant.)
l. **collateral cartilages**
m. **tuber sacrale**
n. **tuber coxae**
o. **tuber ischii**
p. **os coxae and acetabulum**
q. **femur**

r. **patella**
s. **tibia**
t. **fibula**
u. **tarsal bones** (The equine **tarsus** includes the **talus**, or **tibial tarsal bone**; the **calcaneus**, or **fibular tarsal bone**; the **central tarsal bone**; **tarsal bones I and II** fused; **tarsal bones III and IV**; and the **tuber calcanei**)

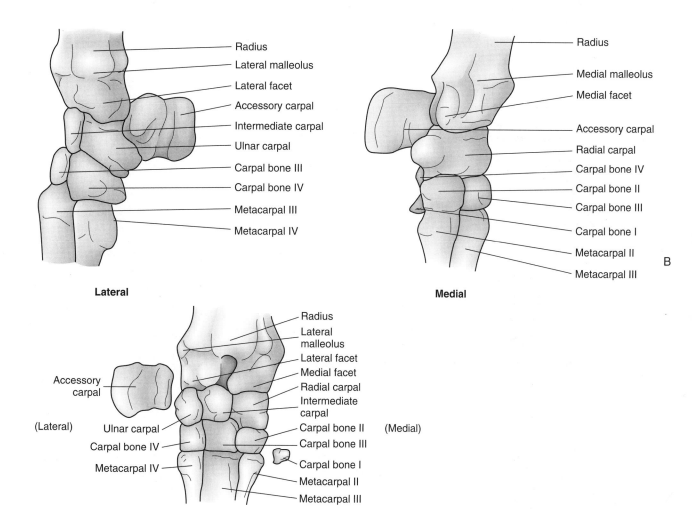

Figure 7.18, cont'd: *B.* Carpus of the horse, lateral, medial, and caudal views.   *Continued*

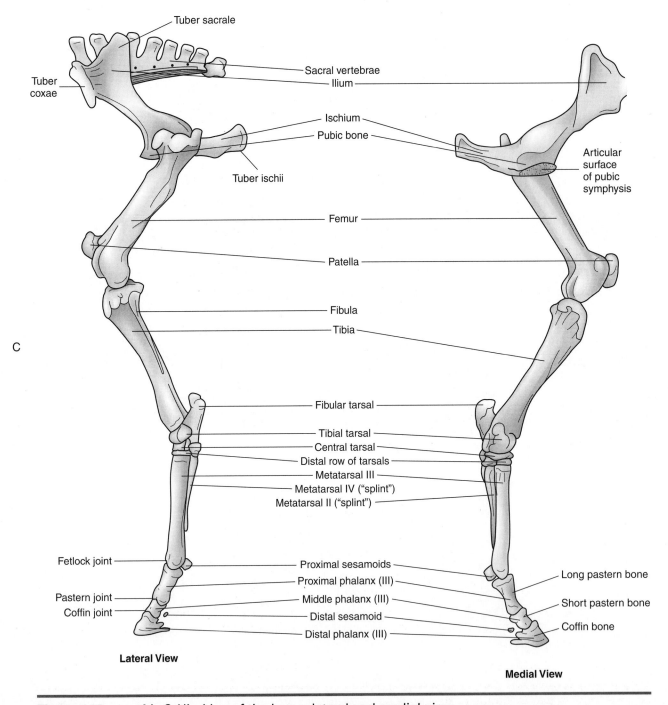

C

Tuber sacrale

Tuber coxae

Sacral vertebrae
Ilium

Ischium
Pubic bone

Tuber ischii

Articular surface of pubic symphysis

Femur

Patella

Fibula
Tibia

Fibular tarsal

Tibial tarsal
Central tarsal
Distal row of tarsals
Metatarsal III
Metatarsal IV ("splint")
Metatarsal II ("splint")

Fetlock joint

Proximal sesamoids
Proximal phalanx (III)

Pastern joint
Coffin joint

Middle phalanx (III)
Distal sesamoid
Distal phalanx (III)

Long pastern bone

Short pastern bone

Coffin bone

Lateral View

Medial View

Figure 7.18, cont'd: *C.* Hind leg of the horse, lateral and medial view.

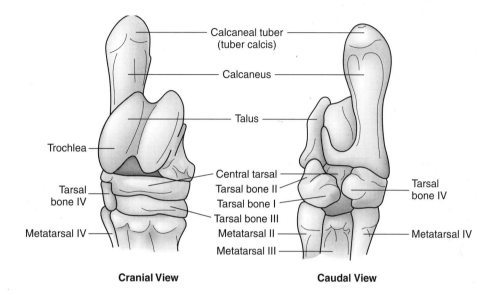

Calcaneal tuber (tuber calcis)

Calcaneus

Talus

Trochlea

Central tarsal
Tarsal bone II
Tarsal bone I
Tarsal bone III

Tarsal bone IV

Metatarsal IV
Metatarsal II
Metatarsal III

Tarsal bone IV

Metatarsal IV

Cranial View

Caudal View

D

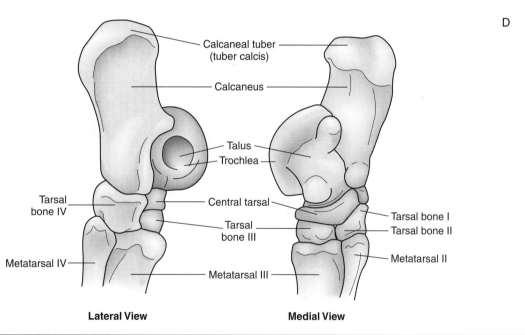

Calcaneal tuber (tuber calcis)

Calcaneus

Talus
Trochlea

Central tarsal

Tarsal bone III

Tarsal bone IV

Metatarsal IV

Metatarsal III

Tarsal bone I
Tarsal bone II

Metatarsal II

Lateral View

Medial View

Figure 7.18, cont'd: *D.* Tarsus of the horse.

Continued

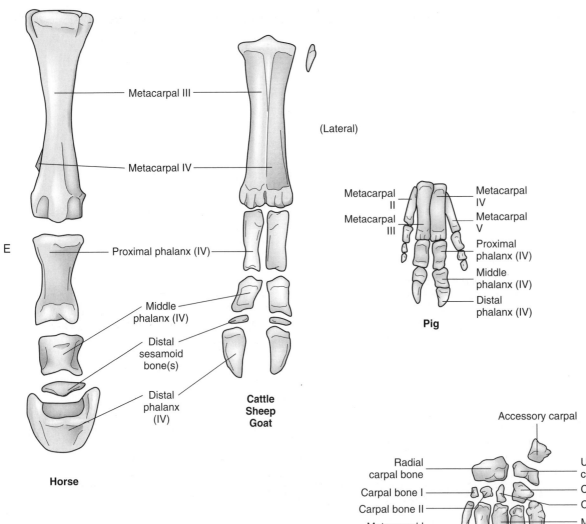

E

Horse

Metacarpal III

Metacarpal IV

Proximal phalanx (IV)

Middle phalanx (IV)

Distal sesamoid bone(s)

Distal phalanx (IV)

(Lateral)

**Cattle
Sheep
Goat**

Metacarpal II

Metacarpal III

Metacarpal IV

Metacarpal V

Proximal phalanx (IV)

Middle phalanx (IV)

Distal phalanx (IV)

Pig

Figure 7.18, cont'd: *E.* Comparative feet of domestic animals. *F.* Carpal bones, metacarpal bones, and phalanges of the dog, dorsal aspect of left forepaw.

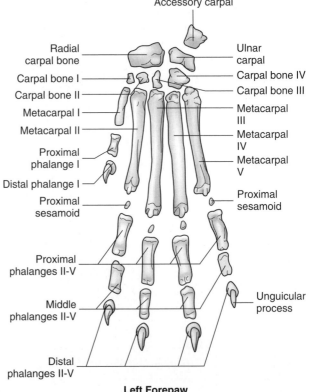

Accessory carpal

Radial carpal bone

Ulnar carpal

Carpal bone I

Carpal bone IV

Carpal bone II

Carpal bone III

Metacarpal I

Metacarpal III

Metacarpal II

Metacarpal IV

Proximal phalange I

Metacarpal V

Distal phalange I

Proximal sesamoid

Proximal sesamoid

Proximal phalanges II-V

Middle phalanges II-V

Unguicular process

Distal phalanges II-V

**Left Forepaw
Dorsal Aspect**

F

3. Using the diagram and the articulated skeleton of a chicken, locate the following bones (Figure 7.19).

a. **axis**
b. **atlas**
c. **cervical vertebrae**
d. **thoracic vertebrae**
e. **synsacrum**
f. **caudal vertebrae**
g. **pyrostyle**
h. **furculum**
i. **coracoid**
j. **humerus**
k. **radius** and **ulna**

l. **phalanges** and **digits** of the wing
m. **keel of the sternum**
n. **ribs**
o. **ilium**
p. **ishium**
q. **pubis**
r. **femur**
s. **patella**
t. **tibia** and **fibula**
u. **tarsometatarsus**
v. **phalanges** and **digits** of the foot

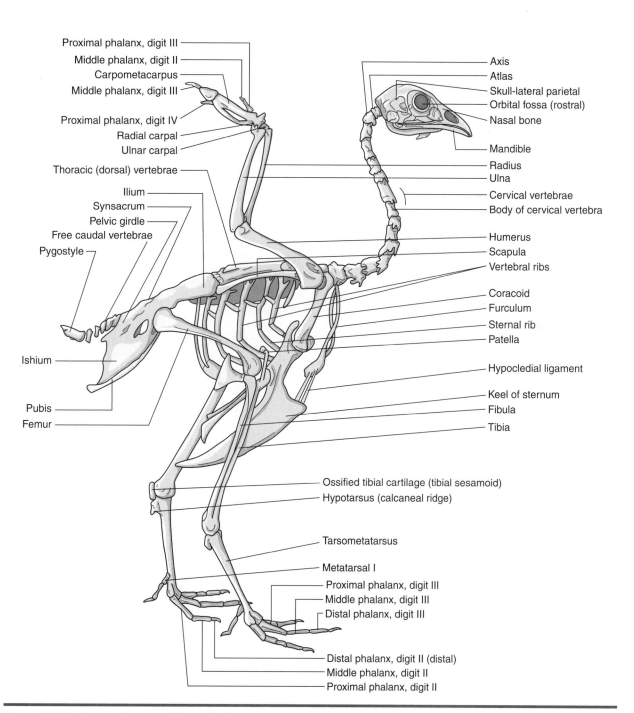

Figure 7.19: The skeleton of the chicken.

EXERCISE 7.4
COMPARATIVE ARTHROLOGY AND DESMOLOGY

1. **Arthrology** is the study of joints. Using the articulated skeleton, models, and Figure 7.18, find the following joints. Items are found in all species except where noted.
   a. **costocondral junction** (where bony rib meets cartilaginous rib)
   b. **temporomandibular joint (T-M joint)**
   c. **shoulder joint (scapulohumeral joint)**
   d. **elbow joint (humeroradioulnar joint)**
   e. **carpal joint** (Found in small and large animals; composed of three joints, each with a separate synovial sac or membrane. Sometimes referred to as a *knee* in the equine.)
      1. **radiocarpal joint**
      2. **intercarpal joint**
      3. **carpometacarpal joint**
   f. **fetlock joint (metacarpophalangeal joint; found in the equine)**
   g. **pastern joint (proximal interphalangeal joint; found in the equine)**
   h. **coffin joint (distal interphalangeal joint; found in the equine)**
   i. **sacroiliac joint** (where the sacrum attaches to the ilium of the pelvis)
   j. **pubic symphysis** (symphysis of the two halves of the pelvis at the pubic bones)
   k. **hip joint (coxofemoral joint)**
   l. **stifle joint (femorotibial joint, true knee joint)**
   m. **tarsal joint (hock joint)** The tarsus of small animals is composed of three separate joints, each with its own joint capsule. Each joint is named based on the names of the bones from proximal to distal, i.e. the tibiotarsal joint is an articulation between the tibia and the tarsus.
      1. **tibiotarsal joint**
      2. **intertarsal joint**
      3. **tarsometatarsal joint**
      The tarsus of large animals is made up of four separate joints, and, like small animals, each has its own joint capsule. As you can see, there is an extra row of tarsal bones in large animals that creates an extra joint when compared to a small animal's tarsus.
      1. **tibiotarsal joint**
      2. **proximal intertarsal joint**
      3. **distal intertarsal joint**
      4. **tarsometatarsal joint**

2. **Desmology** is the study of ligaments. Using the prepared models and diagrams, find the following ligaments.

nuchal ligament: Massive ligament that helps support the head of large animals. It runs from the nuchal crest of the skull to the spinous processes of the thoracic vertebrae. In dogs it runs from the axis to the thoracic vertebrae.

interspinous ligaments: Ligamentous bands between the spinous processes of the vertebrae.

supraspinous ligaments: Runs along the top of the spinous processes from the thoracic vertebrae caudally.

transverse humeral ligament: Holds the tendon of origin of the biceps brachii muscle next to the humerus as it passes the greater tubercle medially.

sacrotuberous ligament: Another support for the pelvis. It runs from the sacrum on both sides (two ligaments) to the dorsal aspect of the tuber ischii (ischial tuberosity) of the pelvis.

round ligament: Helps hold the head of the femur in the acetabulum. It runs from the acetabular bone to the fovea capitis.

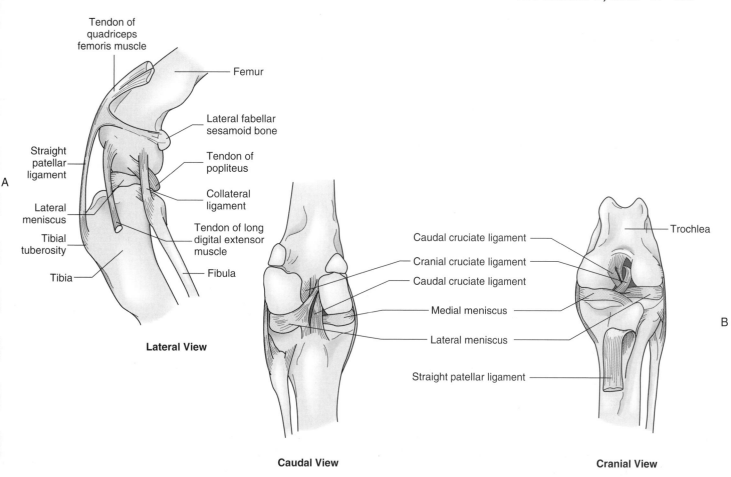

Figure 7.20: *A.* Stifle joint, lateral view of a dog. *B.* Stifle joint, cranial and caudal views of a dog.

patellar ligaments in the horse: Medial, middle, and lateral patellar ligaments. Horses have three patellar ligaments which run from the patella to the tibial tuberosity, one each on both the lateral and medial aspects, and a central ligament in the middle.

straight patellar ligament: In small animals, this runs from the quadriceps femoris muscle group and patella to the tibial tuberosity.

collateral ligaments: These stabilize the stifle medially and laterally, coursing from the distal epicondyle of the femur to the proximal epicondyle of the tibia and fibula (Figure 7.20).

cruciate ligaments: These stabilize the stifle cranially and caudally; the *cranial cruciate* is lateral to the caudal within the knee joint and attached cranially on the central head of the tibia. The *caudal cruciate* attaches caudally on the central head of the tibia, between the medial and lateral menisci (Figure 7.20).

plantar ligament: Mainly observed in the equine. This runs from the caudal aspect of the calcaneus to the fourth metatarsal bone (proximal aspect).

suspensory ligament: Located between the metacarpal bone (or metatarsal in the hindleg) and the deep digital flexor tendon. This attaches at the proximal end of the metacarpal (and metatarsal) and runs distally, dividing the fetlock to attach on both sides to the proximal sesamoid bones. It then sends another branch that passes across the fetlock to join the common extensor tendon and attach to the proximal phalanx. It is a strong support mechanism of the fetlock. In large animals with more than one digit, the suspensory ligament has muscle in it and is called the *interosseus muscle.*

Achilles tendon: This is the combined tendon of the gastrocnemius and superficial digital flexor muscle and attaches at the tuber calcanei.

stay apparatus: Components are muscles, tendons, and ligaments. This allows the horse to stand with little muscular activity.

check apparatus: The portion of the stay apparatus below the fetlock joint.

reciprocal apparatus: A series of muscles and ligaments that permit the stifle and hock to both flex and extend together.

PHYSIOLOGY: CALCIUM MEASUREMENTS

Total serum calcium is approximately 50% ionized; 40% is protein bound, especially to albumin; and 10% is complexed with anions such as citrate or phosphate. It is only the ionized calcium that is biologically active in bone formation, neuromuscular activity, cellular biochemical processes, and blood coagulation.

Alkalosis decreases ionized calcium levels in serum, and acidosis increases it. Ionized calcium is almost always increased in hypercalcemic conditions, and similarly, is almost always decreased in hypocalcemia (<6.5 mg/dl). During hypoalbuminemia (low serum albumin), the amount of calcium bound to protein (and thus the total serum calcium) drops. However, the amount of ionized calcium remains normal, or nearly so. Therefore, in dogs with low serum albumin, low serum calcium would be expected, although clinical signs of hypocalcemia do not occur. Therefore, a correction is necessary.

Procedure

Using the provided serum samples, one from a normal dog (sample A) and another from a hypoalbuminemic dog (sample B), measure the serum calcium, total protein, and albumin in both samples with your clinical chemistry machine.

Use the following formula to calculate the adjusted calcium.

adjusted calcium (mg/dl) = 3.5 − albumin (mg/dl) + measured calcium (mg/dl)

A:_____ = 3.5 − _____ + _____

B:_____ = 3.5 − _____ + _____

Another formula uses total protein to get the adjusted calcium.

adjusted Ca^{++} (mg/dl) = measured Ca^{++} (mg/dl) − [0.4 × serum protein (g/dl)] + 3.3

A:_____ = _____ − [0.4 × _____] + 3.3

B:_____ = _____ − [0.4 × _____] + 3.3

Questions

1. Was there a decrease in measured serum calcium (when compared to the normal dog) associated with a hypoalbuminuria?

2. Was there a decrease in calculated adjusted serum calcium associated with a hypoalbuminuria?

3. How did the two methods of calculating adjusted serum calcium compare, that is, were they close or not?

Discussion

There should have been a decrease in measured calcium in the hypoproteinemic dog because the test is measuring protein-bound calcium as well as free, ionized calcium. With the correction factor, the calculated results should compare well to the normal dog. If the reason for the decrease in serum protein is a drop in the albumin level, there should be a proportional decrease when comparing a decrease in albumin levels to a decrease in total serum protein, and the values should compare well. However, if fibrinogen or globulins are decreased and albumin stays the same, the values will vary.

CLINICAL SIGNIFICANCE

The clinically significant aspects of this chapter could be a book on orthopedic surgery. A common injury in large dogs is a rupture of the cranial cruciate ligament. In Exercise 7.4, the anatomy of the cranial cruciate ligament was discussed. The cruciate ligaments stabilize the stifle joint cranially and caudally. Specifically, the cranial cruciate ligament prevents abnormal displacement of the tibia cranially from its normal articulated position with the femur.

When this ligament ruptures, a pathological movement called cranial drawer movement occurs. The name is presumably because of the movement's similarity to a drawer being opened on a desk, and it is the diagnostic feature of this condition. The veterinarian can assess the degree of movement of the tibia from its normal position to ascertain whether the ligament is strained or completely ruptured. Sometimes there is also a tear of the cartilage of the medial meniscus. Veterinary surgeons can open the joint capsule to observe whether this tear has occurred and trim away any remaining pieces of the torn ligament. If the remaining ligament is not removed, it may calcify and give rise to osteoarthritis in the joint. There are many corrective surgeries developed for repair of this injury.

A recent study, conducted by Sandman and Harariand published in the November 2001 issue of Veterinary Medicine, discussed various methods of repairing cranial cruciate ligament ruptures and the frequency with which veterinarians use the various methods. The most common was the Modified DeAngelis Technique in which a strong, non-absorbent suture material is placed on the outside of the joint from the caudal aspect of the femur to the tibial crest and continued either through the bone, to the attachment of the straight patellar ligament, or a combination of both. This substitutes for the missing ligament, stabilizes the joint, and prevents drawer movement.

Veterinary Vignettes

"Doc, Mary Thompson just called. She's bringing Rocky in right away. He seems to be having difficulty breathing," my receptionist told me.

I was just about to anesthetize my first surgery of the morning. I put the needle cover back on the needle and asked my technician to take the dog back to its cage. This was going to have to wait. I knew that if Mary was in a rush to get her dog here, there was something very wrong with him. Mary was a nurse anesthetist at the local hospital, and unlike some clients, she was not the type to panic unless there was good reason.

When Rocky arrived I could tell immediately he was in great distress. He was open-mouth breathing, and his gums were blue. He was cyanotic; his tissues weren't getting enough oxygen, and his mucous membrane color reflected this state.

"He seemed fine last night," Mary told me. "Then this morning when I got up, he was like this."

"Had he been showing any signs of problems before this, like coughing or tiring easily on exercise?"

"Yes, some coughing occasionally, but not bad. And he has been moving really slow lately, limping a bit on his front left leg, but I thought that was just age. I mean, he's a 10-year-old German Shepard! These dogs don't move fast at this age."

It was hard to understand how this dog had become this ill so quickly without having significant clinical signs previously. The physical examination revealed two major findings. First, Rocky was in congestive heart failure, and his lungs were filled with fluid. This was causing the cyanosis. Second, I found a firm mass on the distal left humerus, just above the elbow joint. This was probably an osteosarcoma, a cancerous bone tumor.

Bone, like other tissues of mesodermal origin, forms *sarcomas* that metastasize via the blood stream to the lungs. Contrast this with *carcinomas,* which arise from tissue of ectodermal or endodermal origin (i.e., epithelial tissue) and metastasize via the lymphatics, then to the lungs. The usual procedure when I find a bony growth is to radiograph both the bone and the lungs. In Rocky's case, the fluid in his lungs would mask any metastatic lesions present.

I gave Rocky an intravenous (IV) injection of furosemide, a diuretic that would help remove the fluid from his lungs and from his body via the kidneys, and put him into an oxygen cage to help relieve the hypoxemia. I then discussed the prognosis and treatment options with Mary. She already knew that the prognosis for both of Rocky's conditions was grave, especially in his current state. Personally, I gave the dog little hope. Even if he did pull through his current crisis, after spending a considerable amount of money on treatment, we could find metastasis to the lungs. It was a difficult decision for Mary to make, especially with the suddenness of the onset of the disease. I thought she would put Rocky to sleep, but she told me to go ahead with my treatment plan.

Back in 1981, when this occurred, we did not have the number of good cardiac drugs that are now available. To increase the strength of the heart muscle's contractions, we put Rocky through a rapid, 48-hour digitalization with digoxin. We monitored him closely with an electrocardiograph so we would know when to back off to a maintenance dose. Rocky gradually improved over the next week. In that time period we took serial radiographs, and I sent an electrocardiogram by phone to a cardiologist with whom I conferred on the case. After one week had passed, the radiographs showed that there was no sign of metastasis. This did not mean that it didn't exist, only that we could not see it. A lesion in the lungs must be greater than 5 mm in diameter to be visible on radiographic examination.

Ten days after entering my hospital, Rocky was stable enough for surgery. The anesthetic protocol I chose was an IV injection of a neuroleptanalgesic, a combination of a narcotic and tranquilizer. I had a narcotic reversal agent drawn up and ready to use, just in case the dog crashed. I was able to intubate Rocky and put him on a gas anesthetic with little stress or struggle. Because this was the distal humerus, not the proximal portion, I amputated the leg at the shoulder joint rather than excising both the scapula and leg.

Having a front leg removed, especially if the dog is large, makes mobility much more difficult than does excision of a rear leg. Because about 70% of the weight is placed on the front legs, the dog must move by hopping on the remaining leg. This places great stress on the joints of that leg, and if there was previous arthritis it would be difficult for the dog to move comfortably. Fortunately, Rocky's good leg was not arthritic.

Rocky went home with a strict exercise regimen, a diet low in sodium, and medications. He adapted to walking on one front leg quite well, I was told. He lived comfortably for another year and died quietly in his sleep. It is said that 5 years of survival after cancer in humans is equivalent to 6 months in dogs. Mary and I figured we had given Rocky the equivalent of an extra 10 years of life.

SUMMARY

In this chapter we covered the comparative anatomy of the skeletal system in great detail. This system has the greatest variation in structure from one species to the next. The shape and size of the bones and joints of the limbs reflect the specific needs of mobility and functionality of each species. Learning the differences in the development of both cancellous and compact bone enables you to understand the process of bone-fracture healing in the young or adult animal. Learning the names of the bones and their processes, depressions, and foramens in the various species covered enables you to become better skilled at animal restraint, blood draws, injections, and especially radiography. All these techniques use bones and their prominences as landmarks for these tasks.

CHAPTER 8

THE MUSCULAR SYSTEM

- identify and differentiate histologically the three different types of muscle tissue
- dissect the muscles of the cat and identify them by name
- gain a general knowledge of the origin, insertion, and action of the muscles dissected

- prepared slides of skeletal, cardiac, and smooth muscle
- compound light microscope
- immersion oil
- colored pencils
- cat cadaver, triple injected (order without skin attached)
- mayo dissecting scissors
- probe
- 1 × 2 thumb forceps or Adson tissue forceps
- #4 scalpel handle with blade
- rubber gloves

Introduction

The **muscular system** is closely tied to the nervous system. This is illustrated by the fact that animals can consciously control the contraction of certain muscles. These muscles are classified as **voluntary** muscles. Other types of muscle, which move independently of conscious thought, are called **involuntary** muscles. **Skeletal** muscle, as you may remember, is *striated* and *voluntary*. **Cardiac** muscle in striated and involuntary, and **smooth** muscle is *non-striated* and *involuntary*.

EXERCISE 8.1

MUSCLE HISTOLOGY

The purpose of this exercise is to demonstrate the microscopic anatomy of the three types of muscle tissue. It also illustrates the complexity of muscle anatomy, especially of skeletal muscle. As with previous chapters, obtain the recommended slide(s) and locate all the things you are instructed to view and that are labeled in the diagrams and photomicrographs. Then, draw and label what you see in the space provided.

Skeletal Muscle

Description: A muscle that is attached to bone is called a *skeletal muscle*. It is composed of multiple bundles of muscle tissue called **fascicles** (Figure 8.1). Each fascicle is surrounded by connective tissue known as the **perimysium,** which separates it from other muscle bundles. Perimysium is composed of reticular and collagenic fibers. The fascicles together make up the entire muscle and are surrounded by **epimysium,** which is composed of dense irregular connective tissue. These connective tissue layers contain blood vessels, nerves, and fat, which enable the cells to obtain nourishment, receive nervous stimulation and contract, and slide past each other with minimal friction. Fat deposits are visible in meat, and their appearance is called *marbling*. The *epimysia* of various muscles blend together to form the **deep fascia,** a coarse sheet of dense connective tissue that binds muscles into functional groups. The **superficial fascia** is less dense connective tissue that connects subcutaneous tissue to the deep fascia.

Individual skeletal muscle cells are **multinucleated** and may be 3 or 4 cm long (depending on the species). The nuclei are located just beneath the cell's plasma membrane (the **sarcolemma**), which is called the **hypolemmal** position. A muscle cell is also called a *muscle fiber*. Inside the cell, or fiber, are numerous tubular bundles called **myofibrils** which nearly fill the **sarcoplasm** (the cytoplasm of a muscle cell). These *myofibrils* are composed of even smaller threadlike structures called **myofilaments.** The myofilaments are composed largely of two varieties of contractile proteins, **actin** and **myosin,** that slide past one another during the contractile phase of muscular activity. Surrounding each muscle cell is the **endomysium,** made up of areolar connective tissue.

Slide: skeletal muscle, longitudinal section and cross-section

On the longitudinal section slide, note the cross-striations; these represent the A, I, H, and M bands of the *sarcomeres* (the contractile unit) of the myofilaments. The letters are just labels for the various bands, they do not refer to any specific name. In Figure 8.2 you can see the various bands that are visible with normal light microscopy.

The depiction of an electron micrograph in Figure 8.3*A* and the drawing in Figure 8.3*B* illustrate the arrangement of lines that gives striated muscle its appearance. The A band represents the thick *myosin* filament; it does not change in length. The thin *actin* filaments slide across the myosin during contraction. Because these fila-

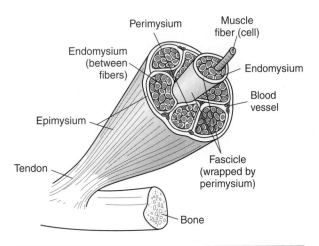

Figure 8.1: Three-dimensional view of muscle and its connective tissue layers.

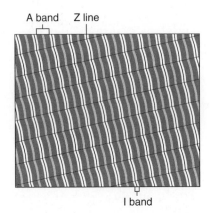

Figure 8.2: Longitudinal section of skeletal muscle from a dog.

ments are anchored at the *Z-line,* the Z to Z distance shortens during contraction (see Figure 4.4). The space between the two Z-lines is a **sarcomere,** which represents the contractile unit of striated muscle. The sarcomeres of each myofilament line up together within each muscle cell, thus each band also lines up together. During contraction, the H-zone and I-bands shorten proportionally as the length of the sarcomere shortens. The M-line is thought to be caused by transversely oriented, slender filaments that hold the myosin in place. All sarcomeres contract simultaneously, causing the entire myofibril to shorten, in turn causing the entire muscle cell to shorten. At each junction of the A and I bands, the sarcolemma indents into the cell, forming a transverse tubule (T tubule) which runs deep into the muscle cell. Surrounding each myofilament is a weblike system of

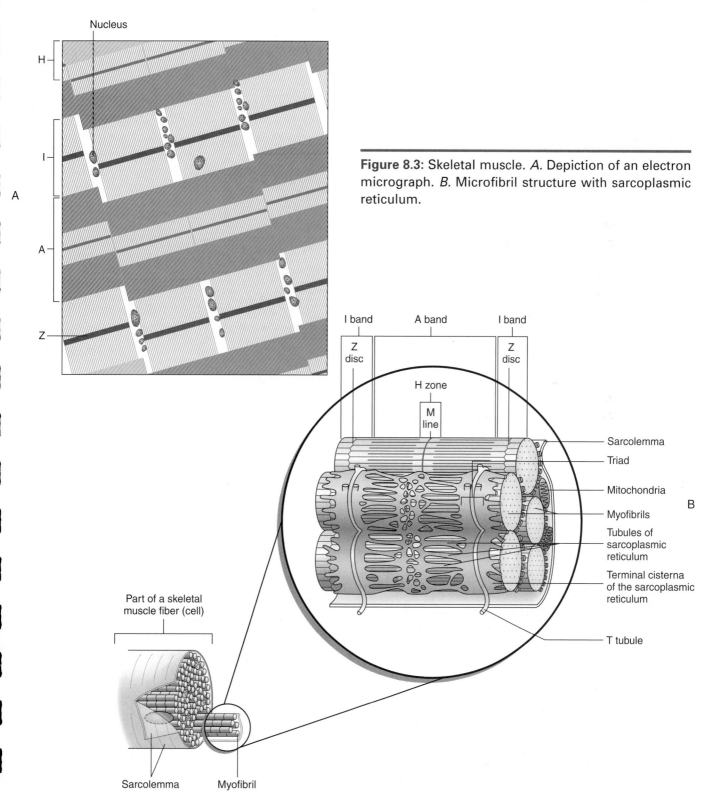

Figure 8.3: Skeletal muscle. *A.* Depiction of an electron micrograph. *B.* Microfibril structure with sarcoplasmic reticulum.

cross channels that make up the smooth endoplasmic reticulum called the **sarcoplasmic reticulum.** It is thickest where it touches the T tubules (Figure 8.4).

On the slide, find the muscle cell/fiber; the nuclei; the endomysium; perimysium (and perhaps a blood vessel within); the A, I, and H bands; and possibly the M line.

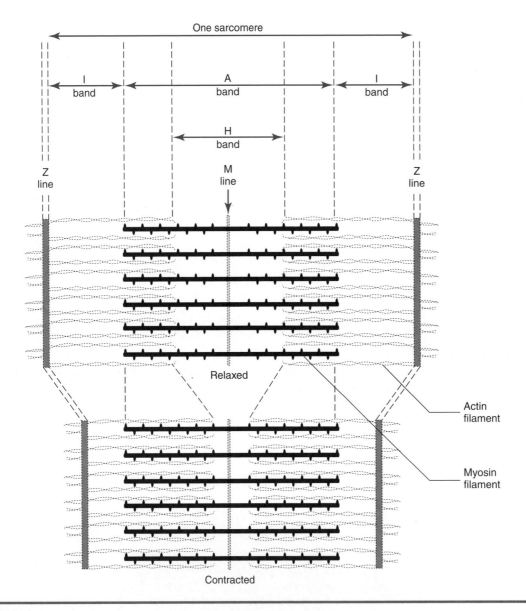

Figure 8.4: A sarcomere.

Draw a section of skeletal muscle in the space provided, and label the parts of the cells.

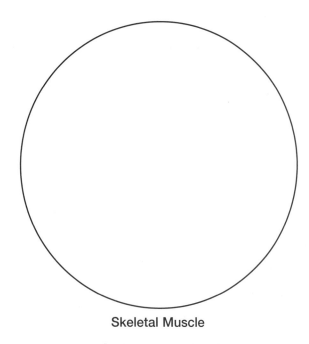

Skeletal Muscle

Cardiac Muscle

Description: *Cardiac muscle* is involuntary and striated. It is different from skeletal muscle in that its cells are generally uninucleated, may branch, and the cells anastomose with one another using special cell-surface modifications called **intercalated discs.**

Slide: cardiac muscle

The striations on the cardiac muscle cells are not as noticeable as they are on skeletal muscle; instead, they appear as multiple, fine-lined cross-striations on each cell. Find the intercalated discs, the nuclei, and an example of a branching cell (Figure 8.5).

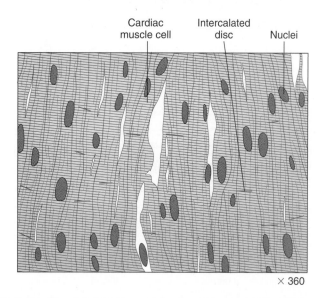

× 360

Figure 8.5: Cardiac muscle of a cat.

Draw a section of cardiac muscle in the space provided, and label the parts of the cells.

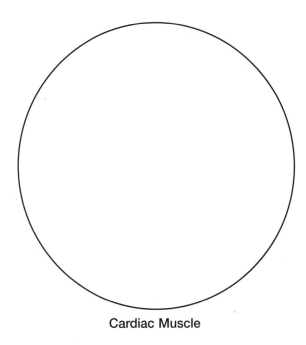

Cardiac Muscle

Smooth Muscle

Description: *Smooth muscle* is involuntary and is found in organs that need to contract to move fluids, or it acts as *sphincter muscle* to control the diameter of the opening in vessels and airways, or between tubular organs. Smooth muscle cells are *not* striated, have *spindle-shaped cells* with a central nucleus, and are arranged closely to form sheets.

Slide: smooth muscle or the intestinal tract

Smooth muscle cells are long and thin, tapered at both ends, and are coursing in the same directions but are not perfectly parallel in configuration. The nuclei may appear rippled if the cell is contracting (Figure 8.6).

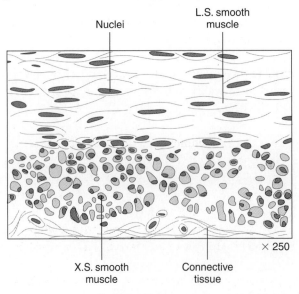

Figure 8.6: Smooth muscle from the jejunum of a sheep.

Draw a section of smooth muscle in the space provided, and label the parts of the cells.

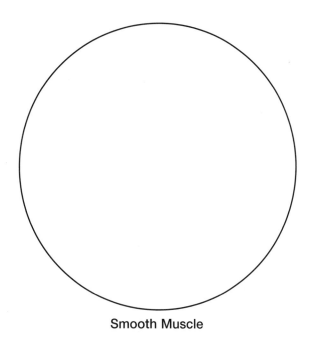

Smooth Muscle

EXERCISE 8.2

DISSECTION OF THE CAT'S SKELETAL MUSCULATURE

The anatomical terms for the musculature of the cat differs from that of humans when naming certain muscles. There are numerous cat-dissection manuals on the market that use the analogous human muscles' names, but these are used to teach human anatomy and physiology in the allied human health professions. The textbook used as the definitive veterinary source for naming these muscles is the *Atlas of Feline Anatomy for Veterinarians* by Hudson and Hamilton, published in 1993.

The important muscles to identify in this exercise are listed in colored bold print. If a muscle is mentioned prior to its dissection, it will be italicized. Also realize that there are many muscles in the cat that you will not dissect out; these are generally the smaller, deeper muscles.

Each muscle has an **origin** (the site of attachment to a fixed bone) and an **insertion** (the attachment to a more freely movable bone). As a muscle contracts, it pulls the *insertion* toward the *origin*. This is known as the **action** of the muscle. Knowing the origin and insertion, in most instances you can determine a muscle's *action*. Muscles may originate directly from the bone or from fascia above the bone, and they are often attached to their insertion by *tendons*. You will remember from your study of histology that tendons, ligaments, and aponeuroses are made of dense regular connective tissue, which are linear parallel bands of collagenous fibers. Some muscles insert by means of an aponeurosis, a broad flattened sheet of tendon.

In the following dissection guide, locate each muscle before making any incisions or attempting to transect the muscle. Clean the surface of each muscle by gently pulling off or cutting away superficial fat. With this method you should be able to see the direction of the muscle fibers. Generally, all the fibers of a muscle will run in one direction, and the fibers of the muscle that is adjacent or deep to it will run in a different direction.

As mentioned previously, muscles are separated from one another by sheets of fascia that envelop each muscle. Discovering these separations will enable the dissector to effectively isolate each muscle. A scalpel or knife can often create false separations; therefore, it is best to use blunt separation whenever possible. This is best accomplished using a probe, the blunt end of a scalpel, your fingers, or scissors (by continually spreading the tips open). Surgeons use scissors (or forceps) to spread tissues apart to minimize bleeding. When separating, leave the fascia that surrounds the muscle attached. This should ensure that the entire muscle is intact and that there are no artificial separations.

To examine the deep muscles, it is frequently necessary to cut through a superficial muscle. Do not *bisect* a muscle unless directed to do so. To bisect (or transect) means to cut into two parts. When bisecting a muscle, cut through it at right angles to the direction of the muscle fibers, halfway between the origin and the insertion, using scissors or a scalpel. To see the underlying muscles, *reflect* the superficial muscle: pull the ends back, one toward the origin and the other toward the insertion. If these directions are followed, the origin and insertion of the muscle will be retained.

Part 1: Superficial Muscles of the Chest

Complete the following steps in the dissection procedure.

1. Clean the fat off the surface of the muscles on the left side of the chest. We will not dissect the right side unless errors have been made on the left.

2. Locate the **pectoralis muscles,** the large group that cover the ventral chest. The group *originates* from the sternum and inserts on the humerus, or close to it. Locate the muscles in this muscle group (discussed in the following steps) by comparing the cadaver's muscles with Figure 8.7. The *action* of this muscle group is to adduct the limb medially.

3. The **pectoralis descendens** muscle (M. pectoralis descendens) (name used in previous cat dissection guides is the pectoantebrachialis) is the most superficial muscle of this group. It is approximately 0.5 in wide and extends from the manubrium of the sternum to insert on the fascia of the forearm. The cranial edge of this muscle is about 0.3-0.5 in from the beginning of the entire pectoralis group. It is a thin muscle and must be carefully dissected from the *pectoralis transversus*. Separate this muscle from the underlying fascia and muscle.

4. The **pectoralis transversus** muscle (M. pectoralis transversus) (previous cat dissection guides use the terms pectoralis major or superficial pectoral) is the portion of the pectoralis group deep to the pectoralis descendens and *cleidobrachialis*. It *originates* on the upper part of the sternum and *inserts* along much of the humerus. It can be difficult to dissect from the more caudal (and caudolateral) *pectoralis profundus* and may appear to have two bodies or sections. The key to this dissection is to note that the fibers of the pectoralis transversus course in the same direction or slight craniolaterally (especially in the caudal part) as those of the pectoralis descendens. There is a slight separation in the fascia where the pectoralis transversus and the pectoralis profundus meet. This muscle is also not very thick, and as you start bluntly dissecting this muscle from the deeper pectoralis profundus, you will see that the deeper muscle's fibers actually run cranially under the pectoralis transversus. Also loosen this muscle away from the cleidobrachialis at its cranial border. Note its wide insertion on the humerus.

5. The **pectoralis profundus** muscle (M. pectoralis profundus) (known in previous guides as the pectoralis minor): lies caudal and deep to the pectoralis transversus. It *originates* on the middle portion of the sternum and *inserts* near the proximal end of the humerus. In the cat, this muscle is larger than the pectoralis transversus.

   a. The caudal part of the pectoralis profundus muscle used to be dissected as a separate muscle and was named the xiphihumeralis. If you want to dissect it separately as depicted in Figure 8.7. Start caudally at its origin and work cranially. Separate it from the cranial part along its caudolateral border. The muscle becomes a tendinous-appearing band as it passes deep to the cranial part of

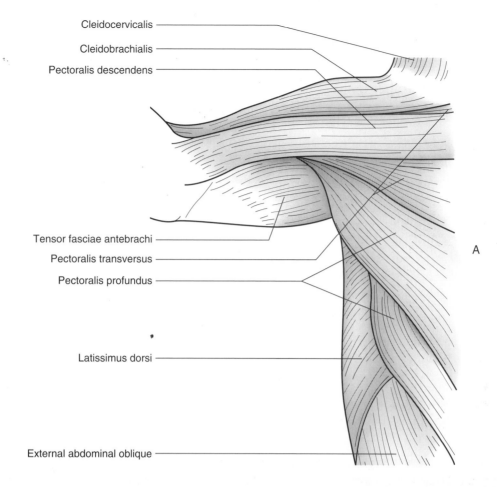

Cleidocervicalis

Cleidobrachialis

Pectoralis descendens

Tensor fasciae antebrachi

Pectoralis transversus

Pectoralis profundus

Latissimus dorsi

External abdominal oblique

A

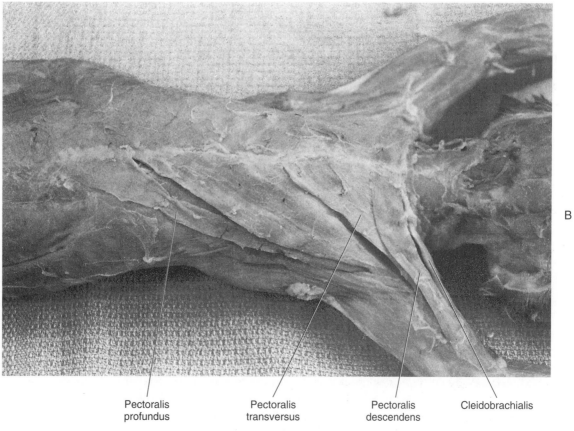

B

Pectoralis profundus

Pectoralis transversus

Pectoralis descendens

Cleidobrachialis

Figure 8.7: *A.* Ventral view of chest musculature. *B.* Ventral view of chest musculature of a cat.

the pectoralis profundus and *inserts* on the humerus. You also will need to separate the lateral edge of the xiphihumeralis from the *latissimus dorsi;* it courses deep to this muscle, too.

6. The muscle on the shoulder cranial to the pectoralis descendens is the **cleidobrachialis** muscle (M. cleidobrachialis) (the clavodeltoid in previous guides). By blunt dissection, separate the cleidobrachialis from the connective tissue deep to it. Palpate the clavicle on the underside of the medial end of this muscle (toward the neck). The cleidobrachialis *originates* on the clavicle and *inserts* on the *ulna* with the pectoralis descendens. It appears as a continuation of the *cleidocervicalis* on the shoulder. Its *action* is to help draw the limb forward.

Part 2: Superficial Muscles of the Neck

1. Remove additional skin from the left side of the neck up to the base of the ear. Be careful not to cut or damage the external jugular or the transverse jugular veins. The external jugular is the large vein that lies on the ventral surface of the neck, and the transverse jugular connects the right and left external jugular veins at the level of the larynx. At this time, free these veins from the underlying muscles. (Refer to Figure 8.8 during this dissection.)

2. Remove any excessive connective tissue or fat from the back of the left shoulder and from the ventral and lateral surfaces of the neck. Do not remove the fascia in the midline of the back because it is part of the origin of the *trapezius* muscle.

3. The **sternocephalicus** muscle (M. sternocephalicus) has two parts, mastoid and occipital, known as the **sternomastoid** and **sterno-occipitalis** muscles. They are difficult to separate by dissection, and it is easier to just identify the main areas of these muscles by their thickness and shape. They *originate* as a unit from the manubrium sterni and course diagonally to *insert* on the mastoid part of the temporal bone and the dorsal nuchal line of the occipital bone of the skull (see Figure 8.8). The sternomastoid muscle is the ventral portion and separates as a thick, elliptical bundle that unites with the cleidomastoid as a strong tendon at its insertion. Both parts of the muscle pass deep to the external jugular vein. The sterno-occipitalis is the broad, thinner, dorsal segment that lies immediately cranial to the cleidocervicalis. The *action* of this muscle is to turn and bend the head. Separate both borders of this muscle.

4. Next, locate the **sternohyoid** muscles (M. sternohyoideus). These are a narrow pair of muscles that are connected together along the midline, just superficial to the trachea in the middle of the neck. The caudal ends are covered by the sternomastoid. The sternohyoids *originate* from the first costal cartilage of the sternum and *insert* onto the hyoid bone. Its *action* is to pull the basihyoid bone and tongue caudally.

5. The **cleidomastoid** muscle (M. cleidomastoideus) is a narrow band of muscle, of which the cranial part passes deep to, and the caudal part courses dorsolateral to the sternomastoid muscle (see Figure 8.8). It *originates* on the clavicle deep to the *cleidocervicalis* and *inserts* on the mastoid region of the temporal bone. Its *action* is to help draw the limb forward and also to fix the neck, holding the head in one position. Locate this muscle by looking between the sternomastoid and cleidocervicalis. Then separate the cleidomastoid along both borders.

6. The **digastric** muscle (M. digastricus) is a thin, superficial muscle that runs adjacent to the inner surface of the mandible (see Figure 8.8). It *originates* on the occipital and temporal bones and *inserts* on the mandible. Its *action* is to help open the mouth.

7. Inserting deep to the digastric and bridging the two sides of the mandible is the **mylohyoid** muscle (M. mylohyoideus). To see its fibers you must carefully remove the thin deep fascia that covers this muscle. The mylohyoid is a thin, sheet-like muscle with transverse fibers; it *originates* on the mandible and *inserts* on the median raphe (the connective tissue along the midline). Its *action* is to raise the floor of the mouth.

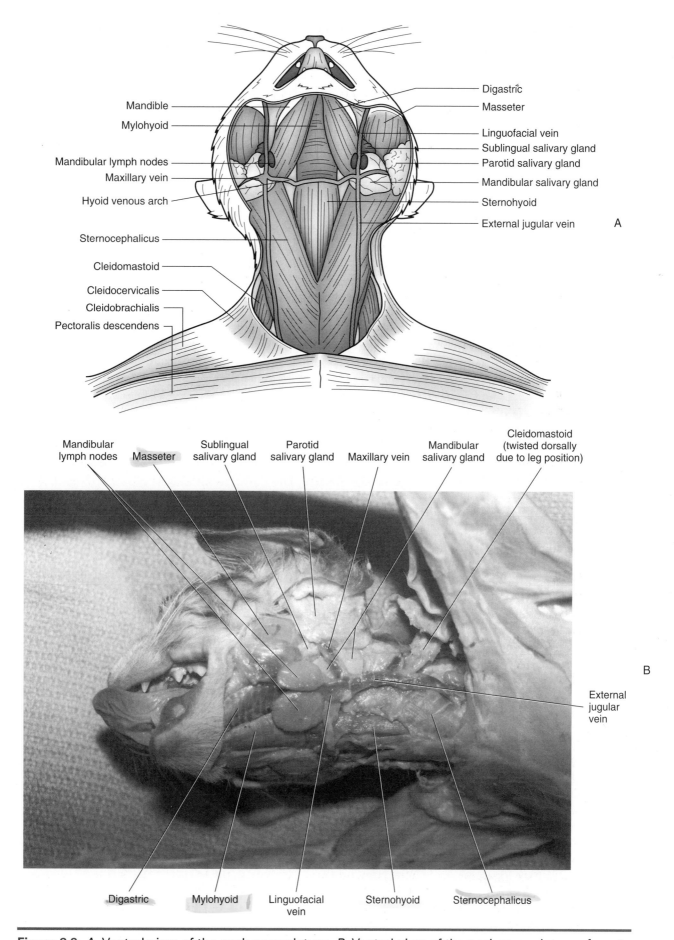

Mandible
Mylohyoid
Mandibular lymph nodes
Maxillary vein
Hyoid venous arch
Sternocephalicus
Cleidomastoid
Cleidocervicalis
Cleidobrachialis
Pectoralis descendens

Digastric
Masseter
Linguofacial vein
Sublingual salivary gland
Parotid salivary gland
Mandibular salivary gland
Sternohyoid
External jugular vein

A

Mandibular lymph nodes
Masseter
Sublingual salivary gland
Parotid salivary gland
Maxillary vein
Mandibular salivary gland
Cleidomastoid (twisted dorsally due to leg position)

B

External jugular vein

Digastric
Mylohyoid
Linguofacial vein
Sternohyoid
Sternocephalicus

Figure 8.8: *A.* Ventral view of the neck musculature. *B.* Ventral view of the neck musculature of a cat.

8. Note the large **parotid salivary gland** located ventral to the ear. The large muscle mass cranial to this gland at the angle of each jaw is the **masseter muscle** (M. masseter). It *originates* from the zygomatic arch and *inserts* onto the mandible. Its *action* is to close the jaw.

9. Other structures to find are the remaining salivary glands and the lymph nodes. Located inferior to the parotid salivary gland is the large **mandibular salivary gland.** Just cranial to this gland is the small **sublingual salivary gland** (monostomatic part). These glands are separated cranially and caudally by the *maxillary vein,* and lie deep and dorsal to the external jugular vein and caudal to the transverse jugular vein. The *linguofacial vein* is the cranial branch that feeds into the jugular vein; on either side are two small **mandibular lymph nodes** (see Figure 8.8).

Part 3: Superficial Muscles of the Shoulder and Back

1. The cranial extension of the cleidobrachialis muscle (dorsal to the clavicle) is the **cleidocervicalis** muscle (M. cleidocervicalis) (in previous guides called the cleidotrapezius). This is the broad muscle located on the back and side of the neck, dorsal to the sternocephalicus muscle. It *originates* from the occipital bone and the first few cervical vertebrae and *inserts* on the clavicle. The cleidocervicalis' *action* is to elevate the clavicle.

2. The **trapezius** muscle covers much of the dorsal surface of the scapula (Figure 8.9). It is divided into two parts: the **cervical part** (in previous guides called the acromiotrapezius) and the **thoracic part** (in previous guides called the spinotrapezius). The cervical part covers the dorsal part of the scapula. It is a fan-shaped muscle that *originates* from fascia along the median dorsal line of the back and *inserts* on the cranial aspect of the scapular spine. The thoracic part of the trapezius group is a triangular sheet of muscle located caudal to the cervical part of the trapezius. It *originates* from the fascia above the spinous processes of the thoracic vertebrae and *inserts* on the caudal aspect of the scapular spine. Separate these muscles from underlying muscles and free their cranial and caudal borders. The *action* of this muscle group is to adduct the scapula.

3. The **latissimus dorsi** muscle (M. latissimus dorsi) is the large muscle caudal to the trapezius group (see Figure 8.9). It *originates* from the spinous processes of the thoracic vertebrae and the lumbodorsal fascia. It courses cranioventrally to *insert* on the proximal end of the humerus with the xiphihumeralis. Its *action* is to draw the trunk forward, depress the vertebral column, draw the limb against the trunk, and draw the limb caudally during flexion of the shoulder joint. Free the cranial and caudal borders of the latissimus dorsi. Part of the cranial border lies beneath the thoracic part of the trapezius.

4. The **omotransversarius** muscle (M. omotransversarius) (in previous guides as the levator scapulae ventralis) in the cat is a strap-like band of muscle that *originates* from the atlas and the occipital area of the skull and passes caudally beneath the cleidocervicalis to *insert* on the acromion process of the scapula. Its *action* is to pull the scapula cranially (see Figure 8.9).

5. The **deltoid** muscle (M. deltoideus) consists of two parts: the **acromial part** (in previous guides as the acromiodeltoid) and the **spinous part** (in previous guides as the spinodeltoid). The spinous part of the deltoid is the most caudal portion. It *originates* along the caudal aspect of the spine of the scapula ventral to the insertion of the thoracic part of the trapezius. It runs almost parallel with the edge of the spine of the scapula and *inserts* on the proximal humerus. Separate and free the borders of this muscle. The acromial part of the deltoid *originates* from the acromion process of the scapula and *inserts* on the proximal end of the humerus. Free the edges of this muscle. The *actions* of the deltoids are to flex, extend, rotate, and abduct the humerus (Figures 8.9*A* and 8.10).

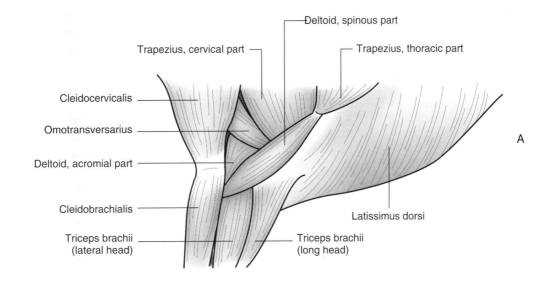

Trapezius, cervical part

Deltoid, spinous part

Trapezius, thoracic part

Cleidocervicalis

Omotransversarius

Deltoid, acromial part

Cleidobrachialis

Triceps brachii
(lateral head)

Triceps brachii
(long head)

Latissimus dorsi

A

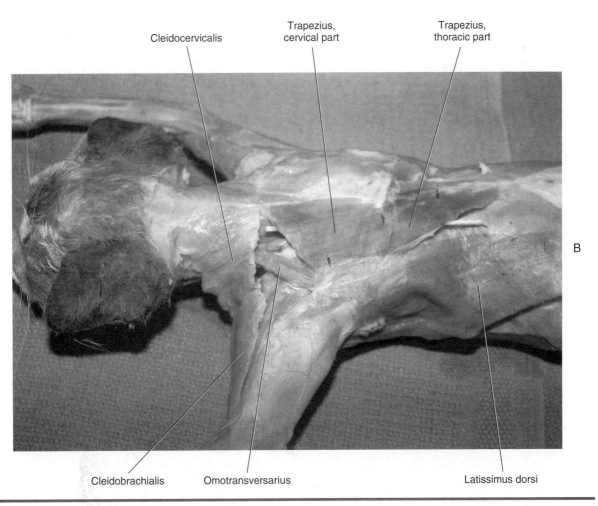

Cleidocervicalis

Trapezius,
cervical part

Trapezius,
thoracic part

B

Cleidobrachialis

Omotransversarius

Latissimus dorsi

Figure 8.9: *A.* Superficial muscles of the shoulder and back. *B.* Superficial muscles of the shoulder and back in a cat.

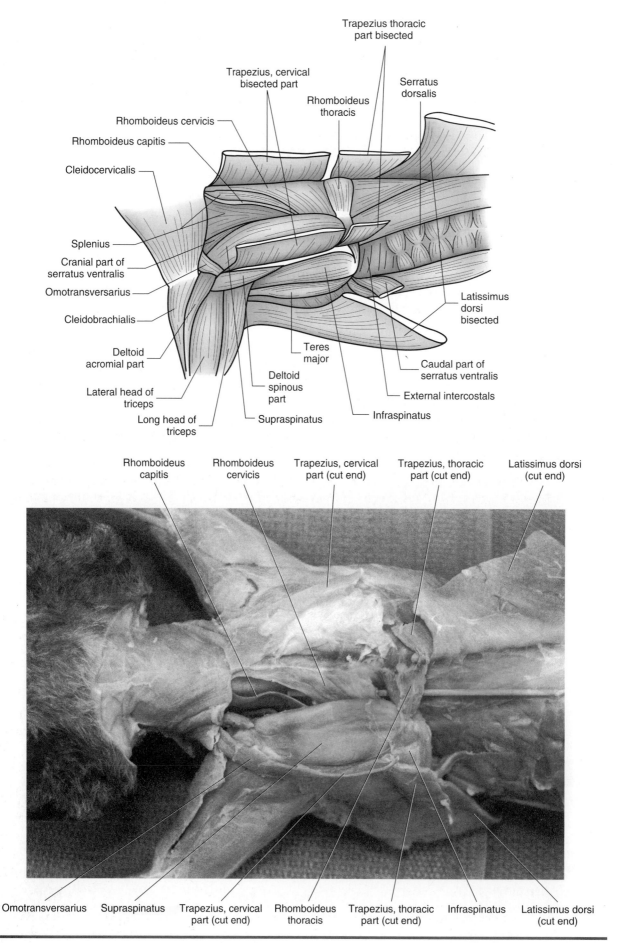

Trapezius thoracic
part bisected

Trapezius, cervical
bisected part

Rhomboideus
thoracis

Serratus
dorsalis

Rhomboideus cervicis

Rhomboideus capitis

Cleidocervicalis

Splenius

Cranial part of
serratus ventralis

Omotransversarius

Cleidobrachialis

Deltoid
acromial part

Lateral head of
triceps

Long head of
triceps

Supraspinatus

Deltoid
spinous
part

Teres
major

Infraspinatus

External intercostals

Caudal part of
serratus ventralis

Latissimus
dorsi
bisected

Rhomboideus
capitis

Rhomboideus
cervicis

Trapezius, cervical
part (cut end)

Trapezius, thoracic
part (cut end)

Latissimus dorsi
(cut end)

Omotransversarius

Supraspinatus

Trapezius, cervical
part (cut end)

Rhomboideus
thoracis

Trapezius, thoracic
part (cut end)

Infraspinatus

Latissimus dorsi
(cut end)

Figure 8.10: Deep muscles of the lower shoulder and superficial muscles of the upper front leg of a cat.

Part 4: Deep Muscles of the Shoulder and Back

1. To expose the deeper muscles of the shoulder and back, carefully bisect the cervical and thoracic parts of the trapezius muscles halfway between their origins and insertions (*not* at the midline of the back). Reflect the bisected muscles dorsally and ventrally.

2. The **supraspinatus** muscle (M. supraspinatus) fills the supraspinous fossa of the scapula cranial to the spine. The **infraspinatus** muscle (M. infraspinatus) fills the infraspinous fossa caudal to the scapula's spine. Both muscles *originate* on the scapula and *insert* on the humerus (see Figure 8.10). The supraspinatus' *action* extends the humerus and the infraspinatus rotates it laterally.

3. The **teres major** muscle (M. teres major) *originates* on the dorsocaudal border of the scapula, caudal to the infraspinatus (see Figure 8.10). It *inserts* on the proximal end of the humerus with the latissimus dorsi. Separate the teres major from the infraspinatus. The *action* of this muscle is to extend and rotate the humerus.

4. Observe the dorsal rim of the scapula. A large muscle group *originates* from the spinous processes of the caudal cervical and cranial thoracic vertebrae. This is the **rhomboideus** group of muscles (M. rhomboideus), which contains three muscles (see Figure 8.10). The most caudal is the **rhomboideus thoracic** (M. rhomboideus thoracic) (in previous guides as the rhomboideus major). It *inserts* on the dorsal caudal angle of the scapula.

   Just cranial to this is the **rhomboideus cervicis** (M. rhomboideus cervicis) (in previous guides as the rhomboideus minor). It inserts along the dorsal rim of the scapula. The division between these two muscles is somewhat obscure (difficult to see). As the rhomboideus thoracic inserts, it partially covers the dorsal aspect of the scapula. Follow the cranial edge of this insertion dorsally toward the spine, and you will find a natural separation in the fibers above the rim of the scapula, and thus the separation between the rhomboideus thoracic and the rhomboideus cervicis. The *action* of these two muscles draws the scapula dorsally and rotates it.

   The **rhomboideus capitis** (M. rhomboideus capitis) is the most cranial muscle of this group. It is a narrow, thin, ribbon-like muscle that *originates* from the occipital bone and *inserts* on the cranial edges of the scapula just ventral to the dorsocranial angle. Its *action* is to draw the scapula cranially. Carefully dissect this muscle from the deeper muscles.

5. Deep to the rhomboideus capitis is a broad, flat muscle called the **splenius** muscle (M. splenius). It covers most of the dorsal and lateral surface of the neck. Its *action* is to turns the head.

6. Bisect the latissimus dorsi muscle at its center and reflect the halves dorsally and ventrally. This will allow you to see the deeper chest muscles. You will note a large, fan-shaped, serrated-appearing muscle called the **serratus ventralis** muscle (M. serratus ventralis). It *originates* from the ventral aspect of the ribs, passes medial to the scapula, and *inserts* on the medial dorsal border of the scapula (Figures 8.10, 8.11, and 8.12*A*). This muscle is also visible on the lateral surface of the body at its origin on the ribs. The *action* of this muscle is to pull the scapula forward and down.

7. The **subscapularis** muscle (M. subscapularis) fills the subscapular fossa of the scapula on its medial side. Thus, it *originates* on the medial scapula and *inserts* on the humerus. Its action is to rotate the scapula medially. Do *not* dissect this muscle.

Part 5: Muscles of the Front Leg

1. Clear away the superficial fat and fascia from the upper leg (also called the *brachium*).

2. The **epitrochlearis** muscle (M. epitroclearis) is a thin, flat muscle that *originates* from the surface of the insertion of the latissimus dorsi, extends along the medial surface of the arm, and *inserts* on the olecranon process of the ulna. The *long head of the triceps brachii* lies deep and caudal to this muscle (Figure 8.13).

3. On the caudal aspect of the front leg is the **triceps brachii** group of muscles (M. triceps brachii) (see Figures 8.10, 8.11, and 8.12). The name *triceps* indicates three muscles in this group. These encompass most of the caudal and lateral surfaces. The **long head** of the triceps is on the caudal

surface and is the largest of the three muscles. It *originates* from the scapula caudal to the glenoid cavity and *inserts* on the olecranon process. The *action* of the triceps group of muscles is to extend the elbow joint. Free both borders of the long head.

4. The **lateral head** is located just cranial to the long head on the lateral aspect of the leg. It *originates* from the proximal end of the humerus and *inserts* on the olecranon process. Free the cranial border first; then free the caudal border, making a linear incision in the fascial attachment at its distal insertion, and continue it for approximately 1 cm down the olecranon process. This will enable you to reflect this muscle cranially. You may choose to bisect the muscle to view the medial head of the triceps. However, some students choose not to do this and instead reflect the muscle cranially to view the medial head. On the medial side of the lateral head is the **accessory head.** Try to separate it along the cranial border, sometimes a distinct separation can be found.

5. The long, narrow **medial head** is located beneath the lateral head and can be viewed by reflecting the lateral head in a cranial direction. It is located just caudal to the humerus. It also *originates* from the proximal end of the humerus and *inserts* on the olecranon process.

6. The **anconeus** muscle (M. anconeus) is a thin, flat, triangular muscle that lies deep to the distal insertion of the lateral head of the triceps. It can be seen once the fascia above the lateral aspect of the olecranon process is opened and the insertion of the lateral head of the triceps is reflected cranially (see Figure 8.12). The anconeus *originates* on the lateral epicondylar area and the distal shaft of the humerus and *inserts* on the olecranon process. Even though this is a small muscle, it is important

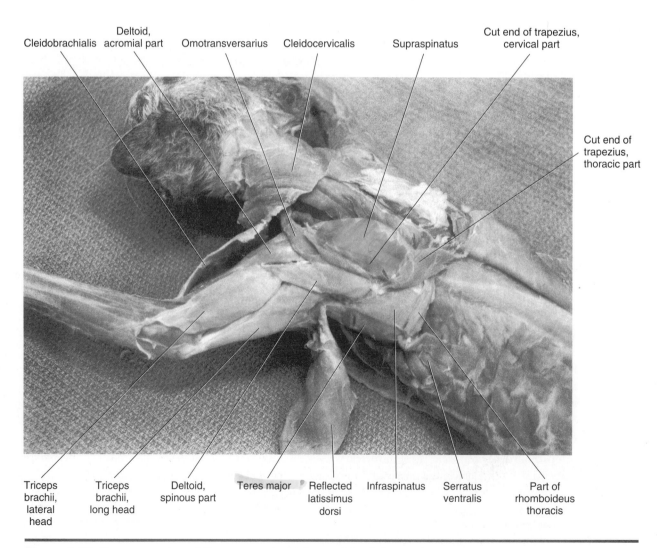

Figure 8.11: Deep muscles of the upper shoulder and back in a cat.

because it must be incised to perform surgery on the anconeal process in a pathological condition known as ununited anconeal process (elbow dysplasia). Just view this muscle; do not attempt to free its edges as it tears easily.

7. The **brachialis** muscle (M. brachialis) is located on the lateral surface of the humerus, cranial to the lateral head of the triceps (see Figure 8.12). It *originates* from the humerus and *inserts* on the proximal end of the ulna. The *action* of the brachialis is to aid in flexion of the elbow joint.

8. The **biceps brachii** muscle (M. biceps brachii) is found on the craniomedial surface of the humerus. Because much of this muscle lies beneath the insertion of the pectorals, it is best viewed by elevating the insertions of the pectoralis transversus, pectoralis profundis, and pectoralis descendens muscles. The biceps brachii *originates* on the scapula and *inserts* on the radius. This muscle's *action* is to aid in flexing the elbow joint (see Figure 8.13*B*).

9. Deep beneath the pectoralis muscles is the **coracobrachialis** muscle (M. coracobrachialis). To locate this muscle, the pectoralis transversus and pectoantebrachialis must be elevated, and the lateral border of the pectoralis profundus must be separated completely from the fascia of the insertion of the latissimus dorsi and epitrochlearis. It should take only a few small snips with your scissors to make this separation. The coracobrachialis is found at the origin of the biceps brachii and is a short band of muscle approximately 4 to 6 mm wide. It runs at a 45° angle obliquely toward the body and away from the insertion of the biceps brachii. It *originates* from

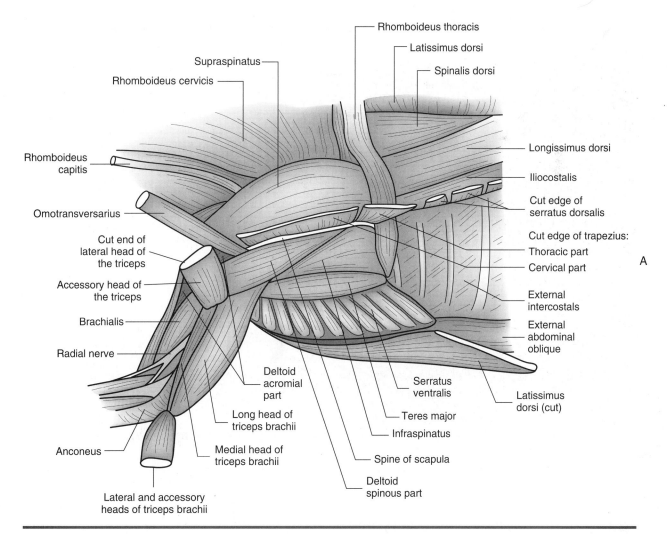

Figure 8.12: *A.* Lateral view of the deep muscles of the upper front leg.   *Continued*

Brachialis | Cut end of triceps brachii, lateral and accessory heads | Deltoid, spinous part

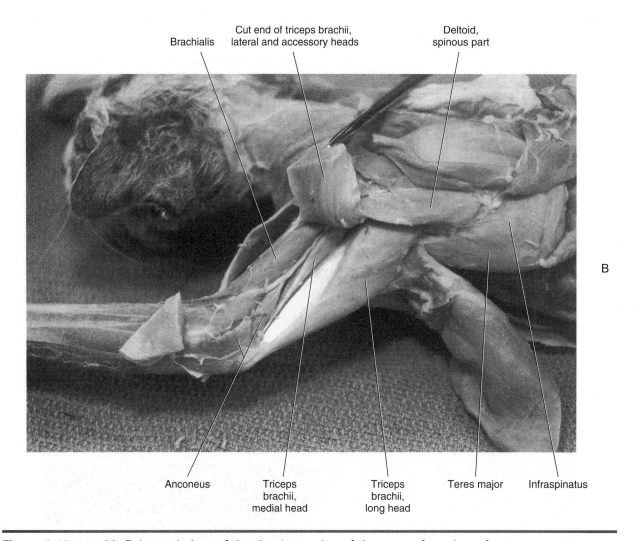

B

Anconeus | Triceps brachii, medial head | Triceps brachii, long head | Teres major | Infraspinatus

Figure 8.12, cont'd: *B.* Lateral view of the deep muscles of the upper front leg of a cat.

the coracoid process of the scapula and *inserts* on the proximal end of the humerus. Its *action* is to adduct the front leg.

The remaining steps (10-12) are optional.

10. Note the thick layer of fascia covering the foreleg muscles. Remove this fascia so that the muscles can be identified. Observe the tendons of these muscles at the carpus, and separate the superficial muscles of the foreleg. Do not bisect any of these muscles. Compare your dissection with Figures 8.14 and 8.15 and identify each of the labeled muscles. The cranial group of muscles, in general, extends and supinates. The caudal group of muscles are flexors and pronators.

11. On the cranial side, the most cranial muscle at the elbow joint is the **brachioradialis** muscle (M. brachioradialis). Proceeding caudally, you will encounter the **extensor carpi radialis** (M. extensor carpi radialis) *longus* and *brevis,* **extensor digitorum communis** (M. extensor digitorum communis), **extensor digitorum lateralis** (M. extensor digitorum lateralis), and the **ulnaris lateralis** muscle (M. ulnaris lateralis). Deep to the tendons of the previously mentioned muscles, running obliquely from caudodorsal to cranioventral, is the **adductor pollicis longus** muscle (M. adductor pollicis longus;) (not shown in Figure 8.14). These muscles, as a group, are called the extensors and are innervated by the **radial nerve.**

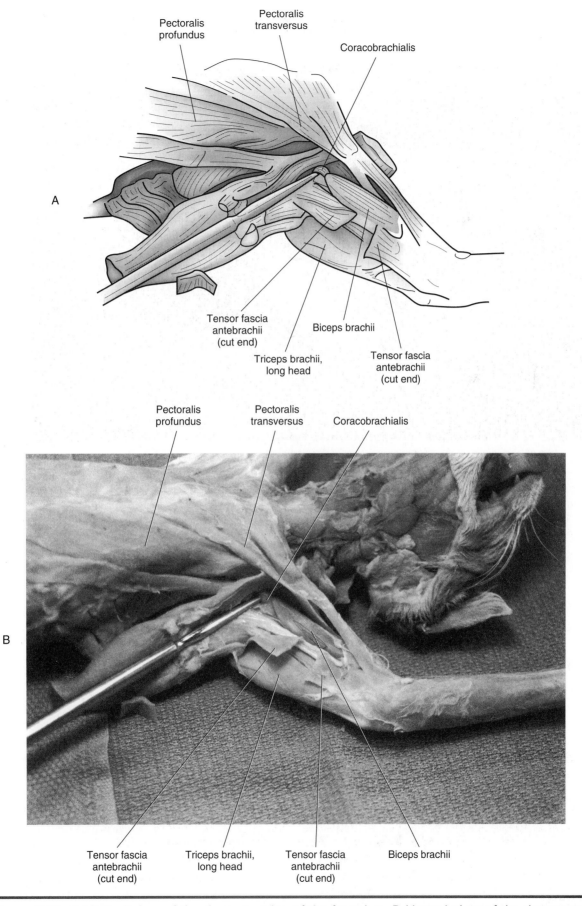

Pectoralis
profundus

Pectoralis
transversus

Coracobrachialis

A

Tensor fascia
antebrachii
(cut end)

Biceps brachii

Triceps brachii,
long head

Tensor fascia
antebrachii
(cut end)

Pectoralis
profundus

Pectoralis
transversus

Coracobrachialis

B

Tensor fascia
antebrachii
(cut end)

Triceps brachii,
long head

Tensor fascia
antebrachii
(cut end)

Biceps brachii

Figure 8.13: *A.* Ventral view of the deep muscles of the front leg. *B.* Ventral view of the deep muscles of the front leg of a cat.

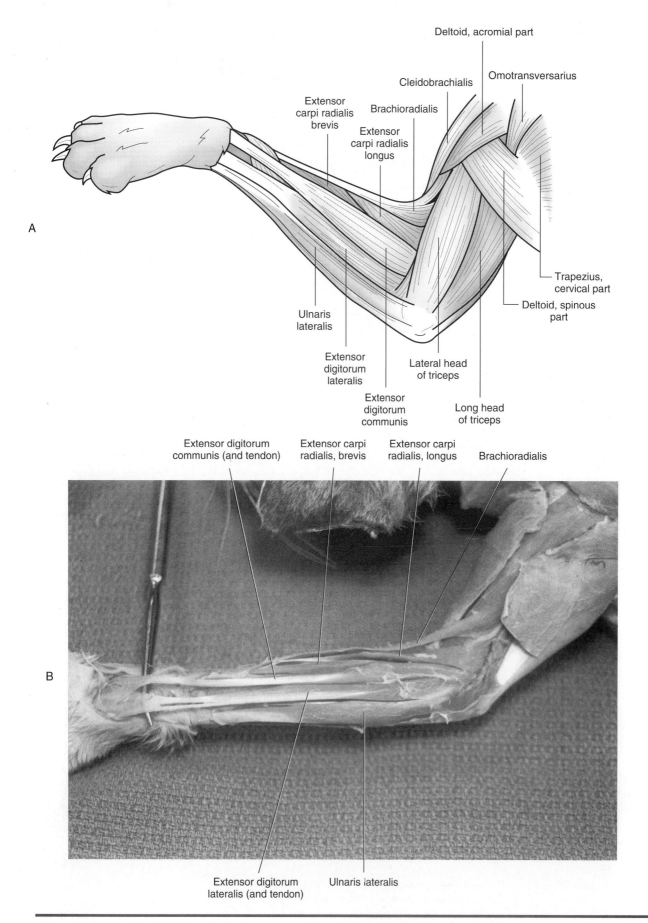

A

Deltoid, acromial part

Cleidobrachialis

Omotransversarius

Extensor
carpi radialis
brevis

Brachioradialis

Extensor
carpi radialis
longus

Trapezius,
cervical part

Deltoid, spinous
part

Ulnaris
lateralis

Extensor
digitorum
lateralis

Lateral head
of triceps

Extensor
digitorum
communis

Long head
of triceps

B

Extensor digitorum
communis (and tendon)

Extensor carpi
radialis, brevis

Extensor carpi
radialis, longus

Brachioradialis

Extensor digitorum
lateralis (and tendon)

Ulnaris lateralis

Figure 8.14: *A.* Cranial view of the muscles of the foreleg. *B.* Cranial view of the muscles of the foreleg of a cat.

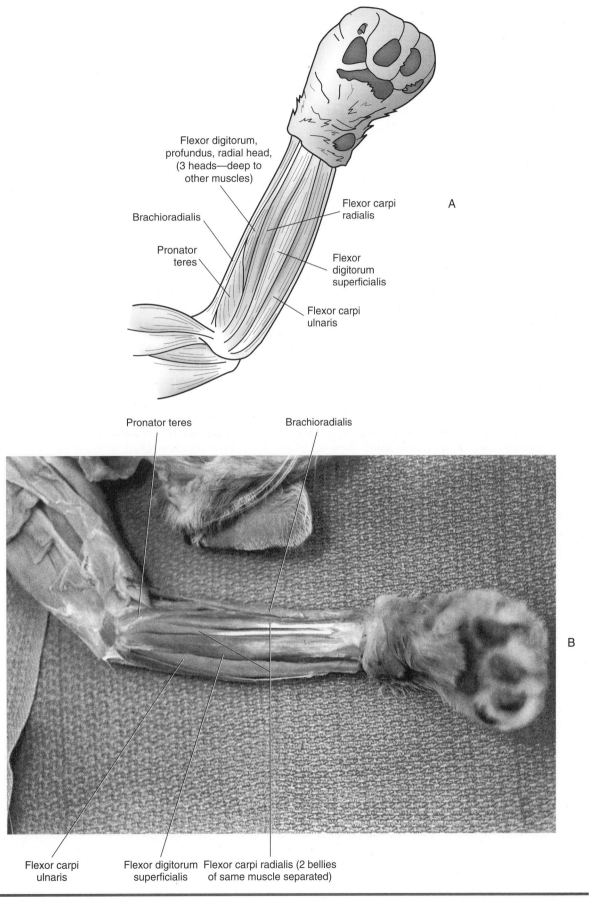

Flexor digitorum, profundus, radial head, (3 heads—deep to other muscles)

Brachioradialis

Pronator teres

Flexor carpi radialis

Flexor digitorum superficialis

Flexor carpi ulnaris

A

Pronator teres

Brachioradialis

B

Flexor carpi ulnaris

Flexor digitorum superficialis

Flexor carpi radialis (2 bellies of same muscle separated)

Figure 8.15: *A.* Medial view of the superficial muscles of the foreleg. *B.* Medial view of the superficial muscles of the foreleg of a cat.

12. On the caudal side is another muscle that angles across the proximal foreleg, the **pronator teres** (M. pronator teres). Just caudal to this muscle is the **flexor carpi radialis** muscle (M. flexor carpi radialis), followed caudally by the **flexor digitorum superficialis** (M. flexor digitorum superficialis) and the **flexor carpi ulnaris** (M. flexor carpi ulnaris). If these three superficial muscles are removed, the **flexor digitorum profundus** (M. flexor digitorum profundus) with its three heads can be viewed (not shown in Figure 8.15). The **radial head** appears most cranially, then the **humeral head,** followed by the **ulnar head.**

Part 6: Abdominal Wall Muscles

1. Remove the fat and superficial fascia from the ventral and lateral surfaces of the left side of the trunk between the xiphihumeralis and the pelvic area. Note the **lumbodorsal fascia,** the wide sheet of white-colored fascia covering the lumbar region of the back.

2. The lateral abdominal wall of the cat is composed of three layers of muscle. These abdominal muscles are very thin, each no greater than 1 mm in thickness; therefore, be extremely conservative in your dissection and cut through only one layer at a time.

3. The *action* of the lateral abdominal wall muscles is to compress the abdomen. This is important in forced expiration, defecation, micturition, and parturition. The muscles are also important in dorsal, ventral, and lateral flexion of the trunk.

4. The **external abdominal oblique** muscle (m. obliquus externus abdominis) forms the most superficial layer of the lateral abdominal muscles (Figure 8.16). It *originates* on the posterior ribs and the lumbodorsal fascia, beneath the caudal edge of the latissimus dorsi muscle, and *inserts* as fascia on the ventral midline. Make a cut in the middle of this muscle parallel to the muscles fibers and separate the opening you have formed. Once this muscle layer is isolated, bluntly dissect between this muscle and the internal oblique. Separate it and extend the opening ventrally to the **ventral fascia.** The top layer of this fascia is a broad sheet of connective tissue called the *aponeurosis* of the external oblique. Notice that the muscle fibers terminate about 1 in before the **linea alba** (also called the *white line;* the mid-ventral line formed by the union of the aponeuroses of the lateral abdominal wall muscles). The aponeurosis is superficial to the *rectus abdominis* muscle.

5. The **internal abdominal oblique** muscle (M. obliquus internus abdominis) lies deep to the external oblique, with its fibers running in the opposite direction (ventral and cranially). It *originates* on the lumbodorsal fascia, *inserts* with an aponeurosis on top of the *transverses abdominus,* and continues to the linea alba. Note where the muscle fibers terminate superficial to the tranversus abdominis before reaching the *rectus abdominis* muscle.

6. The muscle deep to the internal oblique is the **transversus abdominis** muscle (M. transversus abdominis). This may be viewed by making a second cut, in the same direction as the muscle fibers, through the internal oblique muscle. The transverses abdominis *originates* from the vertebral column, the lumbodorsal fascia, and the sacrum, and it *inserts* at the linea alba underneath the *rectus abdominis* muscle. Separate this muscle from the internal oblique and follow it as far ventrally as possible.

7. Carefully separate some of the fibers of the transversus abdominis. The shiny membrane visible beneath the muscle fibers is the parietal layer of the **peritoneum,** which should not be pierced.

8. The longitudinal band of muscle lying lateral and parallel to the linea alba on both sides is **rectus abdominis** muscle (M. rectus abdominis) (see Figure 8.16). To view this muscle, reflect the external oblique. The rectus abdominis *originates* from the cranial aspect of the pubis and *inserts* on the sternum and costal cartilages. Its *action* is to support the abdominal wall and ventrally flex the abdomen. Dissect only the lateral edge of this muscle from the underlying transversus abdominis for a short distance to enable you to see the direction of the muscle fibers.

9. In certain mammals, such as the pig, fat is deposited in the connective tissue layers between the muscles of the lateral abdominal wall so that fat and muscle alternate. This part of the pig is sold commercially as bacon. The small pieces of cartilage often found in bacon are pieces of costal cartilage from the lower ends of the ribs.

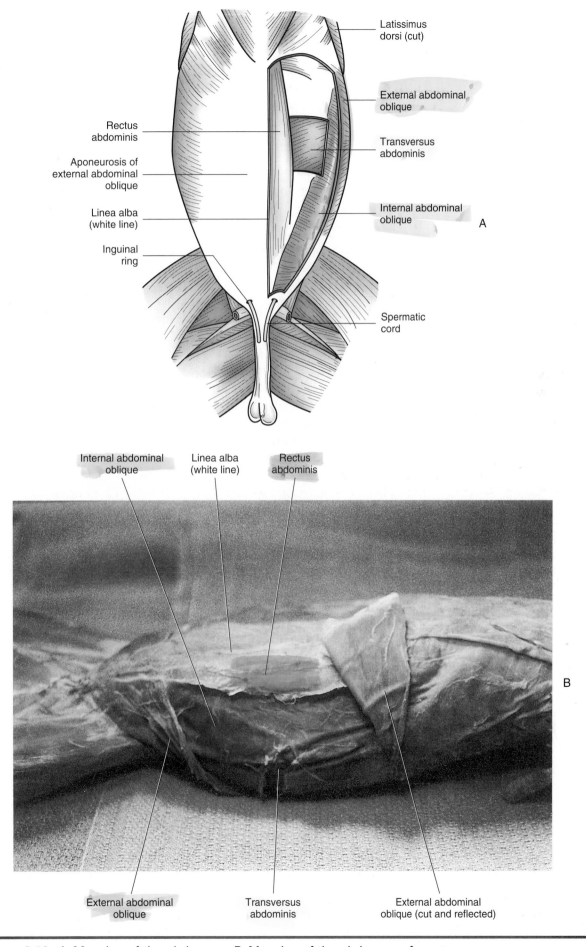

Latissimus
dorsi (cut)

External abdominal
oblique

Rectus
abdominis

Transversus
abdominis

Aponeurosis of
external abdominal
oblique

Internal abdominal
oblique

A

Linea alba
(white line)

Inguinal
ring

Spermatic
cord

Internal abdominal
oblique

Linea alba
(white line)

Rectus
abdominis

B

External abdominal
oblique

Transversus
abdominis

External abdominal
oblique (cut and reflected)

Figure 8.16: *A.* Muscles of the abdomen. *B.* Muscles of the abdomen of a cat.

Part 7: Superficial Muscles of the Hind Legs

1. Remove the fat and superficial fascia from the dorsal surface of the back above the hind legs, around the groin region, and down the legs. Be careful during your dissection not to damage any muscles in the gluteal region, because the fascia adheres tightly to the *gluteal* muscles. Also, be careful not to remove any blood vessels or nerves from the inner thigh.

2. On the superficial craniomedial surface of the thigh (Figure 8.17) is the **sartorius** muscle (M. sartorius). It is a thin, band-like muscle, about 1 to 1.5 in. wide. Free both the cranial and caudal borders from the ilium to the tibia, and bisect this muscle at the middle. The *action* of this muscle is to adduct and rotate the thigh.

3. The **gracilis** muscle (M. gracilis) is the large (but thin), wide, flat muscle covering the caudomedial surface of the thigh, caudal to the sartorius. Free both borders of this muscle and bisect it in the middle. This muscle *originates* on the os coxae bone and *inserts* on the tibia and fascia of the knee area. Its *action* is to adduct the thigh.

4. Now turn the cat over so the dorsal and lateral areas of the rear leg are presented. Caudal to the sartorius on the lateral surface of the thigh is a tough, white sheet of fascia. This is the distal aponeurosis of the **fascia lata** (Figure 8.18). To free the lateral surface of the iliotibial band from the underlying muscles, reflect the bisected sartorius muscle and pass your fingers caudally beneath the tough fascia of the iliotibial band. With your scissors or scalpel, separate the caudal border of the iliotibial band from the *quadriceps femoris* group of muscles. Bisect this band in the middle. When the proximal part of the iliotibial band is elevated, a triangular muscle mass, the **tensor fasciae latae** muscle (M. tensor fasciae latae), can be seen. This is where the aponeurosis originates. The tensor fasciae latae muscle *originates* from the ilium and *inserts* as the fascia lata. It *action* tenses the fascia lata and extends the lower leg. Free the caudal boundary between the tensor fasciae latae and the middle gluteal muscle.

5. The huge muscle caudal to the iliotibial band, covering the entire caudolateral surface of the thigh, is the **biceps femoris** muscle (M. biceps femoris) (see Figure 8.18). This muscle forms the lateral wall of the **popliteal fossa,** which is located caudal to the knee joint. It *originates* on the ischium and *inserts* on the tibia and fascia of the lower hind leg. Its *action* is to abduct the thigh and flex the lower leg. Dissect the fat from the popliteal fossa and locate the *popliteal lymph node* within (Figure 8.19C). Free both cranial and caudal borders of this muscle.

6. When freeing the cranial edge of the biceps femoris, begin near the knee joint and work upward, looking for a small, narrow muscle—the **caudofemoralis** muscle (M. caudofemoralis)—at the cranial edge (Figures 8.18 and 8.19). This muscle *originates* on the coccygeal vertebrae and *inserts* on the patella with a long, narrow tendon visible on the inner surface of the biceps femoris. Its *action* is to help adduct the thigh.

7. The **superficial gluteal** muscle (M. gluteus superficialis) is immediately cranial to the caudofemoralis (Figure 8.19A and B). Its fibers *originate* and extend laterally from the sacrum and coccygeal vertebrae and *insert* on the lateral aspect of the greater trochanter of the femur. Its *action* is to abduct the thigh.

8. The muscle that extends craniad and is located cranial to the superficial gluteal muscle is the **middle gluteal** muscle (M. gluteus medius). It *originates* on the lateral ilium and the sacral and coccygeal vertebrae, and it *inserts* on the greater trochanter of the femur, just craniomedial to the insertion of the superficial gluteal. Its *action* is to abduct the thigh also. The **deep gluteal** muscle is located beneath the middle gluteal and also *inserts* on the greater trochanter. In the Brown Approach to surgery of the hip joint, these tendons of insertion must be cut and the muscles reflected to perform surgery on this joint.

Text continued on page 165

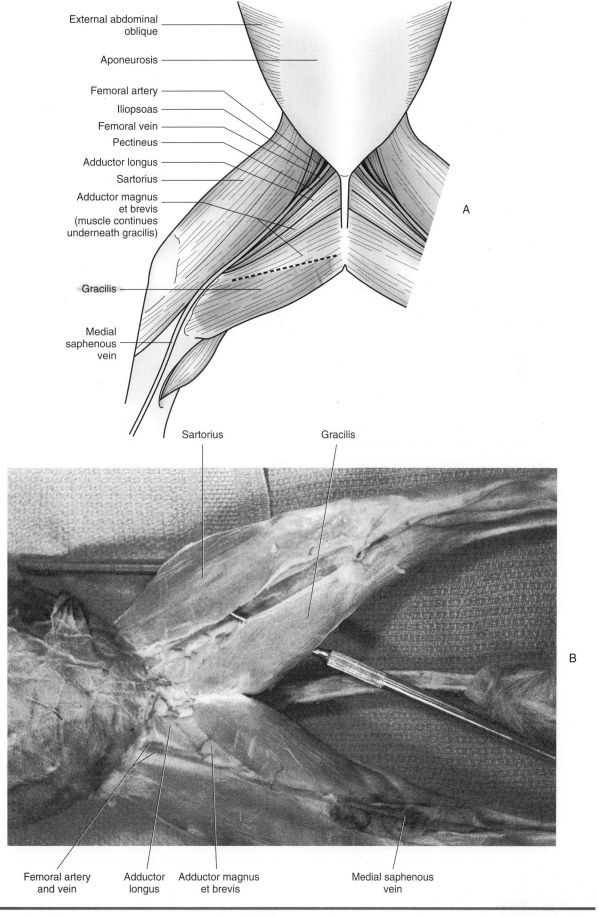

External abdominal oblique

Aponeurosis

Femoral artery

Iliopsoas

Femoral vein

Pectineus

Adductor longus

Sartorius

Adductor magnus et brevis (muscle continues underneath gracilis)

Gracilis

Medial saphenous vein

A

Sartorius

Gracilis

B

Femoral artery and vein

Adductor longus

Adductor magnus et brevis

Medial saphenous vein

Figure 8.17: *A.* Ventral view of the superficial muscles of the upper hind leg. *B.* Ventral view of the superficial muscles of the upper hind leg of a cat.

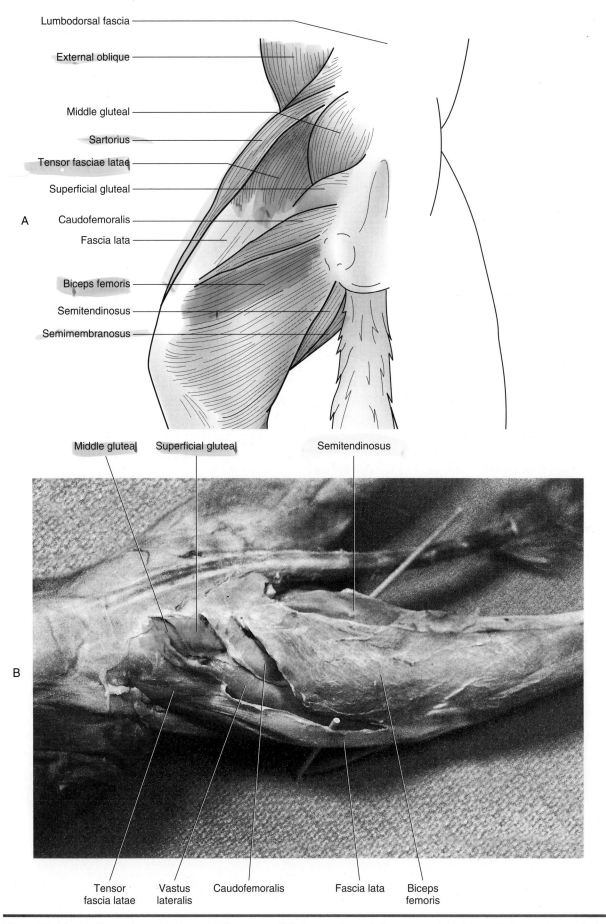

Lumbodorsal fascia

External oblique

Middle gluteal

Sartorius

Tensor fasciae latae

Superficial gluteal

A

Caudofemoralis

Fascia lata

Biceps femoris

Semitendinosus

Semimembranosus

Middle gluteal   Superficial gluteal   Semitendinosus

B

Tensor   Vastus   Caudofemoralis   Fascia lata   Biceps
fascia latae lateralis   femoris

Figure 8.18: *A.* Dorsolateral view of the superficial muscles of the upper hind leg. *B.* Dorsolateral view of the superficial muscles of the upper hind leg of a cat.

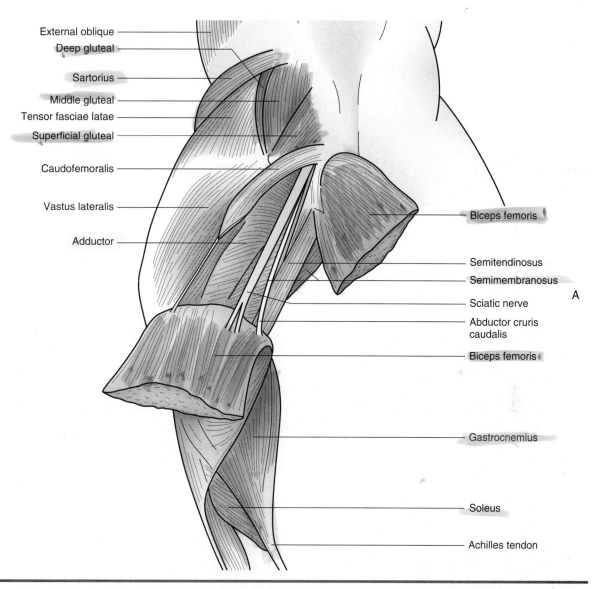

External oblique
Deep gluteal
Sartorius
Middle gluteal
Tensor fasciae latae
Superficial gluteal
Caudofemoralis
Vastus lateralis
Adductor

Biceps femoris
Semitendinosus
Semimembranosus
Sciatic nerve
Abductor cruris caudalis
Biceps femoris

A

Gastrocnemius

Soleus

Achilles tendon

Figure 8.19: *A.* Dorsolateral view of the deep muscles of the upper hind leg.   *Continued*

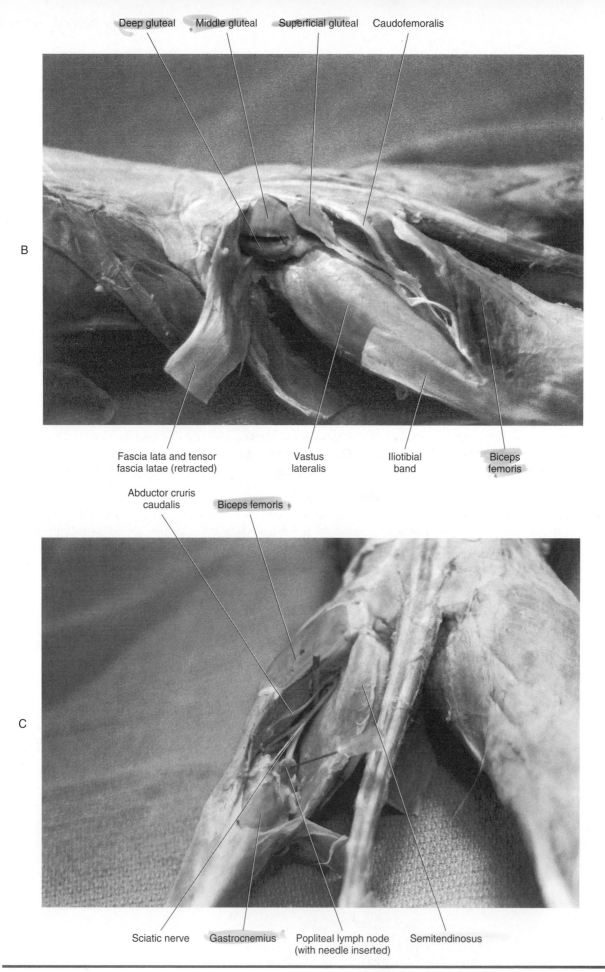

Deep gluteal   Middle gluteal   Superficial gluteal   Caudofemoralis

B

Fascia lata and tensor fascia latae (retracted)   Vastus lateralis   Iliotibial band   Biceps femoris

Abductor cruris caudalis   Biceps femoris

C

Sciatic nerve   Gastrocnemius   Popliteal lymph node (with needle inserted)   Semitendinosus

Figure 8.19, *cont'd: B.* Dorsolateral view of the deep muscles of the upper hind leg of a cat. *C.* Caudal view of the deep muscles of the upper hind leg of a cat. Needle is in the popliteal lymph node.

Part 8: Deeper Muscles of the Thigh

1. On the lateral side of the cat's thigh, the caudal borders of the biceps femoris should be free. Gently lift up the biceps femoris, freeing it from the **sciatic nerve** and the abductor cruris caudalis muscle lying beneath. Carefully remove the fat in the popliteal fossa so the medial aspect of the biceps femoris can be viewed. Be careful not to bisect the sciatic nerve or this thin muscle (see Figure 8.19*A* and *C*).

2. The **abductor cruris caudalis** muscle is an extremely thin, long muscle lying beneath and attached to the biceps femoris and running parallel to the sciatic nerve. This muscle *originates* from the second caudal vertebra and *inserts* on the fascia of the biceps femoris. Its *action* is to abduct the thigh and flex the leg. It is easy to cut accidentally.

3. Turn the cat over so the medial aspect of the thigh is visible. Deep to the gracilis there are four muscles now visible. The most caudal is the **semitendinosus** muscle (M. semitendinosus) (Figure 8.20). It is a thick, strap-like muscle that *originates* from the ischial tuberosity and *inserts* on the tibia, posterior and medial to the biceps femoris. It forms part of the caudomedial border of the popliteal fossa. Its *action* is to flex the leg. Free both borders of this muscle.

4. Just cranial and medial to the semitendinosus is the **semimembranosus** muscle (M. semimembranosus). This *originates* on the ischium and *inserts* on the femur. Its *action* is to adduct the thigh.

5. The large muscle cranial to the semimembranosus is the **adductor** muscle (M. adductor). It can be separated from the much thinner *adductor longus*, which is located cranially. The larger adductor *magnus et brevis* is cranial to the semimembranosus. Be careful not to separate the adductor magnus et brevis in the middle of the muscle belly; there is a slight fascial separation that can be mistaken for the caudal border of the adductor longus. The *origin* of the adductor muscle is the ischium and pubis, and it inserts along most of the length of the femur. Its action is to adduct the thigh (see Figure 8.20).

6. Cranial to the adductor longus muscle is the **pectineus** muscle (M. pectineus). This is a tiny, thin muscle located below the femoral artery and vein. The pectineus *originates* on the pubis and *inserts* on the femur. Its *action* is to extend the thigh (see Figure 8.20). Just proximal and dorsal to the

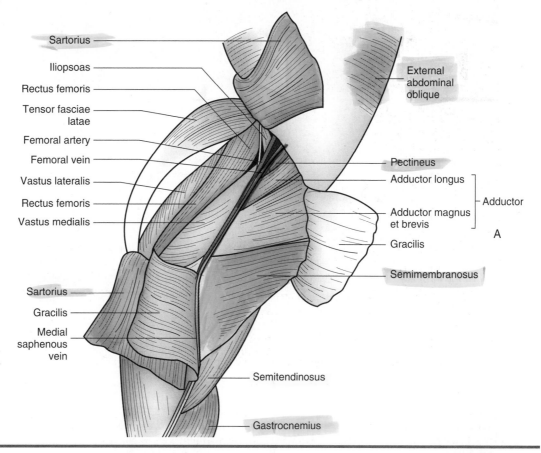

Figure 8.20: *A.* Medial view of the deep muscles of the upper hind leg.

Continued

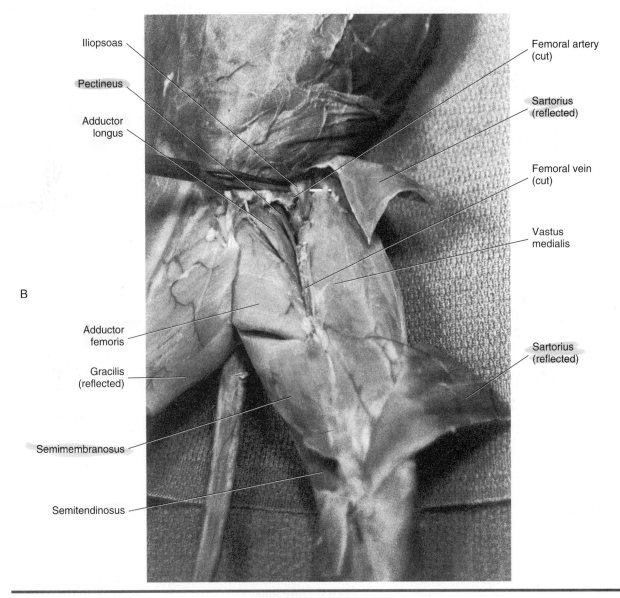

Iliopsoas

Pectineus

Adductor
longus

B

Adductor
femoris

Gracilis
(reflected)

Semimembranosus

Semitendinosus

Femoral artery
(cut)

Sartorius
(reflected)

Femoral vein
(cut)

Vastus
medialis

Sartorius
(reflected)

Figure 8.20, *cont'd: B.* Medial view of the deep muscles of the upper hind leg in the cat.

pectineus is the **iliopsoas** muscle (M. iliopsoas), a compound muscle *originating* from the lumbar vertebrae and ilium and *inserting* on the femur. Its *action* is to rotate and flex the thigh. Only a small, terminal portion of the iliopsoas is visible. It can be seen cranial to the femoral artery and veins. Its fibers run nearly at right angles to the blood vessels.

7. To see the **quadriceps femoris** group of muscles (M. quadriceps femoris), the sartorius and iliotibial band must be bisected (this should already be done). Reflect the sartorius and the tensor fasciae latae muscle. The large muscle now exposed on the craniolateral surface of the thigh is the **vastus lateralis** (Figure 8.21). Craniomedial to the vastus lateralis is a narrow muscle, the **rectus femoris,** that *originates* on the ilium and *inserts* as part of the quadriceps femoris group on the patella and tibial tuberosity. Remove approximately 1 cm from the center of this muscle to view the deeper muscle. The **vastus medialis** muscle is located on the medial surface of the thigh, just caudomedial to the rectus femoris.

The deep muscle beneath the rectus femoris is the **vastus intermedius** muscle. Using blunt dissection, separate the internal margins of the vastus lateralis and medialis from the vastus intermedius. The three vastus muscles *originate* from the upper femur and converge to encompass the

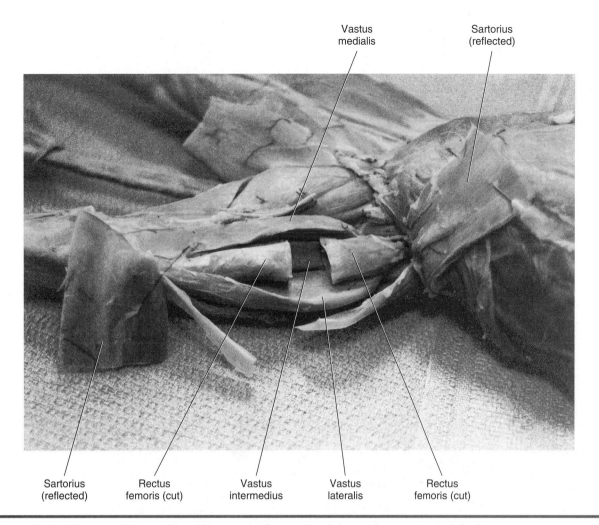

Vastus
medialis

Sartorius
(reflected)

Sartorius
(reflected)

Rectus
femoris (cut)

Vastus
intermedius

Vastus
lateralis

Rectus
femoris (cut)

Figure 8.21: The quadriceps femoris group of muscles in a cat.

patella (a sesamoid bone). They then *insert* as the straight patellar ligament on the tibial tuberosity. This group of muscle's *action* is to extend the stifle joint.

Part 9: Posterior Leg Muscles

1. The **gastrocnemius** muscle (M. gastrocnemius) is the large *calf* muscle (Figure 8.22) visible on the caudal aspect of the tibia, distal to the stifle joint. There are two bellies in this muscle, one medial and the other lateral. The gastrocnemius *originates* primarily on the distal femur and *inserts* as the **tendon of Achilles** *(Achilles tendon)* on the tuber calcis (calcaneal tuberosity) of the fibular tarsal bone (calcaneus). This muscle's *action* is to extend the hock joint and flex the digits of the foot.

2. Ventral to the gastrocnemius, on the lateral aspect of the calf, is the soleus muscle (M. soleus). It *originates* on the fibula and *inserts* by way of the tendon of Achilles on the tuber calcanei of the calcaneus. It *acts* in conjunction with the gastrocnemius to extend the hock joint and flex the digits of the foot. The **superficial digital flexor** lies deep in the gastrocnemius and inserts as part of the tendon of Achilles. It will not be dissected.

Part 10: Deep Muscles of the Hind Leg

This is an optional dissection.

1. On the lateral side of the hind leg, locate the following muscles, starting with the most cranial muscle and proceeding caudally: **cranial tibial muscle** (M. tibialis cranialis), **extensor digitorum longus** (M. extensor digitorum longus), **peroneus longus** (M. peroneus longus), **extensor digitorum**

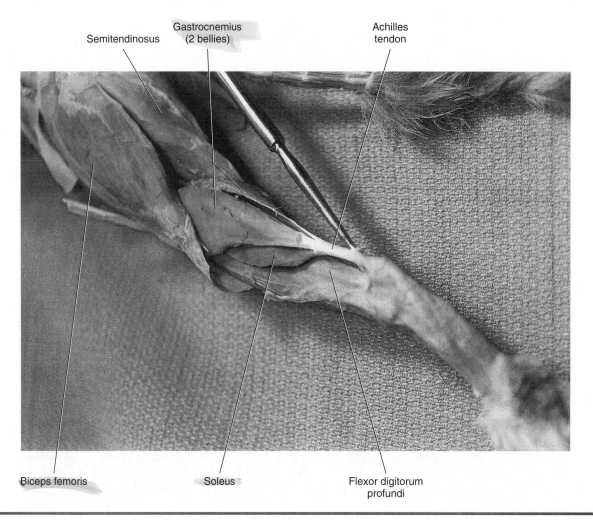

Semitendinosus Gastrocnemius (2 bellies) Achilles tendon

Biceps femoris Soleus Flexor digitorum profundi

Figure 8.22: Caudal view of the superficial muscles of the lower hind leg of a cat.

lateralis (M. extensor digitorum lateralis), and **peroneus brevis** (M. peroneus brevis). Consult Figure 8.23 for help.

2. On the medial side of the lower hind leg, deep to the superficial digital flexor is the **flexor digitorum profundi** (M. flexor digitorum profundi) (see Figure 8.22).

Part 11: Deep Back Muscles

1. To view the **intercostal** muscles (the muscles between the ribs) reflect the caudal border of the origin of the serratus ventralis muscle and expose the ribs.

2. The fibers of the **external intercostals** (Figure 8.24) course in a craniodorsal to caudoventral direction between the ribs. The external intercostals *action* is to lift the ribs in inspiration, thus expanding the thoracic cavity.

3. Remove approximately 1 cm of the external intercostal muscle between two ribs to expose the **internal intercostals,** the fibers of which run at right angles to the external intercostals. These muscles *act* to depress the ribs during active expiration.

4. Dissection of the deep back muscles requires that you reflect the latissimus dorsi toward its origin. Bisect this muscle now if you have not already done so. The **serratus dorsalis** muscle (M. serratus dorsalis) (see Figure 8.24), which is a series of muscles that appear serrated, *originates* from a mid-dorsal aponeurosis and *inserts* ventrally on the ribs. Reflect this muscle ventrally by cutting through the aponeurosis and exposing the *iliocostalis* part of the *sacrospinalis* muscle group.

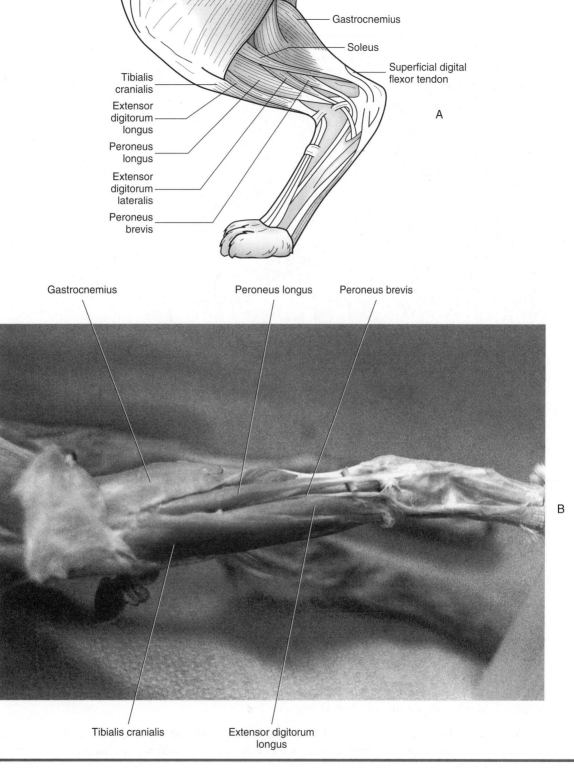

Figure 8.23: *A.* Craniolateral view of the muscles of the lower hind leg. *B.* Craniolateral view of the muscles of the lower hind leg of a cat.

5. The epaxial spinal musculature (dorsal trunk muscles) (see Figure 8.24) is a system of muscles divided into three parts. The m. iliocostalis system, the m. longissimus system, and the m. transversospinalis system. It may be necessary to remove the dorsal fascia to visualize these muscles, we will dissect a muscle from each system. The most ventral, the **iliocostalis thoracis** muscle (M. iliocostalis), lies beneath the serratus dorsalis and *inserts* on the ribs. The middle muscle is the **longissimus thoracis** muscle (M. longissimus thoracis). It occupies the space between the spinous and

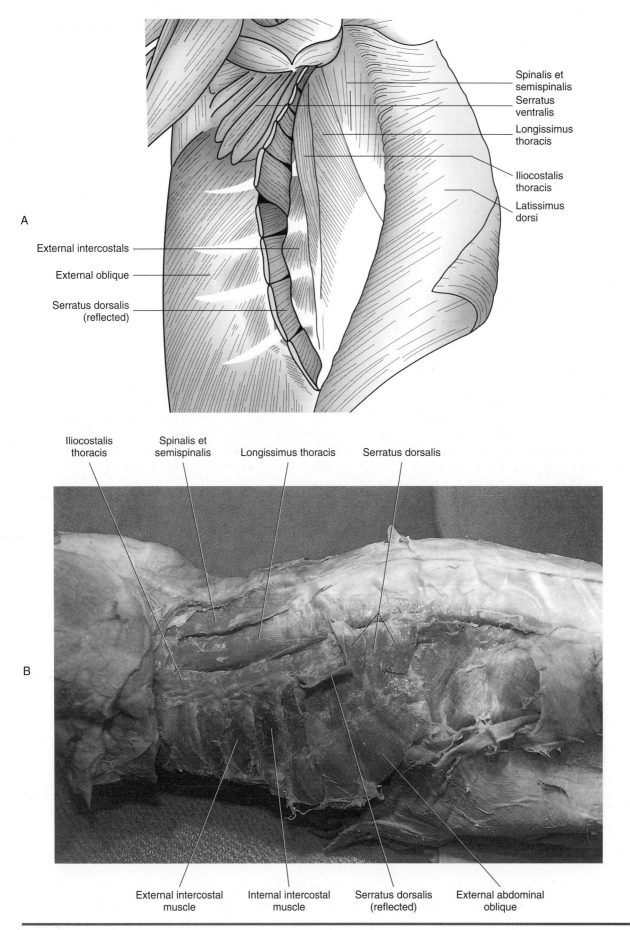

Figure 8.24: *A.* Deep muscles of the back and ribs. *B.* Deep muscles of the back and ribs in a cat.

transverse processes of the caudal thoracic and lumbar vertebrae, and thus runs the length of the back. The dorsal muscle of the transversospinalis system we will dissect is called the **spinalis et semispinalis** muscle (M. spinalis et semispinalis). It is found next to the spinous processes of the thoracic vertebrae. This group of muscle's *action* is to extend the spine and neck. Each system has more muscles cranial and/or caudal to the three dissected, such as the iliocostalis lumborum and longissimus lumborum.

CLINICAL SIGNIFICANCE

The clinical significance of this chapter is illustrated in the following exercise.

EXERCISE 8.3
MUSCLE PHYSIOLOGY AND PATHOLOGY

The classic experiment to illustrate muscle physiology is to sacrifice a frog, remove its gastrocnemius muscle, attach it to a myograph, and stimulate it electrically to show *summation* and *tetany*. Please read about these in your textbook; we will not do this experiment.

Instead, we will demonstrate the importance of an enzyme found within muscle. Creatinine kinase (CK) (previously known as creatinine phosphokinase) catalyzes one of the most important reactions in body chemistry. The concentration of adenosine triphosphate (ATP) in skeletal muscle is relatively low; it supplies only the energy necessary to maintain contraction for a brief period of time. Muscles continue to contract after the initial supply of ATP has been used, leaving adenosine diphosphate (ADP). The ADP is phosphorylated again from another source, creatinine phosphate (CP). There is about five times as much CP stored in the sarcoplasm as there is ATP. When ATP is used to supply energy for muscular contraction, the ADP produced is transphosphorylated by the following equation:

$$CP + ADP \xrightarrow{CK} C + ATP$$

This reaction replenishes the ATP almost as fast as it is being used. Therefore, the ATP level changes little until the concentration of CP gets low.

Horses are susceptible to a condition known as exertional rhabdomyolysis. *Rhabdomyo-* refers to skeletal muscle, and *lysis* means to rupture. In this condition, muscle damage occurs because of the breakdown of skeletal muscle cells. This is actually one of a group of disease conditions that share common symptoms and causes; these are: tying-up syndrome, azoturia, Monday morning disease, and exertion-associated myositis. These conditions most often occur in heavily muscled horses that are working or in training and/or have been rested for one or more days, but were kept on full feed during that period. When returned to exercise, these horses may develop a myopathy.

The resulting pathology is *myonecrosis,* and the severity of the disease is directly related to the extent of the necrosis. When a muscle cell breaks down, it liberates myoglobin and CK, among other chemicals, into the bloodstream. When a sufficient amount of myoglobin has been released, urine will appear coffee-colored because of its presence.

Procedure

1. Three samples of horse serum will be provided: Horse A, normal horse, minimal activity; Horse B, normal horse, exercised vigorously or given an intramuscular injection; Horse C, horse showing much pain, sweating, reluctance to move.

2. Using your blood chemistry machine, measure the CK from each horse.

Horse A = _____

Horse B = _____

Horse C = _____

QUESTIONS

1. Was there elevated CK in Horse B; if so, why?

2. Given that CK is not supposed to determine the extent of muscle damage (despite the fact that it is organ-specific) why were there elevated CK levels in Horse C?

Discussion

CK levels can rise slightly from minor muscular injuries because of the release of CK from the cells. Damage can be as minor as receiving an intramuscular injection, persistent recumbency, bruising, a laceration, or hypothermia. In a situation where there is considerable muscle damage, the CK levels will elevate dramatically, but diagnosing exactly what amount of muscle damage is occurring based on the CK is not possible. The presence of myoglobin in the urine is a better indicator of injury severity than the CK by itself.

Veterinary Vignettes

"Doc, I think Lucas has tetanus," Ann Coles told me over the phone.

"Why do you think that?" I asked. Dogs so rarely get tetanus that we don't vaccinate for it.

"Well, isn't tetanus *lockjaw*?" she asked. "He can't open his mouth."

I assured her it probably wasn't really tetanus, but there were some other things it could be. The first thing a veterinarian must do when an animal can't move a body part is to determine whether it is reluctance or a true inability. The former is usually due to pain, and the latter is often a neurological problem. In Lucas' case, the differential diagnosis was a broken jaw, a tooth problem, a retrobulbar abscess (behind the eye), eosinophilic myositis, and atrophic myositis. The latter two diseases are now combined into a syndrome known as masticatory muscle myositis (MMM).

Ann brought Lucas to the hospital that afternoon. She was very worried; I could tell by the look on her face and the strain in her voice.

Lucas' jaw was really painful; just attempting to open his mouth caused him to cry out loudly. This caused Ann even more distress.

"Ann, the only way I am going to figure out what is going on is by anesthetizing Lucas. I'll open his mouth while he's out and not feeling any pain."

It was senseless to continue my examination of the oral cavity. There was nothing I was going to learn until I got the mouth open. We did check the rest of the dog's systems for abnormalities, but there were none. As I studied Lucas' face, it seemed as though his facial muscles were a little swollen, but this was mostly a subjective impression and was difficult to say with certainty.

After Lucas was anesthetized, I was able to open his mouth. Because there was no muscle atrophy and the jaw was not locked shut, he did not have the atrophic form of this disease. The teeth looked fine, as did both mandibles. There did not appear to be an abscess behind either eye, and that left only an inflammatory or autoimmune cause on my list of diseases to rule out. I made an incision, took a small piece of muscle tissue from the masseter muscle, and sent it to the lab. We also ran blood chemistries and found the CK level mildly elevated. Eosinophils were normal, but this wasn't an unexpected finding. The lab confirmed the diagnosis of MMM.

Because this is an immune-mediated disease, immunosuppressive doses of corticosteroids were indicated, and the dog responded nicely. In talking to Ann I soon realized that her fear of giving her dog cortisone would hamper its needed therapy. I tried to convince her that Lucas would need to stay on the cortisone for six months, even though he would appear to be doing well and be back to normal. Just to be safe, I decided to educate her into giving treatment. I found some good articles on the disease, complete with photos from my veterinary journals, copied them, and gave them to her. The photos of both the atrophic and chronic forms of the disease were very graphic. I don't think the dog missed a pill the entire six months he was on the medication.

SUMMARY

This chapter has been mainly a detailed dissection of the muscular anatomy of the cat. We started at the chest, moved to the neck, upper back, front leg, abdomen, hind leg, and then to the deep muscles of the back and ribs. We also studied the histology of the three types of muscle tissue and discussed muscle physiology. The muscles of other animals are similar to that of the cat, so it serves as a good basis for other domestic animals.

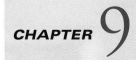

CHAPTER 9

THE DIGESTIVE SYSTEM

OBJECTIVES:

- state the overall purpose and function of the digestive system
- using diagrams, identify the structural anatomy of the alimentary canal wall
- dissect the structures and organs of the digestive system and identify them by name
- describe and identify the teeth of a cat or dog, including their dental formula and eruption table
- identify the parts of a tooth
- know the methods used in dental charting and the Triadan System of numbering teeth
- using a model or diagram, identify the parts of the equine large intestine
- using a model or diagram, identify the parts of the ruminant digestive system, including rumen, reticulum, omasum, abomasum, and the large intestine
- differentiate between the ansa spiralis of the ruminant and the pig
- know the parts of the fowl digestive system

MATERIALS:

- cat cadaver, triple injected (order without skin attached)
- Mayo dissecting scissors
- probe
- 1 × 2 thumb forceps or Adson tissue forceps
- #4 scalpel handle with blade
- bone-cutting forceps
- rubber gloves
- models of the equine colon and ruminant forestomachs and stomach
- three dogs for blood analysis, and two of the three for fecal analysis
- 10-ml graduated cylinders
- blood chemistry machine, amylase and lipase tests

Introduction

The **digestive system** is made up of the structures of the **alimentary tract** and the **accessory digestive organs.** Its purpose is to ingest, digest, and absorb food and to eliminate the undigested remains as feces. It provides the body with the nutrients, water, and electrolytes that are essential for health and growth.

The *alimentary tract* or *canal* is a hollow tube extending from the mouth to the anus. The interior of the canal is considered technically *outside* the body because the contents within only make contact with the cells that line the canal. These cells are mostly the simple columnar epithelial cells studied previously. They protect the inside of the body from the bacteria and by-products of digestion found within the alimentary canal. For the ingested food to be available, it first must be broken down into smaller, diffusible molecules. The steps in this process are *prehension*, or using the mouth and lips to grasp the food; *mastication,* or chewing, which mixes food with salivary enzymes to both physically and chemically break it down; *swallowing,* moving the food through the esophagus to the stomach; *digestion* and *absorption,* which take place in the stomach and small intestines; and *elimination* via the large intestines and anus.

The *accessory digestive organs* empty their secretions into the alimentary canal to aid digestion and absorption. These structures include the **salivary glands, gallbladder,** and **pancreas.** The **liver** is also considered—among other things—an accessory digestive organ. It is the largest gland in the body and performs many metabolic roles, such as formation of plasma proteins, vitamin storage, removal of foreign substances via the **Kupffer cells** (phagocytic cells, also called *stellate reticuloendothelial* cells), detoxification, carbohydrate metabolism, and storage of glycogen. The liver's role in digestion is to produce **bile,** which leaves it via the **hepatic ducts** and is carried to the **duodenum** through the **common bile duct.** When bile is not being used, it backs up through the **cystic duct** into a small reservoir sac known as the **gallbladder.** Bile emulsifies fats, or, in other words, it breaks the large fat particles into smaller ones. This allows the **lipase** released from the pancreas to break down the fat molecules into fatty acids for absorption.

The Structural Anatomy of the Alimentary Canal

The alimentary canal walls have four layers, or **tunics.** From the lumen outward, these are: the *mucosa, submucosa, muscularis (muscularis externa),* and *serosa (adventitia).* Each of these layers is composed of a predominant type of tissue and has a specific function in the digestive process (Figure 9.1).

The **mucosa** *(mucous membrane)* consists of surface **epithelium** (which is in most cases is simple columnar epithelium), a layer of **lamina propria** (consisting of areolar connective tissue), and the **muscularis mucosae** (a thin layer of smooth muscle that enables local movement of the mucosa). The major function of the mucosa is *secretion* (of enzymes, mucus, and hormones), *absorption* (of digested foodstuffs), and *protection* (against bacterial and endotoxin invasion). Each layer of the alimentary canal walls may be involved in one or more of these three functions.

The **submucosa** is composed of moderately dense irregular connective tissue and contains blood and lymph vessels, scattered lymph nodules, and nerve fibers. The layer's intrinsic nerve supply is called the **submucosal plexus.** The submucosa's major functions are movement of absorbed of nutri-

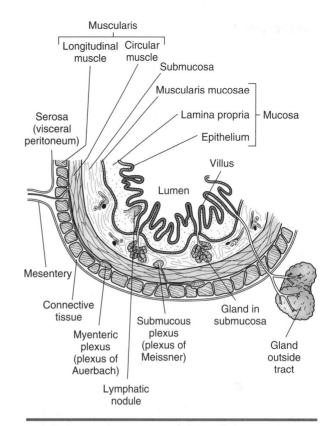

Figure 9.1: Basic structural pattern of the alimentary canal wall.

ents into the blood, structural support for the canal, and protection. When a veterinarian places sutures in the wall of the intestinal tract, this is the layer that has the holding power.

The **muscularis externa** consists of two layers: an inner circular and an outer longitudinal, both of smooth muscle. Between the two layers is another intrinsic nerve plexus, the **myenteric plexus.** The muscularis externa and its nerve plexus are responsible for the gastrointestinal (GI) motility of the organ.

The **serosa** is composed of visceral peritoneum, which histologically is epithelial (simple squamous) tissue, and a thin layer of areolar connective tissue. In areas outside the abdominal cavity, the serosa is replaced by **adventitia,** a layer of coarse fibrous connective tissue that binds organs, such as the **esophagus** and **rectum,** to the surrounding tissues. The serosa, with its smooth surface and lubricating peritoneal fluid, reduces friction during contractions of the GI tract. The adventitia anchors and protects the surrounded organs of the GI tract.

Teeth

There are four types of teeth, classified according to their shape, function, and time of eruption. The front teeth are called **incisors** (I) and are used cutting; the fang teeth are the **canines** (C), which are used for grasping; the next set of teeth are the **premolars** (P); and caudal to the premolars are the **molars** (M). The premolars and molars are used for chewing, and since the molars have a flatter occlusal surface they have a grinding function. The first set of teeth an animal grows are the **deciduous teeth;** these fall out and are replaced by the adult **permanent teeth.** As you can see from the following **dental formulas,** only the incisors, canines, and premolars are deciduous.

Dental Formulas

Students should know at least the dental formulas for the dog and cat, which are listed in the following paragraphs. The dental formulas for the other main domestic animals are listed at the end of this section.

Dental formulas can be written in two ways. The first is with the numbers of teeth of each type on the upper arcade written over the numbers of teeth of each type on the lower arcade. For example, the adult dental formula for dogs is as follows.

$$2\left(I\frac{3}{3}, C\frac{1}{1}, P\frac{4}{4}, M\frac{2}{3}\right)$$

The second method of writing this is to know that the order written is incisors, canines, premolars, then molars. Then you simply list the numbers of teeth of each type.

$$\frac{3142}{3143} \times 2$$

We multiply by two because you must consider both the left and right sides of the mouth to calculate the correct total number of teeth.

Using this same method, the dog deciduous formula is:

$$\frac{313}{313} \times 2$$

Hint for remembering these formulas: It is easier to remember one number than many (i.e., it is easier to remember 313 than 3, 1, 3 in the dog deciduous formula). Try remembering the formulas this way:

puppy	313 over 313	top same as bottom
adult	3142 over 3143	bottom one more than top

This is the formula for a cat's deciduous teeth:

$$\frac{313}{312} \times 2$$

Similarly, the adult cat formula is as follows.

$$\frac{3131}{3121} \times 2$$

Hint for remembering these formulas:

kitten	313 over 312	top same as puppy, bottom is one less
adult	3131 over 3121	bottom is 10 less than top, or just add one molar to kitten formula on top and bottom

Dental formulas of other domestic species

horse:

$$\frac{313\text{-}43}{3133} \times 2$$

ox:

$$\frac{0033}{3133} \times 2$$

pig:

$$\frac{3143}{3143} \times 2$$

Aging a Dog or Cat

A student should learn to *age* a dog or cat using the eruption table shown in Table 9.1. Note that this is a simplified version showing the median age for the incisors. The figure in parentheses is the documented range.

The cat has approximately the same eruption table as the dog, except it has no upper P1, no lower P2 and P3, and no M2 or M3.

After the permanent teeth come in, aging an animal is an estimate based on tartar on the teeth, wear-

Table 9.1	**Permanent Teeth Eruption Ages in the Dog**
Tooth	**Age**
I1	3 mos. (2-5 mos.)
I2	4 mos. (2-5 mos.)
I3	5 mos. (4-5 mos.)
C	6 mos. (5-6 mos.)
P1	4-5 mos.
P2	5-6 mos.
P3	5-6 mos.
P4	5-6 mos.
M1	5-6 mos.
M2	6-7 mos.
M3	6-7 mos.

ing of the teeth, graying of the muzzle, and the overall condition of the dog or cat.

Dental Charting

Currently there is no standardized method of tooth identification in veterinary medicine. Human dental insurance carriers have helped to standardize nomenclature for human teeth by using a numerical system. Veterinarians have been using a form of dental shorthand for years to identify a specific tooth and its corresponding quadrant. The letters indicating the type of tooth are the same as those used in the previously discussed dental formulas. Adult teeth are designated by capital letters and deciduous teeth by lower-case letters. A number in superscript or subscript to the left or right of the letter indicates the quadrant the tooth occupies, and this is always

based on the veterinarian viewing the animal from the rear.

For example:

I_2 = lower right second incisor

^1c = deciduous upper left canine

P^4 = upper right fourth premolar

Triadan System

Most veterinary dentists are now using the **Triadan system** in which each tooth has a three-digit number. The first number represents the quadrant, starting on the animal's upper right side (1) and continuing clock-wise (as viewed from the front of the animal's head) to the lower right (4). See Figure 9.2 for additional information. Quadrants for deciduous teeth are numbered 5 through 8.

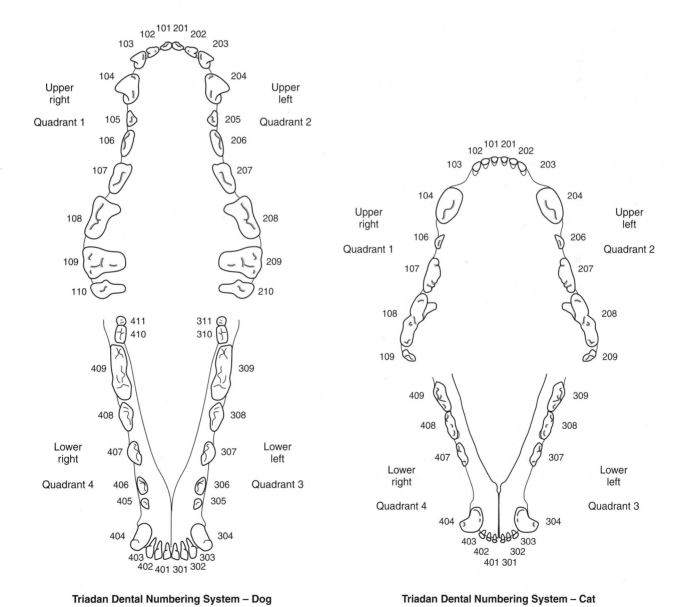

Triadan Dental Numbering System – Dog

Triadan Dental Numbering System – Cat

Figure 9.2: Triadan System of tooth identification.

Individual teeth are represented by the second and third digits. For example, all canines are numbered 04 and all first molars are numbered 09. Thus, the four canine teeth would be 104 to 404, starting on the upper right (104), upper left (204), lower left (304), and lower right (404). The same numerical pattern is used for each first molar, 109 to 409. This system works for both dogs and cats and takes into consideration that cats have fewer teeth than dogs. Cats have three premolars on the upper jaw and two premolars on the lower jaw, therefore in the Triadan system tooth 05 on top and 05 and 06 on the bottom are not present. If you look at the shape of the cat's teeth in comparison to the dog's, it makes sense to identify the teeth by this method.

The Structure of a Tooth

Each tooth consists of a *crown, neck,* and *root* (Figure 9.3). The **crown** is the portion of the tooth that is visible above the **gum,** or **gingiva.** This portion of the tooth also can be called the **clinical crown.** The entire area covered by enamel is called the **anatomical crown,** and it extends a short distance into the neck area. The **neck** is the portion of the tooth between the gum line and where the gum attaches to the tooth, and it is associated with a slight constriction in the shape of the tooth. The **root** is the area below the neck that extends into the bone.

The crown is covered by **enamel,** which is the hardest substance in the body, yet it is also fairly brittle. It is 95 to 97% inorganic calcium salts (chiefly $CaPO_4$). The crevice between the upper margin of the gingiva and the limit of the anatomical crown (i.e., the enamel) is called the **gingival sulcus,** and its border is the **gingival margin.** The outermost surface of the root is covered by **cementum,** which is similar to bone in composition and less brittle than enamel. The cementum attaches the tooth to the **periodontal ligament,** which is the connective tissue that holds the tooth in the **alveolar bone.** The periodontal ligament also has a cushioning effect on the tooth.

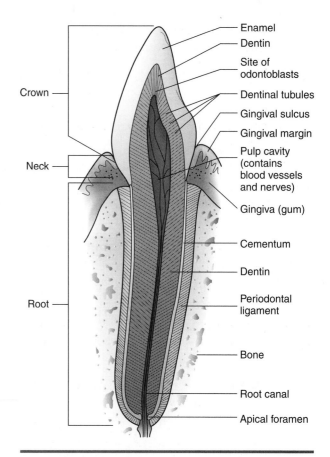

Figure 9.3: The structure of the tooth in a dog or cat.

The inner part, which is the bulk of the tooth, is called the **dentin.** It also is composed of a bone-like material. In the center of the tooth is the **pulp cavity,** which contains the **pulp,** a connective tissue containing blood vessels, nerves, and lymphatics. This provides the nutrients and sensation to the tooth. The *odontoblasts,* located on the outer margins of the pulp cavity, produce the dentin. The pulp cavity extends into the deep portions of the root and becomes the **root canal.** An opening exists at the root apex, called the **apical foramen,** which provides a route of entry into the tooth for the structures found in the pulp cavity.

EXERCISE 9.1

DISSECTION OF THE DIGESTIVE SYSTEM

Complete the following steps in the dissection procedure. The important organs and tissues to identify are listed in colored bold print. If a structure is mentioned before its dissection or its definition, it is italicized.

1. In the previous chapter you dissected the left side of the head and neck to expose the major salivary glands and lymph nodes of the cat. If any of the glands were destroyed during the dissection of the musculature, remove the skin on the right side of the neck and dissect out these structures. The sali-

vary glands produce saliva, which is carried to the mouth via **salivary ducts** that pass through the head and neck musculature.

2. The largest salivary gland in the cat is the **parotid salivary gland,** which is located ventral to the pinna (Figure 9.4). This gland can be recognized by its lobular texture. The duct of the parotid crosses the masseter muscle and enters the oral cavity opposite the last upper premolar tooth. This is one of the reasons this tooth often has a lot of tartar build-up in older dogs. Two branches of the facial nerve also cross the masseter, dorsal and ventral to the parotid duct.

3. The **mandibular salivary gland,** which has the same lobular texture as the parotid gland, lies posterior to the angle of the jaw, ventral to the parotid gland (see Figure 9.4). The maxillary vein passes over this gland. This gland's duct opens into the floor of the mouth near the lower incisors.

4. The smaller **sublingual salivary gland** (monostomatic part) is located rostral to the mandibular gland and the maxillary vein, and it is just superior to the mandibular lymph nodes. This gland also has a lobular appearance. Do not confuse this gland with the **mandibular lymph nodes** (which are smooth and elliptical in shape) located on either side of the linguofacial vein (which feeds directly into the jugular vein).

5. The cat has two additional salivary glands, the **molar salivary gland,** which lies near the angle of the mouth, and the **infraorbital salivary gland** on the floor of the orbit beneath the eye. They are hard to locate, and both, especially the latter, require difficult dissection.

6. On the side opposite the dissection of the salivary glands and lymph nodes (should be the animal's right side), use a scalpel to incise through the masseter muscle from the corner of the mouth to just ventral to the parotid salivary gland. Then, using bone cutters, cut through the ramus of the

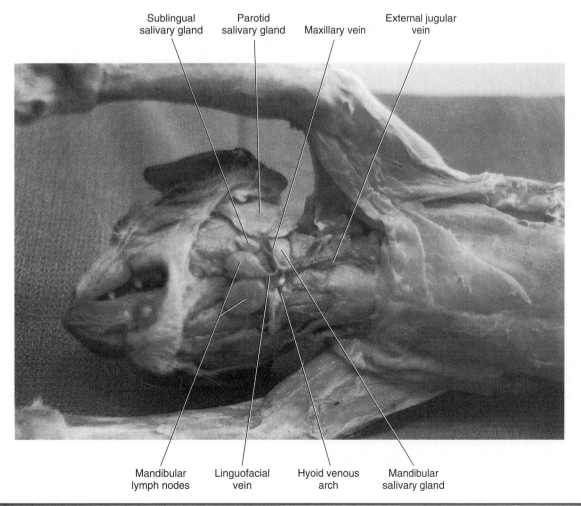

Figure 9.4: Salivary glands of a cat.

mandible. As you start to pry open the oral cavity, use scissors to cut through the caudal tissue on the same (right) side all the way to the epiglottis. When you are done, the entire mouth and throat of the cat will be exposed, yet the structures of one side will be completely intact (Figure 9.5).

Locate the following structures in the oral cavity.

a. The **vestibule** is the area between the teeth and cheeks.
b. The **hard palate** is the bony structure containing transverse ridges called the **palatine rugae,** and it makes up the roof of the mouth.
c. The **soft palate** extends caudally as a membranous continuation of the hard palate.
d. The **fauces** (a "space" bordered on each side by the **palatoglossal glossopalatine arches,** ventrally by the base of the tongue, and dorsally by where the soft palate starts) is the opening between the **oral cavity** and the **oropharynx.**
e. The two small **palatine tonsils** are masses of lymphatic tissue located on dorsolateral wall of the oropharynx, about 1 cm caudal to the cranial edge of the glossopalatine arches.
f. The **lips** can be found on both the dorsal and ventral portions of the mouth.
g. The **labial frenulum** is the mucous membrane attached to each lip, upper and lower, at the midline.

7. The **tongue** lies on the floor of the oral cavity. Underneath the tongue is the **lingual frenulum,** the tissue covered by mucous membrane that anchors the tongue to the bottom of the mouth.

8. The dorsum of the tongue's surface is covered with many papillae. There are four different types (Figure 9.6). The **filiform papillae** are the most numerous; they are hard, pointed, spinous pro-

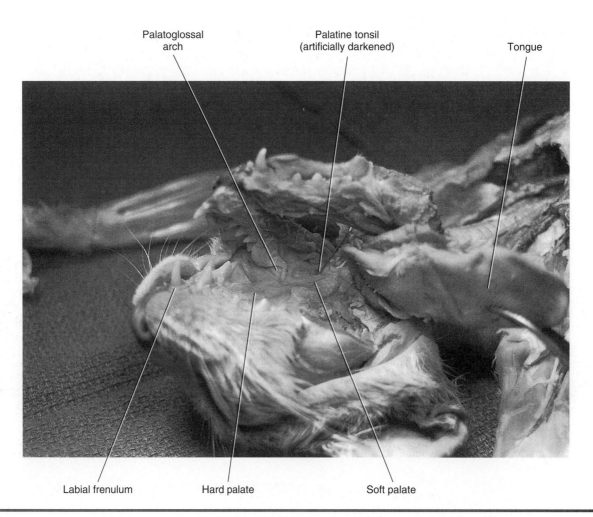

Palatoglossal arch   Palatine tonsil (artificially darkened)   Tongue

Labial frenulum   Hard palate   Soft palate

Figure 9.5: Dissected oral cavity of a cat.

jections with which the cat grooms its fur. The filiform papillae are located on the front half of the tongue. The **fungiform papillae** are small, more rounded, mushroom-shaped, and are located between and behind the filiform papillae. The **vallate papillae** are located on the top surface of the tongue laterally toward the base. There are only two or three of these on each side. Each is large and rounded and is surrounded by a circular groove (like a moat around a castle). The circumvallate papillae are difficult to see in a preserved cat. The last type are the **foliate papillae,** which are leaf-like projections located both at the base of the tongue and along the lateral side. Microscopic taste buds are located on all the papillae except the filiform type.

9. Examine the teeth of the cat and number them using the Triadan System.

10. To expose the **nasopharynx,** make a small slit in the middle of the soft palate (see Figure 11.2 in chapter 11). If you remember your skull anatomy, there is a depression just behind the internal nares bordered by the pterygoid bones. Inhaled air goes through the nasal cavity, exits through the internal nares, and enters the nasopharynx before proceeding to the larynx and into the trachea. In the dorsolateral wall of the nasopharynx there are two small holes; these are the nasal openings of the **auditory** or **Eustachian tubes.** These tubes go to the middle ear and help maintain the correct pressure within.

11. Thus, the oropharynx and nasopharynx are separated by the soft palate. Next, locate the **laryngopharynx,** or the part of the pharynx cranial to the larynx and dorsal to an open *epiglottis*. This part of the pharynx opens into the larynx and esophagus.

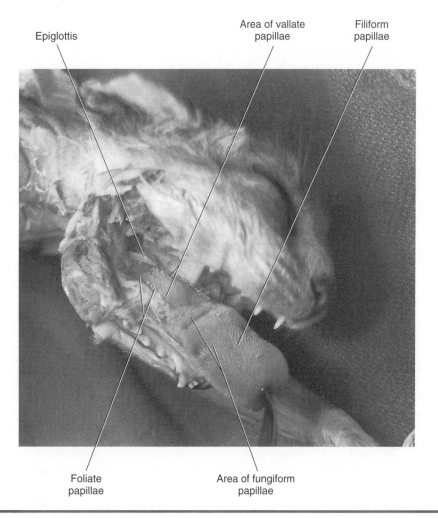

Figure 9.6: Tongue and epiglottis of a cat.

12. Locate the **epiglottis,** the pointed cartilage covered with mucous membrane, just in front of and below the larynx, caudal to the base of the tongue (see Figure 9.6). It closes the larynx and prevents fluids or food from entering during the swallowing process.

13. To study the remaining digestive organs, you must open the abdominal and thoracic cavities. First, locate the center of the pubic bone ventrally and make a small incision (large enough to permit entry of the blunt end of your scissors so you do not accidentally cut any organs). Open the abdominal cavity by cutting along the *linea alba* to the xiphoid process, just caudal to the sternum on the ventral midline. At this point, continue the cut cranially just to the left of the sternum (the animal's left side), cutting through the costal cartilages up to the manubrium. Do not sever the blood vessels supplying the ventral body wall.

Observe the thin, transparent membrane dividing the chest cavity into two halves; this is the **mediastinum.** On the right side, locate where the diaphragm (the large, flat, muscular membrane that separates the abdomen from the thorax) meets the ventral chest wall. Using scissors, make two cuts about 2 cm apart: First feel with your index finger where the diaphragm attaches to the wall, then

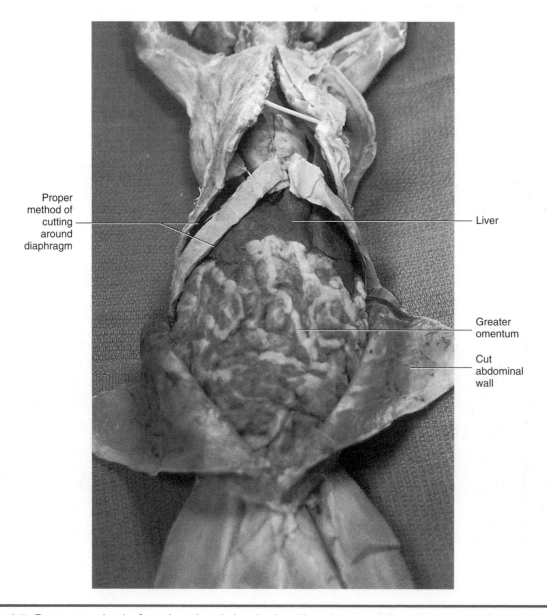

Figure 9.7: Proper method of cutting the abdominal wall and around the diaphragm of a cat.

cut caudodorsally at an angle past your finger, making certain you are slightly caudal to the attachment of the diaphragm the entire distance of the cut. Make a second cut parallel the first, only this time cut on the cranial side of the diaphragm (Figure 9.7). When you are finished, make these cuts on the left side in the same way. Now the abdomen and caudal thorax are visible, and both sides of the diaphragmatic attachments have been preserved.

14. Spread the chest wall so the heart and lungs are visible. Move the caudal lobe of the left lung out of the way and locate the **esophagus,** which runs along the dorsal wall of the thoracic cavity. It is best seen caudal to and behind the heart, just before it penetrates the diaphragm (Figure 9.8). The hole through the diaphragm where the esophagus penetrates is called the **esophageal hiatus.** The esophagus is the muscular tube connecting the laryngopharynx to the stomach. Try to visualize where the esophagus passes through the diaphragm.

15. Observe the **greater omentum,** the lacy structure containing adipose tissue, which covers and protects the ventral abdominal organs (see Figure 9.7).

16. The **liver** is the largest organ in the abdominal cavity. It is the reddish-brown structure located immediately caudal to the diaphragm. Observe the ligaments of peritoneum attaching the liver to the diaphragm and ventral body wall. The ventral ligament is called the **falciform ligament** (which may have been partially cut while opening the body wall), and the cranial one is the **coronary ligament** (Figures 9.9 and 9.10). The coronary ligament can be seen by moving the liver caudally. This ligament is short and attaches the liver to the central tendon of the diaphragm (a sheet of connective tissue in the dorso-central area of the diaphragm).

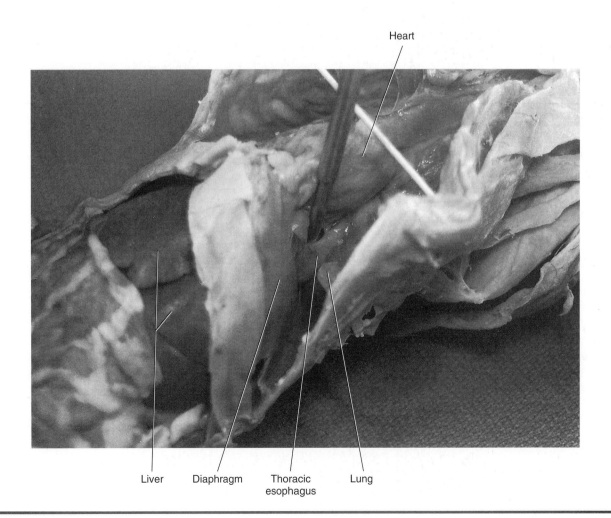

Figure 9.8: Thoracic esophagus, cranial to the esophageal hiatus, in a cat.

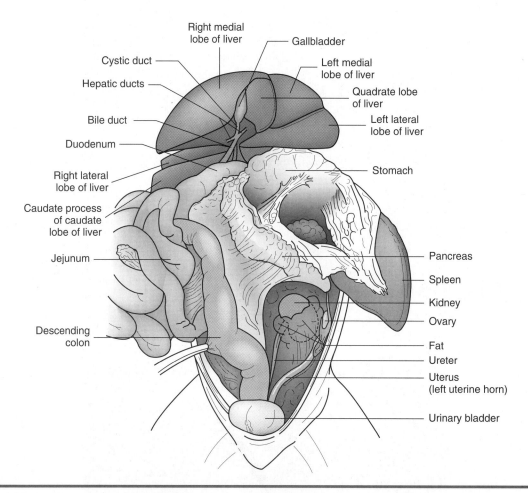

Figure 9.9: A cat's abdominal contents.

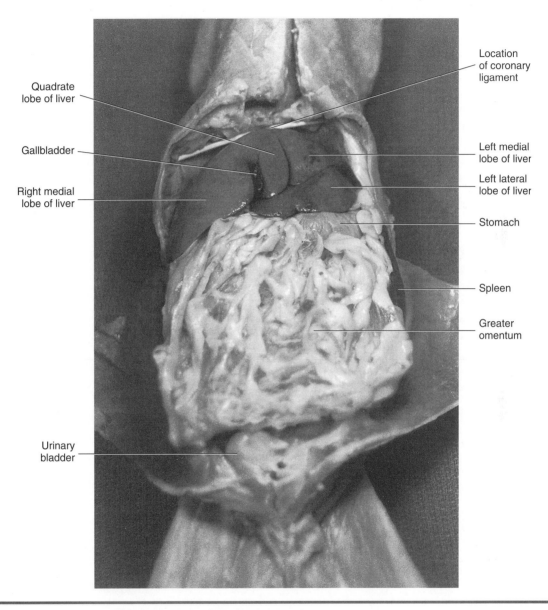

Quadrate
lobe of liver

Gallbladder

Right medial
lobe of liver

Urinary
bladder

Location
of coronary
ligament

Left medial
lobe of liver

Left lateral
lobe of liver

Stomach

Spleen

Greater
omentum

Figure 9.10: Ventral view of the lobes of the liver in a cat.

17. The falciform ligament divides the liver into right and left halves. Each half is subdivided into multiple lobes. These are named the **right lateral** and **right medial lobes,** the **left lateral** and **left medial lobes,** and the **quadrate lobe** (see Figures 9.9, 9.10, and 9.11). There is also a **caudate lobe** which has two processes, the largest, located caudal to the right lateral lobe, is called the *caudate process,* and a smaller *papillary process* can be found below the lesser omentum (Figure 9.12).

18. Observe the dark green **gallbladder** in a depression in the large right medial lobe (see Figures 9.9 and 9.10).

19. The largest part of the **stomach** lies mainly on the left side of the abdominal cavity. Separate the stomach and liver to observe the esophagus as it courses through the esophageal hiatus. There are four regions of the stomach (see Figure 9.11). The **cardiac sphincter,** or **cardia,** is the muscular valve that prevents esophageal reflux. The **cardiac region** is caudal to the cardiac sphincter and surrounds it for a short distance. The **fundic region** (or **fundus**) is the dome-like part of the stomach that projects dorsal to the cardia. The *body* of the stomach is the largest region and includes the majority of the *greater curvature* and *lesser curvature.* The **pyloric region** is distal to the body and extends from the level of the angular notch of the lesser curvature to the **pylorus.**

As noted previously, the shape of the stomach is such that it has two significant curvatures: the **greater curvature,** which is the larger outer curve located on the left lateral margin, and the **lesser curvature,** the smaller curve located on the inner medial margin of the stomach (Figures 9.12 and 9.13). The **greater omentum** is a double-membraned sac attached to the greater curvature, and the **lesser omentum** is attached to the lesser curvature. Because the greater omentum is a double-layered struc-

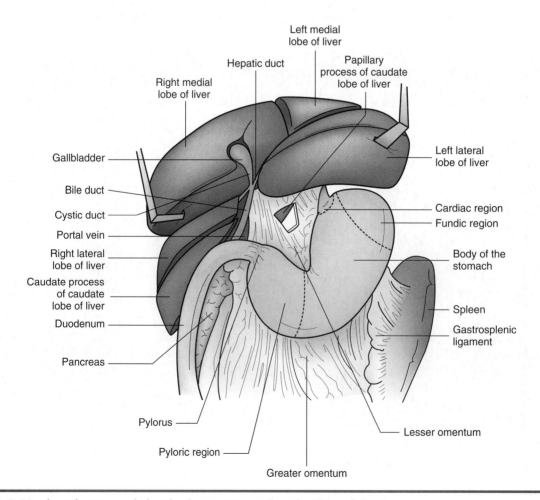

Figure 9.11: A cat's upper abdominal contents and underside of the liver.

ture, there is a cavity present, called the **omental bursa,** between the superficial and deep walls. To verify this, make a small, shallow tear in the omentum and insert your finger between the two layers.

20. Make a longitudinal incision through the stomach wall, beginning in the esophageal region and continuing to the duodenum. Wash out the stomach's contents. From the interior, observe the cardiac sphincter at the opening between the esophagus and stomach. Also look at the pyloric sphincter, which is between the stomach and duodenum. Notice the longitudinal gastric folds of the wall of the stomach.

21. Locate the **spleen,** the large reddish-brown organ to the left of and attached via the **gastrosplenic ligament** to the stomach (see Figure 9.13).

22. Lift up the greater omentum and fold it back over the stomach so that the small intestines are visible. There are three parts that make up the small intestines. The first part is the **duodenum.** It courses from the pylorus caudally on the right side (Figure 9.14). It ceases to be called the duodenum as it turns sharply around the **root of the mesentery,** which is the mesenteric extension of the parietal peritoneum from which the intestines hang when the cat is in its normal standing position (Figure 9.15).

23. Just medial to the duodenum and lying within its mesentery is the **pancreas** (see Figures 9.14 and 9.15). There are three parts of the pancreas: The *body* of the pancreas lies within the mesoduodenum and is attached to the duodenum; the *left lobe* is thicker and traverses across the abdomen between the layers of the deep wall of the greater omentum; and the *right* is the free tip of the pancreas that separates from the duodenum. It can be recognized by its lobular appearance (similar to that of the parotid salivary gland, although darker in color).

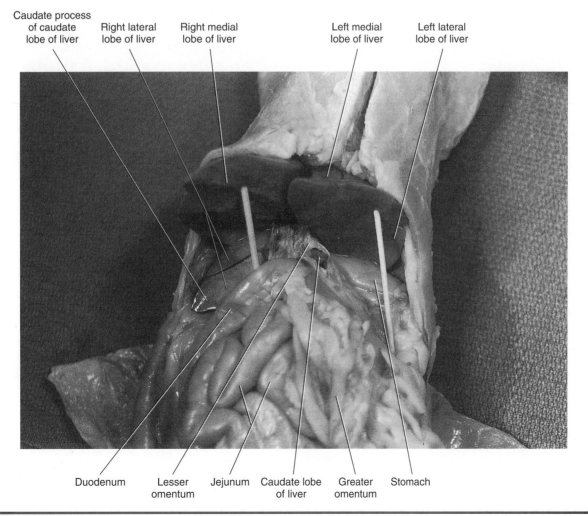

Figure 9.12: Underside of the liver in a cat.

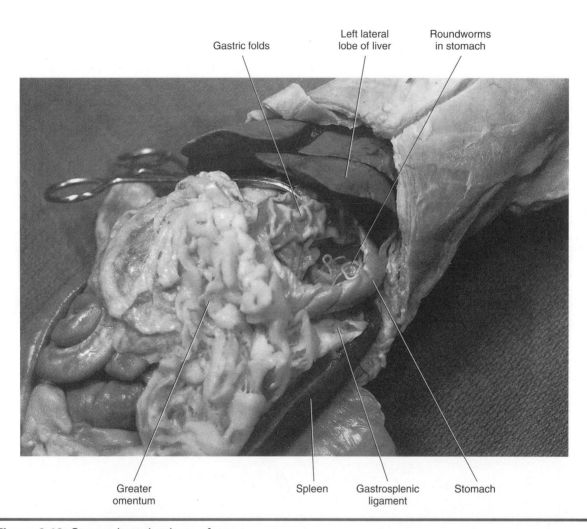

Figure 9.13: Stomach and spleen of a cat.

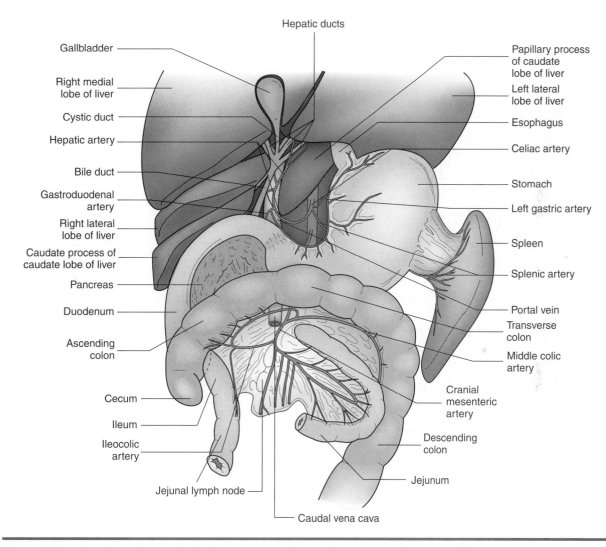

Figure 9.14: Ducts and vessels in the small and large intestines of the cat (most of the jejunum and part of the ileum have been removed).

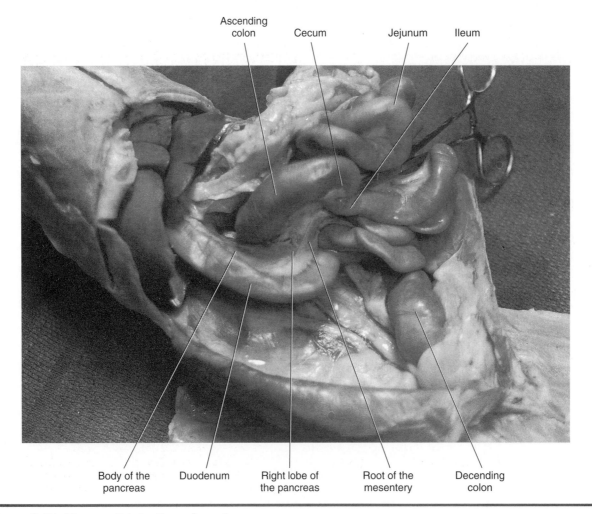

Ascending colon Cecum Jejunum Ileum

Body of the pancreas Duodenum Right lobe of the pancreas Root of the mesentery Decending colon

Figure 9.15: The small intestine of a cat.

Locate the sharp cranial flexure of the duodenum, found a short distance from the pylorus, where it curves caudally. Lift it up, view the dorsal side, and carefully dissect away the most cranial part of the body of the pancreas, just caudal to where the common bile duct enters (the greenish-looking duct connecting the liver to the duodenum. You will dissect this out in Step 30). By teasing apart the tissues, locate the **pancreatic duct** in the interior of the pancreas. This is a white, thread-like duct carrying pancreatic juice to the duodenum. It unites with the *common bile duct* next to the duodenum to form the **hepatopancreatic ampulla** (Figure 9.16). An **accessory pancreatic duct** can be found a short distance (approximately 1 cm) distal to the main duct. These ducts carry the enzymes produced by the exocrine glands of the pancreas. (See Exercise 9.5 for more information on these enzymes.)

24. The second and largest part of the small intestine, the **jejunum** (Figure 9.17), is the next portion encountered; it makes up about half of the length of this organ. The **ileum** is the last part of the small intestine; it opens into the large intestine. There is no gross demarcation between the jejunum and ileum except for a thin mesenteric ileocecal fold. Open the jejunum or ileum. The velvety appearance in the interior is due to the presence of numerous microscopic villi, which are finger-like projections that aid in the absorption of food.

25. Trace the ileum to where it enters the large intestine. This is approximately 0.5 cm from the beginning of the large intestine. Now locate the **ileocolic junction,** a sphincter at the junction between the small and large intestines, by making a longitudinal incision through the wall of the colon in the region of the juncture of the two intestines (see Figure 9.17).

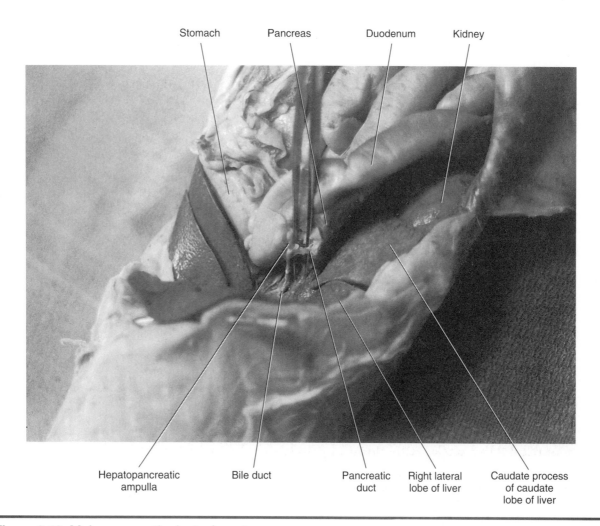

Stomach   Pancreas   Duodenum   Kidney

Hepatopancreatic   Bile duct   Pancreatic   Right lateral   Caudate process
ampulla   duct   lobe of liver of caudate
                                                          lobe of liver

Figure 9.16: Main pancreatic duct of a cat.

26. The large intestine starts as a blind pouch known as the **cecum** (see Figure 9.17). The large intestine continues cranially on the right side as the **ascending colon,** arches transversely as the **transverse colon,** and continues caudally as the **descending colon** (Figure 9.18). There is no distinctive division between the ascending, transverse, and descending colon. The next part of the large intestine, the **rectum,** is located within the pelvic canal and opens to the exterior. This external opening of the rectum is known as the **anus.**

27. Examine the interior surface of the abdominal wall. The shiny membrane over the muscles is called the **parietal peritoneum.** The **visceral peritoneum** is the shiny covering on the abdominal organs inside the abdominal cavity. These two layers are connected by the *mesentery.* The visceral peritoneum is also called the *serosa* or *serosal surface* of the intestinal organs. The serosa is made up of visceral peritoneum and areolar connective tissue, which attaches it to the underlying intestinal smooth muscles.

28. The visceral peritoneum also covers other organs within the abdominal cavity. Identify the **urinary bladder** (see Figure 9.18), the small ventral organ located just cranial to the pelvic cavity, and the **uterus,** if your specimen is a female.

29. The portion of the abdominal cavity between the urinary bladder and the uterus is called the **vesicogenital pouch** in the female. The **genital pouch** extends between the uterus and rectum. In males, the above pouches still exist due to a genital fold, containing the ductus deferens, which lies between the urinary bladder and rectum. These pouches are lined by visceral peritoneum abdominally.

Jejunum   Mesojejunum

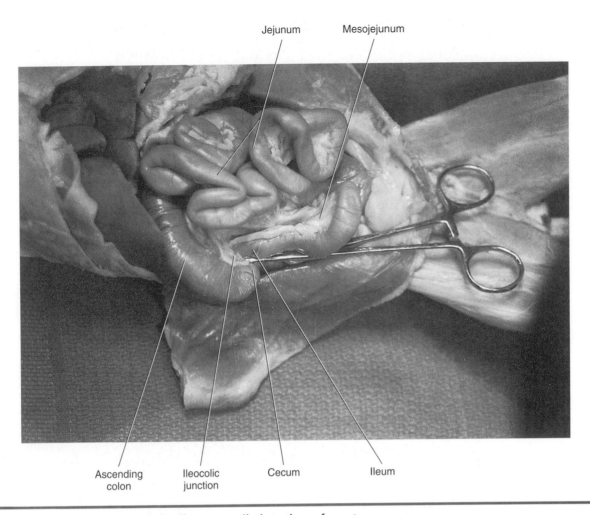

Ascending   Ileocolic   Cecum   Ileum
colon   junction

Figure 9.17: The ileum and the ileocecocolic junction of a cat.

30. The *lesser omentum* (or **gastrohepatoduodenal ligament**) can be seen extending from the left lateral lobe of the liver to the stomach and duodenum. There are three important structures located in the lesser omentum: the *bile duct,* the *hepatic artery,* and the *portal vein.* The **bile duct** is formed by a convergence of the **hepatic ducts** from each liver lobe with the **cystic duct** coming from the gallbladder. If the ducts contain bile, they will be green in color; otherwise, they are colorless and difficult to see. The common bile duct enters the duodenum about 3 or 4 cm from the pylorus.

Carefully dissect away the lesser omentum and locate the **hepatic artery** dorsal and to the left of the common bile duct. This vessel should be injected with red latex. The **portal vein** is located under the duodenum and beneath the common bile duct as the cat is viewed on the table. Lift the duodenum and observe the dorsal surface of the lesser omentum and the duodenum to find this vein (Figure 9.19). It is a large vessel entering the liver. If it is not injected with yellow latex, it will be dark brown from the presence of coagulated blood. The portal vein is thin-walled and may be damaged during the dissection.

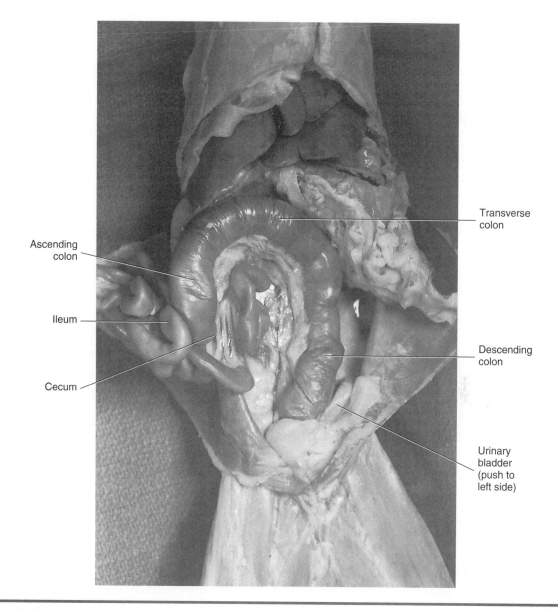

Figure 9.18: Large intestine, cecum, and parts of the colon of a cat.

31. Lift a loop of the jejunum and find the largest **mesenteric lymph node,** called the **jejunae lymph node,** located within the *mesojejunum* (see Step 32). Also note the arteries, veins, lymphatic vessels, adipose tissue, and other mesenteric lymph nodes in the mesentery.

32. Each mesentery is named for the organ it supports; for example, the **mesocolon** is the mesentery attached to the colon, and the **mesoduodenum** is the mesentery attached to the duodenum. Locate and name these structures for each abdominal organ.

33. The kidneys and ureters are located dorsal to, or behind, the peritoneum. For this reason they are considered **retroperitoneal.**

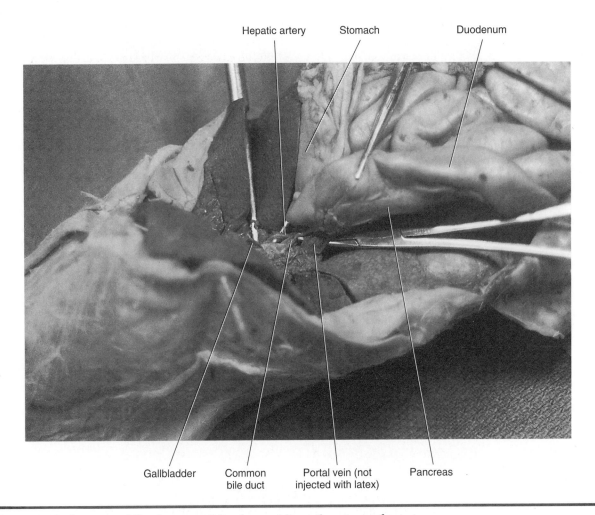

Figure 9.19: Common bile duct, portal vein, and hepatic artery of a cat.

ANATOMY OF THE RUMINANT STOMACH

The process of digestion in the ruminant is classified as anterior fermentation; in other words, fermentation takes place anterior to the small intestine. The **rumen** is a fermentation vat in which bacteria and protozoa process cellulose and hemicellulose from ingested roughage and convert them to volatile fatty acids. These are absorbed by the ruminal papillae to produce energy. The volatile fatty acids provide more than 70% of the mature ruminant's daily energy supply.

The *rumen,* **reticulum,** and **omasum** are considered the **forestomachs,** which are non-glandular and lined with stratified squamous epithelium. The **abomasum** is the true, or glandular, stomach. The reticulum and rumen are located on the left side of the animal, whereas the omasum and abomasum are located on the right. This is important when considering where the incision line will be so that the technician can clip the animal appropriately for an abdominal surgical procedure.

Procedure

Using Figure 9.20, find the following structures. Note that on the exterior, indentations are called *grooves*, and on the interior, these grooves are called *pillars*.

1. **rumen**
   a. **esophageal sphincter**
   b. **rumino-reticular orifice**
   c. **rumino-reticular fold or groove**
   d. **ventral sac**
   e. **dorsal sac**
   f. **caudal dorsal** and **ventral blind sacs**
   g. **caudal groove**
   h. **right and left longitudinal pillar** and/or **groove**
   i. **ruminal papillae**

2. **reticulum**
   a. **reticular groove** or **esophageal groove**
   b. **reticulo-omasal orifice**
   c. **honeycombed interior mucosal surface**

3. **omasum**
   a. **studded laminae** (designed to move food to the abomasum)
   b. **omaso-abomasal orifice**

4. **abomasum**
   a. **cardiac region**
   b. **fundic region**
   c. **pyloric region**

5. **pylorus**

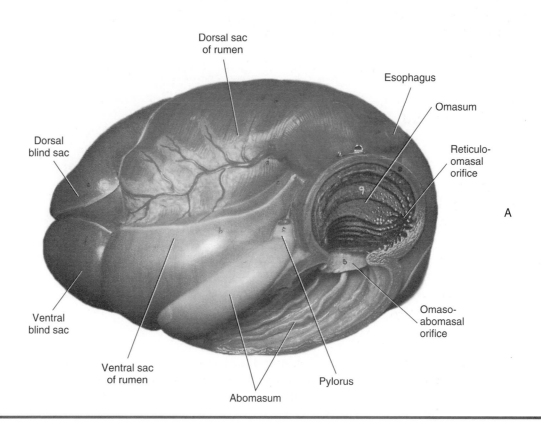

Figure 9.20: Ruminant stomach. *A.* Right side. (Courtesy of SOMSO Modelle, www.somso.de.)

Continued

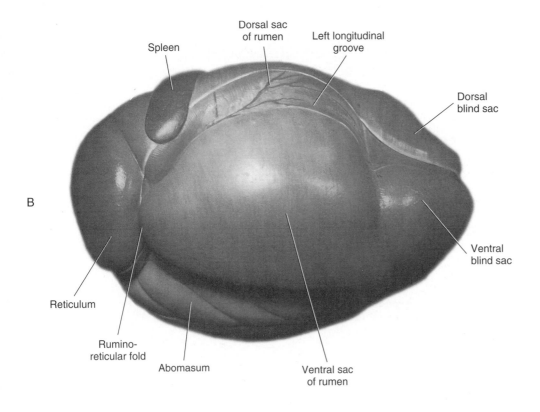

B

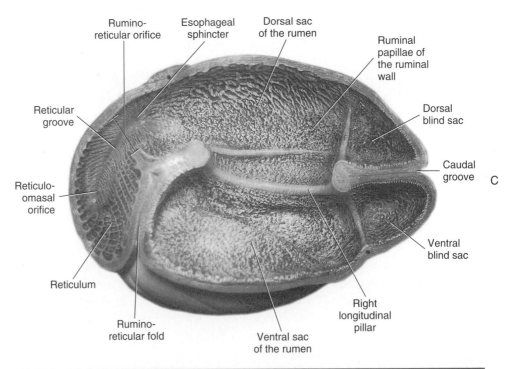

C

Figure 9.20, *cont'd: B.* Left side. *C.* Interior view of the rumen and reticulum. (Courtesy of SOMSO Modelle, www.somso.de.)

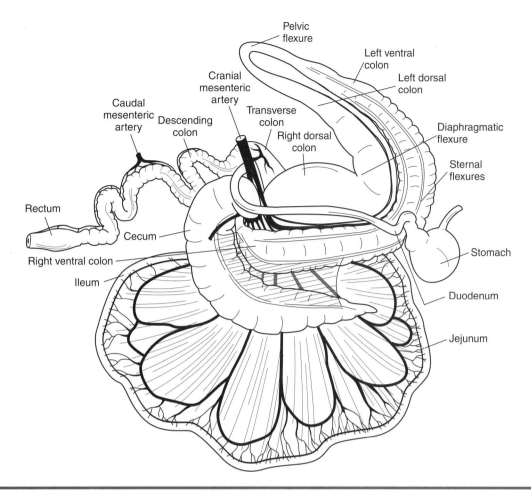

Figure 9.21: The gastrointestinal tract of the horse.

COMPARATIVE ANATOMY OF THE LARGE INTESTINE

Locate the following structures (in bold or italic type) in the figures that follow.

Equines

Unlike the ruminant, the equine species is a **posterior fermenter** (i.e., fermentation takes place posterior to the small intestine). Because of this, the horse has the largest and most complex large intestine of any domestic animal. For a large-animal veterinarian, knowledge of the horse's extensive intestinal anatomy is crucial for the performance of a competent rectal palpation, and thus for an accurate diagnosis in a case of equine colic.

The *cecum* is large and comma-shaped and extends from the pelvic inlet to the floor of the abdominal cavity behind the diaphragm near the xiphoid cartilage. The *ileum* empties into the cecum, and from there the intestinal contents move to the **great colon.** This has four parts, and food moves through them in the following order: (1) **right ventral colon** (bends at the **sternal flexure**); (2) **left ventral colon** (bends at the **pelvic flexure**); (3) **left dorsal colon** (bends at the **diaphragmatic flexure**); (4) and **right dorsal** colon (Figure 9.21). This then attaches to the **transverse colon,** leading to the **descending colon,** then to the *rectum*. At the flexures and the entry to the relatively narrow transverse colon, foreign bodies may become trapped and cause problems.

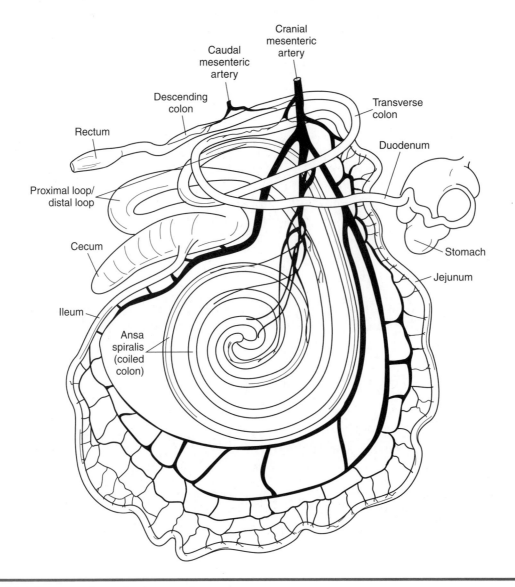

Figure 9.22: The gastrointestinal tract of an ox.

Ruminants

In the ruminant, as in the horse, the *ileum* empties its contents into the *cecum*. From there the material moves into the **proximal loop,** then to the **distal loop** of the colon, and then to the **ansa spiralis** (the **coiled colon**), which makes up the bulk of the colon (Figure 9.22). From the ansa spiralis it enters the *transverse,* then *descending,* colon and finally goes on to the *rectum.* In contrast to the large intestine of small animals, in ruminants the ascending colon is absent and is replaced by the proximal and distal loops and ansa spiralis.

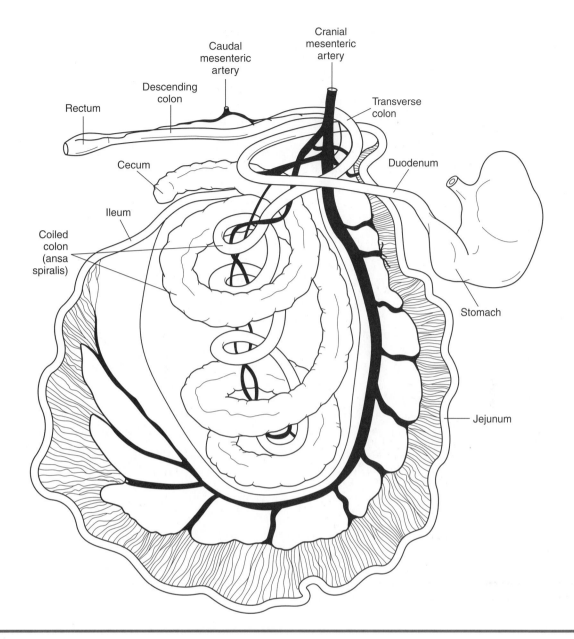

Figure 9.23: The gastrointestinal tract of a pig.

Pigs

Like the ruminant, the ascending colon is absent in the pig. The *ileum* empties into the *cecum*, and from there the contents move directly into an *ansa spiralis*. However, in pigs this *coiled colon* is not nearly as symmetrical in coiling as in the ruminant (Figure 9.23). Locate the same parts as listed previously for the ruminant (except the proximal and distal loops).

EXERCISE 9.4

EXERCISE 9.4

DIGESTIVE ANATOMY OF THE FOWL

Locate the structures in bold type in Figure 9.24.

1. *esophagus*

2. The **crop** is a temporary food storage area where food waits until the proventriculus and gizzard empty. Carbohydrate digestive enzymes are secreted by the crop's wall. *Peristalsis* moves food out of the crop and to the proventriculus.

3. The **proventriculus** has no storage function but mixes foodstuffs with pepsin, a proteolytic enzyme released from the cells within. Little acid is produced by this organ.

4. The **gizzard** macerates the food.

5. The **small intestine** has the same function as in other animals.

6. The **large intestine** also has the same function as in other animals.

7. The **cloaca** is the common opening for the gastrointestinal, urinary, and genital systems. In birds, the transit time for foodstuffs through the digestive system is very rapid. Feces begin to be formed 2 to 2.5 hours after food is ingested, and half the waste has been excreted within 5 hours.

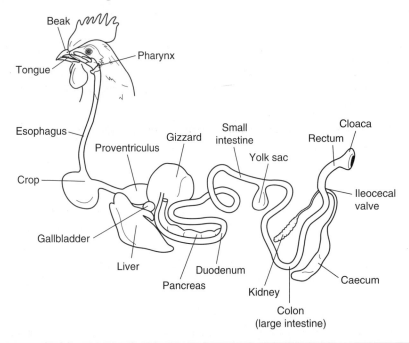

Figure 9.24: Digestive system of a chicken.

EXERCISE 9.5

EXOCRINE PANCREATIC ENZYME ANALYSIS

The pancreas is a mixed gland producing both enzymes and hormones. Enzymes are produced by the exocrine glands of the pancreas, and hormones are produced by the endocrine glands. As you may remember from chapter 5, exocrine glands secrete their product onto an epithelial surface, whereas endocrine glands secrete directly into the bloodstream.

Two pancreatic enzymes are routinely assayed in dogs because their levels rise in cases of *acute pancreatitis*. These are **amylase** and **lipase.** A third enzyme, **trypsin,** can be assayed quantitatively or qualitatively in suspected cases of exocrine pancreatic insufficiency. The following list indicates the process by which these enzymes are produced by the pancreas and the nutrients they digest.

Precursor		Enzyme	Substrate
trypsinogen	$\xrightarrow{\text{H}^+,\text{trypsin}}$	trypsin	proteins
none		amylase	carbohydrates
none		lipase	fats

Assays for Lipase and Amylase

The object is to quantitatively determine whether an animal's pancreas is responding to the presence of fat and carbohydrates in the diet.

1. After three dogs (A, B, and C) have fasted overnight, draw blood from all three and perform the following:
   a. Administer 3 ml/kg of corn oil to dog A.
   b. Administer 3 ml/kg of 50% dextrose to dog B.
   c. Keep dog C NPO (non per os or no food orally); this will be the control.

2. Draw blood from A and B again one hour after administration of the corn oil and dextrose, respectively, and again at 3 hours post-administration from all three dogs.

3. Using a blood chemistry machine, measure amylase and lipase in each sample. Make a graph of the results, using time on the x axis and mg% (milligrams percent) of amylase and lipase on the y axis.

QUESTIONS

1. Should there be a rise in serum levels of these enzymes if they are being secreted from the pancreas into the intestinal tract?

2. Was there a change in the appearance of the serum of either dog A or B at one hour or three hours post administration?

3. If any of the dogs were exhibiting signs of pancreatitis (painful abdomen, vomiting), what would you expect to see as a change in the enzyme levels of the pancreas?

Discussion

There should be no significant absorption of lipase or amylase back into the bloodstream from the intestinal tract after these enzymes are released by the pancreas in a normal animal. However, in an animal with an inflamed pancreas, the administration of a large amount of fat or dextrose could further stress the pancreas and cause release of these enzymes into the blood, and thus both lipase and amylase would be expected to rise. The serum of Dog A should have become lipemic at either 1 hour or 3 hours post administration. (This occurs both in normal and in clinically ill dogs.) There should have been no change in the appearance of the serum from Dog B.

Qualitative Assay for the Presence of Trypsin Using the Fecal Film Test

The object is to diagnose which dog is trypsin deficient. This is an older test that has been replaced by more modern assay methods, such as the trypsin-like immunoreactivity (TLI) test. However, it is a simple test to run and will demonstrate the presence or absence of trypsin.

1. Your instructor will provide samples of feces from two dogs. Label them dog A and dog B.

2. For each dog's sample, using a graduated cylinder, bring 9 ml of 5% sodium bicarbonate to 10 ml total volume by adding feces. Mix well with a stirring rod.

3. Place a ¾-inch wide strip of undeveloped x-ray film into the test tube (cut the strip so that it sticks out the top of the test tube about 1 inch).

4. Incubate the test tubes at 37 degrees centigrade for 30 minutes.

5. Remove the film strip, rinse under running water, and examine whether the gelatin on the end of the strip has been removed (by enzymatic action). If it has, the end of the strip will be transparent.

6. If the gelatin has been removed, this indicates the presence of trypsin.

QUESTIONS

1. In dogs with acute pancreatitis, feces have an increased fat content. Why might this occur?

2. What might be a clinical sign of exocrine pancreatic insufficiency, in which the digestive enzymes of the pancreas are not produced?

Discussion

In acute pancreatitis, the lipase is not breaking down the pieces of fat in the duodenum. Therefore it cannot be absorbed, and the fat is excreted in the feces. Animals that cannot produce sufficient enzymes for digestion will have weight loss and will appear malnourished after a long period of time, in spite of having a good appetite. They may also have chronic diarrhea.

CLINICAL SIGNIFICANCE

Topics that could be discussed in this section include crop burns after eating food that is too hot in hand-reared birds, differentiating between malabsorption syndrome and exocrine pancreatic insufficiency in hungry dogs that are losing weight, and periodontal disease in dogs and cats. However, placement of Cesarian section incisions will be our focus.

When considering where to make an incision for a Cesarian section on a cow, ewe, or doe, knowledge of the abdominal anatomy and the arrangement of abdominal organs is crucial. The gravid uterus lies just above, or dorsal to, the mid-to-caudal ventral belly wall. To access this structure, there are generally two approaches: a paramedian incision between the midline and left mammary vein, and a linear incision starting in the left paralumbar fossa and extending ventrally. The former is done with the animal in dorsal recumbency and the latter with the animal standing or lying on its right side.

Each incision has its advantages and disadvantages, both during the operation and during the post-op recovery period. If the approach were to be made on the right side, the small intestine and ansa spiralis make access to the uterus impossible, for this reason it is not done. On the left side, there is an area just caudal to the rumen where the uterus is accessible. As a technician, knowing how a particular veterinarian likes to do this surgery and knowing the animal's internal anatomy will help you clip and prep the animal for surgery correctly.

Veterinary Vignettes

I really hated this type of emergency call. Geoff Spencer had called telling me his horse had severe colic. He lived about halfway between my hospital and Willamette Ski Bowl—about 70 miles up a winding, two-lane highway. It was the dead of winter, with the temperature dropping about as fast as my enthusiasm for my chosen profession. I might not have minded it so much if they had night skiing; after I cured this horse I could relax with a few runs on the slopes. But it was already getting dark, the roads were icy, and my right defroster wasn't doing the job it should have.

I don't know; maybe I was just nervous, or scared, or both . . . about the drive and about whether I could cure this animal. The horse was reported to be rolling, sweating, kicking at its side, and getting up and down: classic symptoms of a very painful abdomen. I told Geoff to start walking the horse and not to let it roll. We didn't want a twisted intestine to complicate matters.

Veterinarians classify colic in two categories: of gastrointestinal origin and of non-gastrointestinal origin. Causes of non-gastrointestinal colic are conditions such as peritonitis, hepatitis, uterine torsion, urinary tract disease, or parturition (giving birth). The clinical signs of a colicky horse are classified as mild or severe.

A horse with mild colic is restless, anorexic, has an anxious expression in its eyes, and may be sweating and looking at its side. The gastrointestinal signs are yawning, stretching the upper lip, grinding the teeth, and abnormal feces. Behavioral signs include groaning, pawing the ground, getting up and down, stretching, a desire to roll, and tail twitching. Pain is classified as slight and intermittent.

In severe colic all the previously listed signs are present but are more severe. In addition, the horse may be depressed, may want to vomit, and may have profuse diarrhea. Other more severe behavioral signs include rolling, kicking at the belly, disregard for self and handlers, walking in short circles, and falling or suddenly dropping to the ground in pain. We also see labored breathing and profuse sweating. Pain is classified as continuous.

The diagnosis of the nature and cause of the colic is made based on The Eight Ps of Colic. These are: pain (how severe), pulse (how high), peristalsis (increased or decreased), perfusion (of the mucus membranes), passage of a stomach tube (to note ability to pass the tube and the stomach's contents), palpation (by rectum), peritoneal fluid assessment (normal or not), and packed cell volume (the percentage of red blood cells).

In this case the diagnosis was not difficult, and I had it soon after starting my physical exam. The horse's colic was severe. After taking the horse's pulse (78 beats per minute, which is high), temperature (102°F, also increased), and listening to its heart, lungs, and gut sounds, I performed a rectal palpation. There was a huge, solid, hard mass caught at the entry to the pelvic flexure, and there was no way, short of surgery, to get to it. I recommended transferring the horse to an equine surgeon, but Geoff did not want to do this and requested that I put the horse down. The pain this horse was exhibiting I am sure weighed heavily in his decision. We were also facing the prospect of hauling a colicky horse on icy roads in the dark, and that was provided he could have borrowed his neighbor's horse trailer. Under the circumstances, he did what he thought was best, and I euthanized his horse.

"Dr. Cochran, how much will it cost to find out what caused this problem?" Geoff asked.

If I don't know the cause of a problem, my natural curiosity compels me to find the source of the physical symptoms of my patient by performing a necropsy. I told him I was about to suggest that. If we didn't find out what the foreign body was that caused this horse's problems, one of his other horses might ingest something similar.

I opened the horse's left side, starting my cut at the dorsal aspect of the paralumbar fossa. I immediately found the pelvic flexure, where the left ventral colon becomes the left dorsal colon. The object was stuck solidly in this flexure. I cut open the wall of the colon, pulled out the object, and washed it off. At first I didn't recognize what it was, but as the fecal material washed away, its shape became uncommonly clear to me. It looked like a white, ankle-length gym sock—wrinkled at the ankle as if it had been sitting in an old locker for years. This sock had become completely mineralized. It must have lain in the bottom of the cecum for years.

This horse had been a range horse for a period of time in its life, eating off the ground and picking up some mineralized dirt with its food. Why the horse had eaten this sock and how it made it through the small intestine to the cecum is a mystery, but it happened, and then something caused it to be expelled from the cecum. It was even surprising that the sock had made it past the sternal flexure, considering its size and shape.

Now we knew what had caused this problem, and chances are, this one I will never see again.

SUMMARY

The dissection of the gastrointestinal system of the cat should provide you with a good knowledge of the anatomy of the *monogastric* animal (having one stomach). We examined the structures of the mouth, esophagus, stomach, small intestine, large intestine, and accessory digestive organs. If time permitted and the specimens were available, a detailed study of the equine, bovine, porcine, and avian digestive systems would be included rather than just diagrams of their structures. By matching specimens to the diagrams provided, a study of the digestive system structures of the aforementioned animals and the bird would be greatly enhanced.

CHAPTER 10

THE CARDIOVASCULAR SYSTEM

OBJECTIVES:

- name and describe the layers of the heart and the layers of an artery
- understand how the heart is positioned in the chest and how that relates to the sites used to auscultate the valves
- understand and describe the flow of blood through the heart
- name and locate the major anatomical structures of the heart using models, diagrams, and a sheep's heart
- explain the operation of the atrioventricular valves and semilunar valves, and their relationship to the heart sounds
- dissect and name the arteries and veins of the cat; also identify the pulmonary, systemic, and portal systems of blood flow
- understand the pathway of electrical conductivity through the heart and how this produces the electrocardiogram
- know the various intervals, segments, and deflection waves produced during the electrocardiogram and their relationship to the electrical conductivity of the heart
- understand arterial blood pressure and how to measure it

MATERIALS:

- cat cadaver (triple injected) without skin attached
- sheep heart
- Mayo dissecting scissors
- probe
- 1 × 2 thumb forceps or Adson tissue forceps
- #4 scalpel handle with blade
- bone cutting forceps
- rubber gloves
- model of the heart
- electrocardiograph
- doppler and sphygmomanometer
- electric clipper and #40 blade
- rubbing alcohol
- electrocardiogram gel

Introduction

The study of the **cardiovascular system** can be divided into two areas: the **heart** and the **blood-vascular system.** The major function of the cardiovascular system is transportation, with blood as the transport vehicle. The blood carries oxygen, nutrients, cellular wastes, electrolytes, and many other substances and chemicals that are essential to life and to maintaining homeostasis in the body. The heart is a muscular pump that propels the blood through the vascular system. The vascular system is composed of a number of circulatory paths. The **systemic circulation** is the blood flow to and from most of the parts of the body (Figure 10.1).

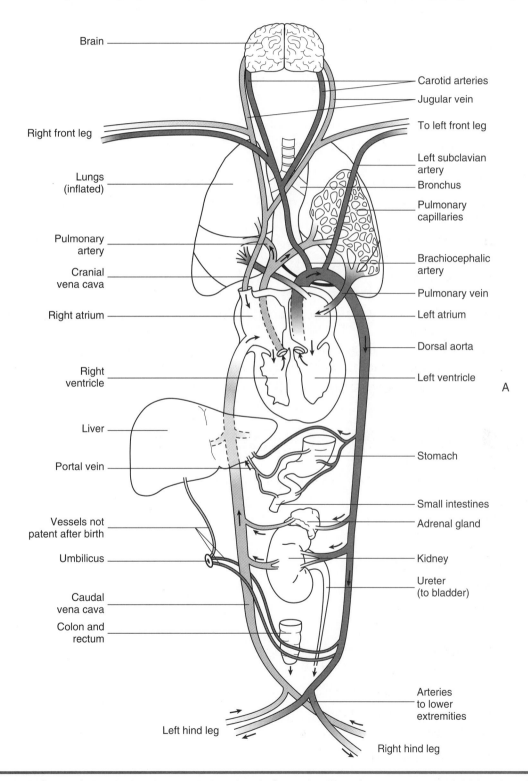

Figure 10.1: *A.* Plan of circulation in an adult animal. Gray shading shows oxygenated blood and color shading shows unoxygenated blood.   *Continued*

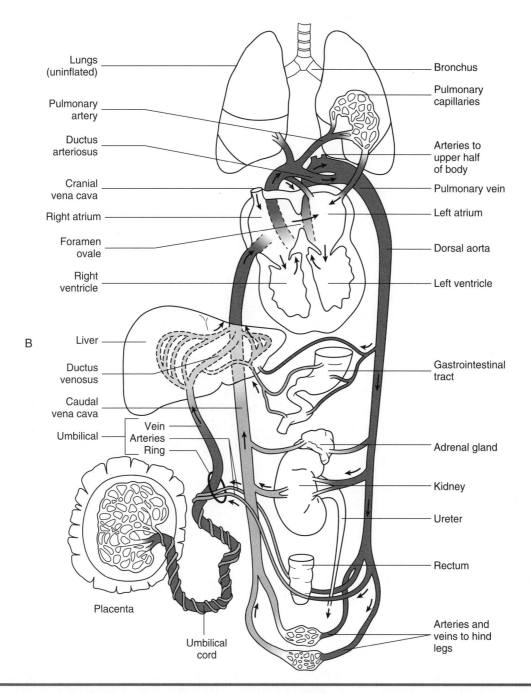

Lungs (uninflated)

Pulmonary artery

Ductus arteriosus

Cranial vena cava

Right atrium

Foramen ovale

Right ventricle

B

Liver

Ductus venosus

Caudal vena cava

Umbilical — Vein / Arteries / Ring

Placenta

Umbilical cord

Bronchus

Pulmonary capillaries

Arteries to upper half of body

Pulmonary vein

Left atrium

Dorsal aorta

Left ventricle

Gastrointestinal tract

Adrenal gland

Kidney

Ureter

Rectum

Arteries and veins to hind legs

Figure 10.1, *cont'd: B.* **Plan of circulation in a fetal animal. Gray shading shows oxygenated blood and color shading shows unoxygenated blood.**

The **pulmonary circulation** is the blood flow to and from the lungs, during which it picks up oxygen and returns it to the heart. The **coronary circulation** is the blood flow to the heart muscle itself, which provides it with the oxygen and nutrition it needs to do its job as a pump. The **hepatic portal system** is a venous system in which blood returns from the intestines and proceeds to the liver before returning to the heart. A portal system, by definition, is a series of vessels between two capillary beds. Also

by definition, the **venous system** returns blood to the heart and the **arterial system** carries blood away from the heart.

Part One: The Heart

The heart has two phases: *contraction* and *relaxation*. Contraction, or **systole,** is the active phase when energy is expended, and relaxation, or **diastole,** is the resting phase. There are four chambers

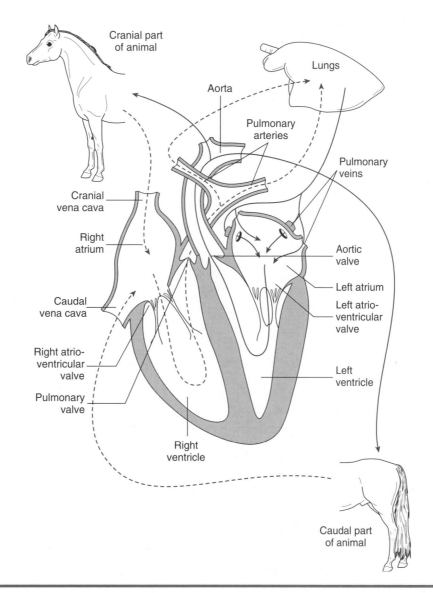

Figure 10.2: Diagram of the heart and the blood flow through it.

of the heart: the upper chambers are the **right** and **left atria,** and the lower chambers are the **right** and **left ventricles** (Figure 10.2).

Both atria contract at virtually the same time, as do the ventricles. The cardiac cycle is illustrated in Figure 10.3. The blood enters the atria while they are relaxed. The **atrioventricular valves** (between the atria and ventricles), also known as the **AV valves,** are open, and so blood flows rapidly into the ventricles. Approximately 70% of the filling of the ventricles occurs during this phase. The atria then contract, which is called **atrial systole,** and the ventricles fill completely. Next, both ventricles begin to contract, and the atrioventricular valves are forced to close, producing an audible sound; this is called the **first heart sound** (the *lub* of the *lub-dub* of the heartbeat).

This phase is a period of isometric contraction. As contraction continues, the pressure within the ventricles overcomes the closed **semilunar valvules** (the valves to the large arteries that exit off the base of the heart); the valves open, and the blood is ejected into the **pulmonary artery** and **aorta** (the large artery that begins systemic circulation). This contraction is called **ventricular systole.** As blood moves into the two arteries, they stretch, and because of their elasticity, the pressure within these vessels becomes sufficient to cause the semilunar valves to snap closed. This is the **second heart sound.** The cycle is repeated, starting with the period of relaxation of the heart called *diastole.* Figure 10.3 shows only the right side of the heart—the same cycle is occurring simultaneously on the left side. The third heart sound is

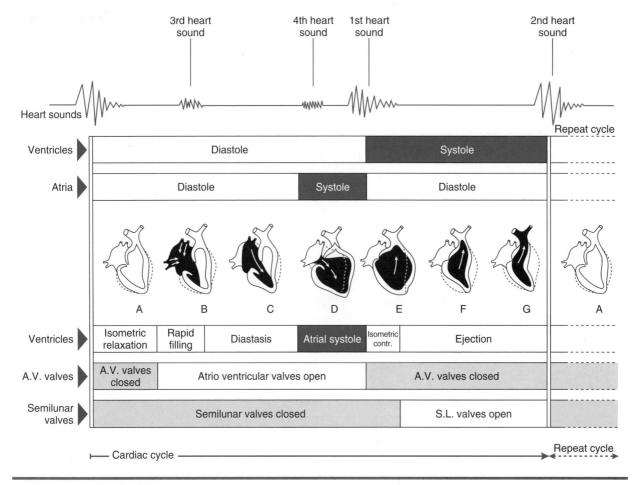

Figure 10.3: Events of the cardiac cycle.

rapid ventricular filling, and the atrial systole causes the fourth heart sound. The third and fourth heart sounds can be heard normally with an esophageal stethoscope (a tubular device placed within the esophagus, usually during anesthesia and surgery, that amplifies the heart beat and respiratory sounds). If these sounds are heard with a normal stethoscope,

there may be some cardiac problems, and the veterinarian should be alerted.

In the exercises in this chapter, the important structures are listed in colored bold print. If a structure is mentioned before its dissection, it is italicized. Structures discussed before their dissection may also be in bold print for special emphasis.

EXERCISE 10.1

ANATOMY OF THE HEART

The heart is positioned in the chest so the **base** (the top part) is located just inferior to the tracheal *carina* (tracheal bifurcation), and it angles caudoventrally until the **apex** (the pointed end) is resting just superior to the last few sternabrae (in most species—in the horse the apex is located mid-sternum). Because the base of the heart is oriented more cranially than the apex, the valves of the **pulmonary artery** and **aorta** are more cranial than the valves that separate the atria from the ventricles.

The heart is enclosed within a double-walled sac called the **pericardium,** or **pericardial sac.** The inner surface of this sac attached to the heart is the **visceral pericardium,** which is also the **epicardium** of the heart. The outer surface of the sac is the **parietal pericardium.** It is intimately attached to the **pericardial pleura** by a thin layer of connective tissue. The pericardial pleura is an extension of the **mediastinal pleura,** which was first noted when the chest was cut open in the dissection of the digestive system. The space between the visceral and parietal pericardia is the lumen of the pericardial sac, which contains fluid that acts as a lubricant, allowing the heart to beat freely (Figure 10.4).

Complete the following steps in the dissection procedure using a cat's and sheep's heart as indicated.

1. Using the cat, observe the **pericardial sac** surrounding the heart. Make a tiny incision in the sac at the apex of the heart. Using your scissors, cut up through the pericardial sac to the heart's base and fold the sac above the heart and out of the way. Find on the cranial surface of the heart a red **paraconal artery,** which is a branch of the **left coronary artery** traversing the heart in the

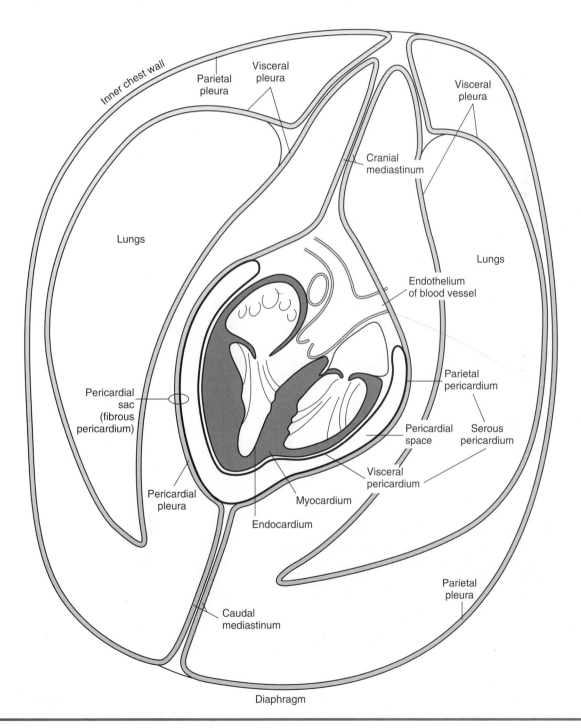

Figure 10.4: The pericardial sac and thoracic pleura.

paraconal interventricular groove. We look for these arteries in the cat's heart because they are not injected with latex in the sheep (Figure 10.5).

2. Rinse the sheep heart in water to remove as much preservative as possible. Observe the pericardium around the sheep heart in Figure 10.6. The sheep heart will be used for the remaining steps in this exercise.

3. If the pericardium has not been removed, remove it from the heart, being especially careful to dissect it from the vessels at the base of the heart. If it has been removed, look for the remnants of this membrane attached to the large blood vessels above the heart.

4. There are three layers to the heart: the **epicardium, myocardium,** and **endocardium** (Figure 10.6). Separate a small portion of the *epicardium* (the **visceral pericardium**) from the *myocardium* (the muscle layer) by careful dissection with a scalpel. The third layer of the heart, the *endocardium,* will be visible when the heart is opened.

5. Start the dissection by locating the **pulmonary artery** on the cranioventral left surface of the heart (Figure 10.7*A*). This artery emerges from the cranial surface of the heart at the base, medial to the *left auricle.* The simplest method of determining the ventral left surface of the heart is to look for the **paraconal interventricular groove,** which separates the right ventricle from the left ventricle. In this groove lies the **paraconal artery,** a branch of the *left coronary artery.*

Turn the heart over and look at the caudal surface. Another coronary artery runs down the middle and angles to the right just above the apex. This is the **subsinuosal artery.** This is where the **coronary vascular system** in domestic animals differs from species to species. In carnivores and rumi-

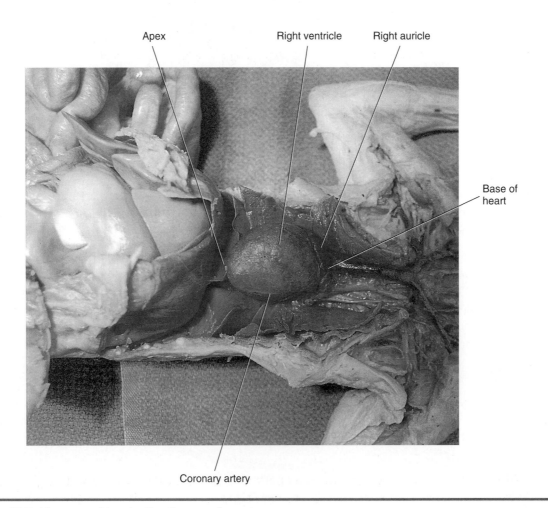

Figure 10.5: Heart position in the thorax of a cat.

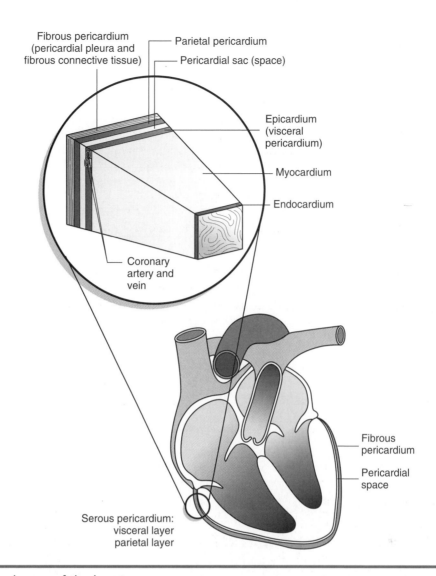

Figure 10.6: The layers of the heart.

nants, the subsinuosal artery is a branch of the left coronary artery (see Figure 10.7*B*). In the horse and pig, the subsinuosal artery arises from the **right coronary artery** (see Figure 10.7*C*). The right coronary artery can be located beneath the *right auricle* as it circles the heart.

The coronary venous system of the heart on the left cranial side includes the **paraconal vein,** which joins the **great cardiac vein** to enter the **coronary sinus** as it courses within the coronary groove beneath the left auricle. On the right lateral side is the **subsinuosal vein.** It lies in the **subsinuosal interventricular groove** and joins the coronary sinus as it courses toward the base of the heart. It does this in all species. Adjacent to the right coronary artery beneath the right auricle is the **small cardiac vein.**

6. Insert the handle end of a probe through the pulmonary artery and down into the **right ventricle.** Note that the distance to the bottom of this ventricle is approximately half the distance to the apex. There are several methods to know you are in the right ventricle: the chamber of the right ventricle does not go all the way to the apex; the myocardial wall is thinner when compared to the left ventricle; and within the chamber there is a cord that attaches the inner wall to the outer, called the moderator band (see Step 15).

7. Using scissors, make a cut through the ventral wall of the *pulmonary artery* into the right ventricle. Continue the cut parallel to and 1 cm to the right of the *paraconal interventricular groove* (Figure 10.8). If you are too far from this groove, you will cut structures within the heart.

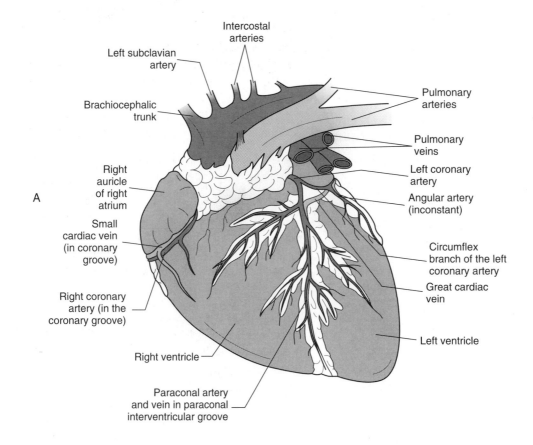

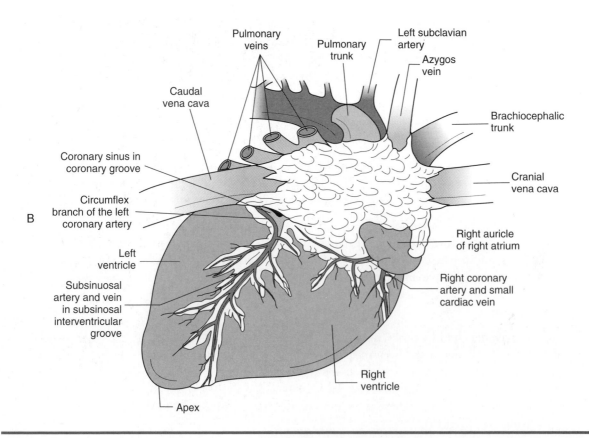

Figure 10.7: *A.* Left lateral view of the coronary circulation of the heart, all species. *B.* Right lateral view of the coronary circulation of the heart in carnivores and ruminants. The subsinuosal artery is a continuation of the circumflex artery (a branch of the left coronary artery).

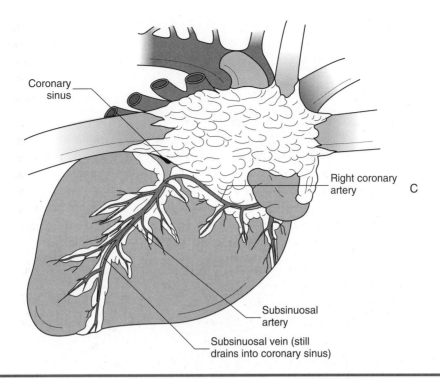

Coronary
sinus

Right coronary
artery

C

Subsinuosal
artery

Subsinuosal vein (still
drains into coronary sinus)

Figure 10.7, *cont'd: C.* Right lateral view of the coronary circulation of the heart in horses and pigs. The subsinosal artery is a continuation of the right coronary artery rather than the left coronary artery.

8. Continue the cut around the bottom of the right ventricle and up the opposite wall (still staying 1 cm from the edge of the wall) through the opening of the *caudal vena cava* (Figure 10.9).

9. Pry open the *pulmonary artery* and observe the **pulmonary valve.** It should have two crescent-shaped pouches intact, with the third incised by the cut you just made (see Figure 10.8).

10. From the caudal opening of the **caudal vena cava,** insert your finger or probe into the **right atrium** dorsally and caudally and find the opening of the **cranial vena cava.** These two large vessels are the routes for blood returning to the heart from the cranial and caudal halves of the body (see Figure 10.9).

11. Open the right ventricle as shown in Figure 10.9, and look into the area of the *right atrium* and *auricle.* First observe the internal structure of the **right auricle,** the flap-like or ear-like (*auricle* means ear) structure seen from the exterior of the heart. The web-like arrangement of muscles on the interior of the auricle is made of multiple **pectinate muscles,** so called because they resemble a comb (pecten).

12. Locate the orifice of the **coronary sinus,** just below the caudal vena cava. It is a small canal-like tube that courses around the heart between the atria and the ventricles in the groove just under the auricles.

13. Examine the **tricuspid valve,** or **right atrioventricular (AV) valve,** between the right atrium and right ventricle. Count the number of *cusps,* or flaps, that make up this valve. From the name, you should determine that it has three cusps.

14. Locate the **papillary muscles** and the attached **chordae tendineae** in the wall of the right ventricle. The papillary muscles exert tension on the cusps of the valve during the contraction of the ventricles, thus preventing eversion into the atria.

15. Find the trabecula septomarginalis crossing the lumen of the right ventricle. This is thought to prevent overdistension of the ventricle. Also note, the trabeculae carneae, myocardial ridges projecting mainly from the outer wall and run toward the apex (in both ventricles).

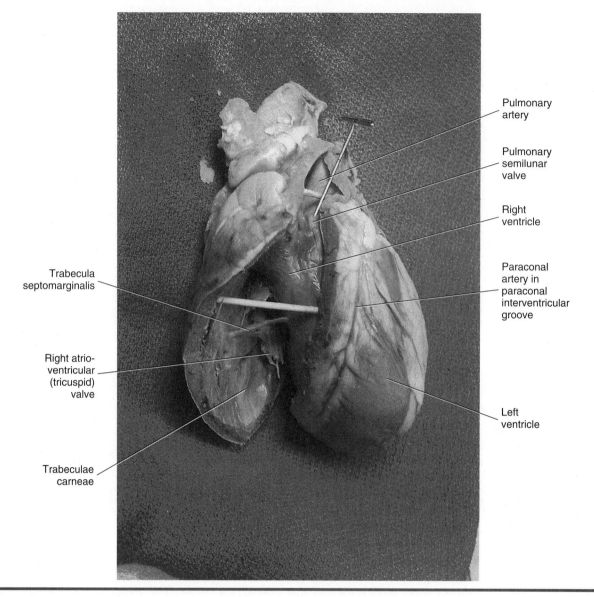

Pulmonary artery

Pulmonary semilunar valve

Right ventricle

Paraconal artery in paraconal interventricular groove

Trabecula septomarginalis

Right atrio-ventricular (tricuspid) valve

Trabeculae carneae

Left ventricle

Figure 10.8: Proper cut through the pulmonary artery and right ventricular wall in a sheep.

16. Find the **aorta.** This is the large, thick-walled vessel coming from the **left ventricle.** It can be found by inserting the handle of the probe into the remaining large vessel at the base of the heart and on into the left ventricle. Insert one side of the scissors deep into the aorta and down into the ventricle. If you encounter resistance, pull out slightly and reposition the scissors (because you probably have slipped into one of the cusps of the aortic valve). With your opposite hand, rotate the pulmonary artery out of the way, position the visible side of the scissors under this artery, and make a longitudinal cut through the aorta and ventricle to the apex (Figure 10.10).

17. Try to find the **pulmonary veins** entering the **left atrium.** This may be possible in hearts that have the pericardium still attached, but if it has been removed, often there is just a large hole in the dorsum, or top, of the atrium where these vessels once entered the heart.

18. Locate the most ventral and lateral of the pulmonary veins, or the most ventrolateral point on the opening into the left ventricle. Make a longitudinal incision through this pulmonary vein or opening, staying at least 2 cm from the previous cut (which was actually through the left lateral wall of the heart). Continue this incision through the wall of the left atrium and the left ventricle to the apex, where it should join the previous cut (Figure 10.11).

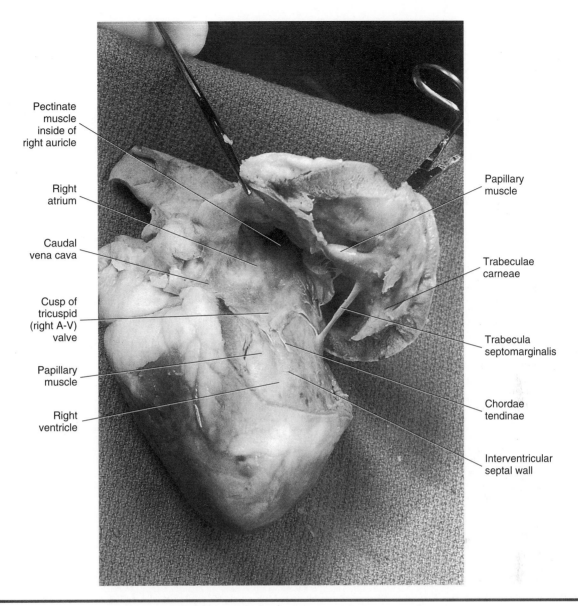

Pectinate muscle inside of right auricle

Right atrium

Caudal vena cava

Cusp of tricuspid (right A-V) valve

Papillary muscle

Right ventricle

Papillary muscle

Trabeculae carneae

Trabecula septomarginalis

Chordae tendinae

Interventricular septal wall

Figure 10.9: Right atrium and ventricle. The previous cut is continued around the ventricle and dorsally to join the entrance of the caudal vena cava.

19. Spread open the left side of the heart (see Figure 10.11). Compare the wall thickness of the left ventricle with that of the right. Observe the **bicuspid valve,** or **left atrioventricular (AV) valve.** This valve has two major cusps, hence the name bicuspid valve. Veterinarians often call it the **mitral valve.** Note that this ventricle also has chordae tendineae and papillary muscles, but no moderator band.

20. Look into the *left atrium* and note that it also has *pectinate muscles* within the **left auricle.** Locate the **interatrial septum,** the wall that separates the two atria. This is best accomplished by grasping the heart with your thumb in the right atrium and your index finger in the left atrium (or vice versa); if you pinch your thumb and finger together, that will be where the interatrial septum is located. Now examine this septum from the interior of the right atrium. Locate the **fossa ovalis,** the oval-shaped depression ventral to the entrance of the cranial vena cava. This was a portal at one time through which the fetal blood (containing oxygen from the placenta) by-passed the pulmonary circulation and flowed into the left side of the heart to be pumped to the body (see Figure 10.1*B*).

21. Spread apart the cut made through the aorta into the left ventricle (see Figure 10.10). Note the valve between these two structures; like the pulmonary valve it also has three moon-like, crescent-shaped valvules and is therefore called the **aortic valve.**

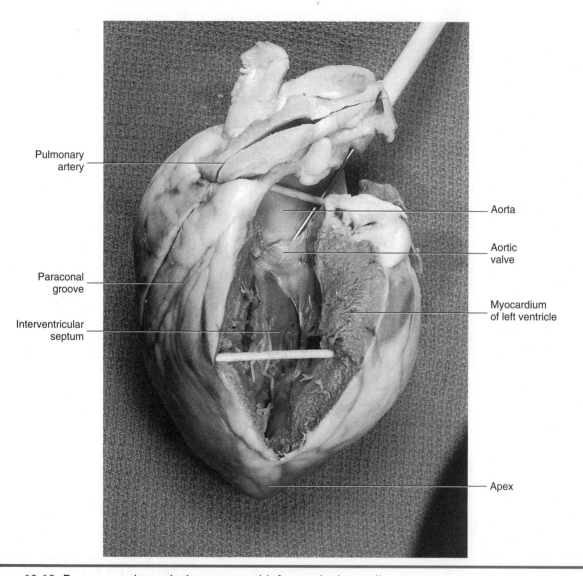

Pulmonary artery

Paraconal groove

Interventricular septum

Aorta

Aortic valve

Myocardium of left ventricle

Apex

Figure 10.10: Proper cut through the aorta and left ventricular wall.

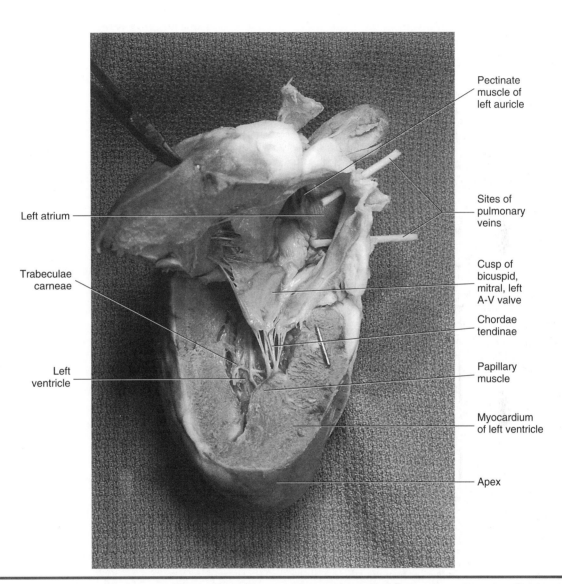

Pectinate muscle of left auricle

Sites of pulmonary veins

Cusp of bicuspid, mitral, left A-V valve

Chordae tendinae

Papillary muscle

Myocardium of left ventricle

Apex

Left atrium

Trabeculae carneae

Left ventricle

Figure 10.11: Left atrium and ventricle. The cut is made toward the apex through one pulmonary vein hole on the lateral side and connects with the previous cut at the apex of the heart.

Part Two: The Vascular System

The **blood-vascular system** is a closed transport system; blood leaves the heart in the **arteries** and returns via the **veins.** Generally, veins contain venous blood, which is deoxygenated, and arteries contain arterial blood, which is oxygenated. There are two exceptions to this general rule for both the venous system and arterial system. In the venous system, the pulmonary vein(s) returning from the lungs contain oxygenated blood, as does the umbilical vein of the fetus, which returns oxygenated blood to the fetal heart. This oxygen is picked up from the mother as blood courses through the placenta (see Figure 10.1*B*). Conversely, in the arterial system, the pulmonary arteries taking blood to the lungs, and the umbilical artery taking blood to the placenta, contain deoxygenated blood.

The flow of blood through the *vascular system* starts with the *arteries.* As the arteries branch and enter tissues they get smaller and become **arterioles,** which are the smallest arteries. From there they become **capillaries,** and a network of capillaries within tissue is called a **capillary bed.** This is the site of **internal respiration,** where the nutrients and oxygen enter the cells of tissue, and carbon dioxide (CO_2) is removed to the blood according to the following formula.

$$CO_2 + H_2O \rightleftharpoons H_2CO_3 \rightleftharpoons H^+ + CO_3^-$$

Most of the CO_2 combines with water to produce carbonic acid, which rapidly dissociates into hydrogen ions and bicarbonate ions. The hydrogen ions combine with the blood's hemoglobin, which has released its oxygen to tissue, thus forming *reduced hemoglobin* that acts as a buffer. Most of the bicarbonate ions are carried inside the red blood cells, but some diffuses out into plasma, and chloride diffuses in. Approximately 21% of the carbon dioxide combines with the protein amino groups on hemoglobin to form carbaminohemoglobin. Another 7% is dissolved in plasma. Once the blood reaches the lungs, the reaction reverses itself, and CO_2 passes from the capillary blood into the alveoli and then is exhaled. Simultaneously, oxygen enters the blood and is picked up by the red blood cells and hemoglobin within. This shows the close relationship between the blood-vascular system and the respiratory system.

Capillaries are tiny, microscopic tubes composed almost entirely of simple squamous epithelium. These were examined in the histology section of chapter 5, Figure 5.3. Blood exits the capillaries and enters the smallest veins, called **venules,** which then unite to form larger *veins* and return the blood to the heart.

The walls of blood vessels, except for the capillaries, have three layers, or **tunics** (Figure 10.12).

The **tunica interna** (or tunica intima) lines the lumen of the vessel and is composed of a thin layer of simple squamous epithelial cells, also called the **endothelium** (squamous cells underlain by a scant basal lamina). It is continuous with the endocardium of the heart. The cells of this layer fit closely together, forming an extremely smooth inner lining that helps to decrease resistance to blood flow.

The **tunica media** is the thicker middle layer of blood vessel walls and is composed primarily of smooth muscle and elastin. The smooth muscle is under the control of the *sympathetic nervous system,* a component of the *autonomic nervous system.* It plays an active role in regulating the diameter of the blood vessels, which controls the peripheral resistance and thus blood pressure.

The **tunica externa** (or *adventitia*) is the outermost tunic and is composed of areolar, or fibrous, connective tissue. Its function is basically support and protection.

In general, the walls of the arteries are thicker than those of the veins because their tunica media have heavier smooth muscle and elastin. Arteries and arterioles in tissues appear round, whereas the venules and veins are larger and can take on a variety of elliptical shapes during the sectioning of tissues. The larger veins have valves, usually two paired cusps placed irregularly along the vessels. They direct blood to flow only toward the heart and prevent backflow. Horse's legs do not have valves in their veins—instead they depend on the pumping action of the horse's weight and movement on the hooves to move the blood up the legs.

Arteries, because they are closer to the pumping action of the heart, must expand as blood is propelled into them and then recoil passively as the blood continues on its path into circulation during diastole. The **systolic blood pressure** is the pressure within the arterial system at the peak of systole and is controlled by two factors: (1) peripheral resistance (based on the size and elasticity of the vessel) and (2) stroke volume of the heart (how much blood is pumped out per stroke). **Diastolic blood pressure** (when the heart is relaxed) is controlled only by peripheral resistance. If arteries did not have elasticity (no elastin), were built similar to veins, and could not passively recoil, the blood pressure would drop precipitously during diastole. Conversely, if the arteries lose their elasticity (via sclerosis or hardening) and cannot stretch, blood pressure rises during both systole and diastole. This condition is known as **hypertension.** Excessively low blood pressure is known as **hypotension.**

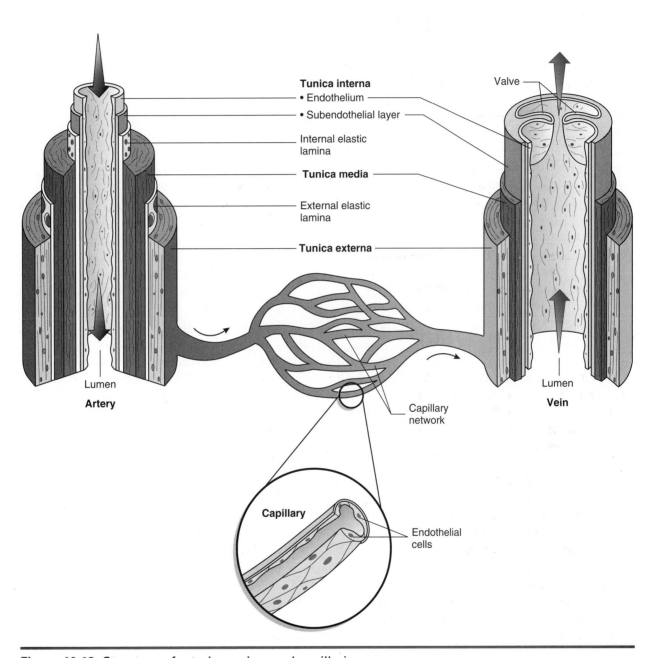

Figure 10.12: Structure of arteries, veins, and capillaries.

DISSECTION OF THE ARTERIES AND VEINS

During this dissection, be careful to not damage the kidneys, ureters, uterus, ovaries, ductus deferens, or the vessels to these structures. Also take care not to damage any nerves. Use both the diagrams and the photos shown in the figures to determine the locations of the veins and arteries during your dissection.

PART A: DISSECTION OF THE VEINS

1. Using the cat, look at the *pericardium* surrounding the heart. After determining the location of its attachment, remove the parietal layer, the thymus gland, and any lymph nodes in the cranial thorax. Remember that the visceral layer of the pericardium forms the epicardium of the heart. Figure 10.13 illustrates the important veins which will be identified during dissection.

2. Find the **cranial vena cava,** the large blue vessel entering the cranial aspect of the right atrium. This vein returns blood from the head, neck, and front legs.

3. Lift up the heart and cranial vena cava and carefully dissect away any tissue found beneath the cranial vena cava. The **azygos vein** can be seen entering the dorsal surface of the cranial vena cava immediately cranial to the heart (Figure 10.14). Lift up the right lung; the azygos vein is now visible on the right side against the vertebral column. The tributaries of the azygos vein are the **intercostal veins** from the body wall, the **esophageal veins,** and the **bronchial veins.** It is difficult to find the esophageal and bronchial veins if they are not injected with latex, which they may not be.

4. Trace the cranial vena cava forward; note that it is formed by the union of the two **brachiocephalic veins** (see Figures 10.13 and 10.14).

5. The cranial vena cava also receives the **internal thoracic vein** from the ventral chest wall. The right and left internal thoracic veins unite shortly before they empty into the cranial vena cava. The **right vertebral vein** from the brain in some cats might enter the dorsal surface of the cranial vena cava, but in most cats it enters the right brachiocephalic vein. (see Figures 10.13 and 10.14).

6. Trace both brachiocephalic veins forward. The left brachiocephalic vein receives the **left vertebral vein.** The vertebral veins come from the vertebrae and thus can only be traced a short distance dorsally in your dissection. Both brachiocephalic veins are formed by the union of the external jugular veins, which drain the head, and the short subclavian veins, which drain the front legs (see Figures 10.13 and 10.14).

7. Start tracing the **external jugular vein** toward the head. The first branch is the smaller **internal jugular vein,** and it joins immediately above the point of union of the external jugular with the subclavian. The internal jugular vein, which drains the brain, lies next to the left common carotid artery near the trachea. Another vessel, which is often difficult to locate, is the **thoracic duct.** It is a large lymphatic vessel that empties into the external jugular at the point of union of the external jugular with the subclavian. Sometimes this duct may appear blue, if latex was forced into it, or brown and beaded (because of its valves) if it is empty. Continue tracing the external jugular vein toward the head. The large **superficial cervical vein** (formerly known as the transverse scapular vein) empties into the external jugular (see Figures 10.13 and 10.14).

8. The external jugular vein is formed by the union of the **maxillary vein** (the dorsal branch) and the **linguofacial vein** (the ventral branch) just caudal to the point of the mandible. The **transverse jugular vein** can be seen connecting the two external jugular veins at this point (see Figures 10.13 and 10.14).

9. We will use the left side of the cat to dissect the vascular system and the right side for the nerves. If both the arteries and veins on the right side have superior latex injection (as was the situation in the photo of the cat in Figure 10.14), use that side. However, if on one side the arteries are better injected and on the other the veins are better injected, use both sides for the dissection. Just be sure not to damage any nerves as you do this.

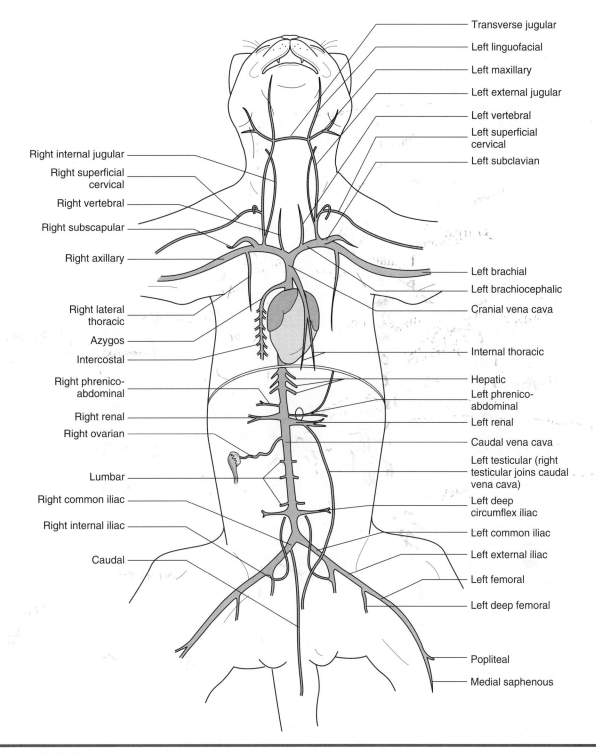

Figure 10.13: Venous system of the cat.

Trace the left subclavian vein through the chest wall. It receives the **subscapular vein** from the shoulder a short distance from the site where it joins with the external jugular. Distal to where the subscapular vein joins, the continuation of the subclavian is called the **axillary vein.** When the axillary vein attaches to the muscles of the front leg, it becomes the **brachial vein.** The **cephalic vein** lies on the cranial surface of the foreleg, coursing up the leg and dividing at the elbow joint (see Figure 10.14). It continues up the cranial surface of the upper leg as the cephalic vein, whereas the other branch, the superior brachial vein, passes medial to the humerus to course adjacent to the

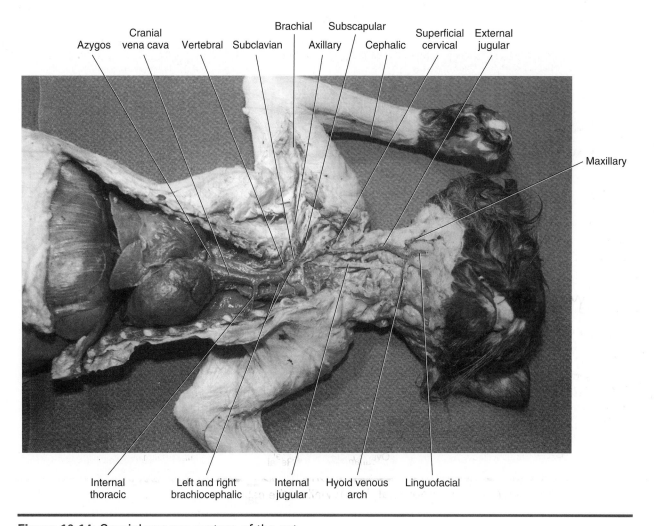

Figure 10.14: Cranial venous system of the cat.

superficial brachial artery and join the brachial vein (see Figures 10.13 and 10.14). The cephalic vein may join the superior cervical vein or enter the external jugular on its own.

10. Raise the apex of the heart and spread the lung lobes apart so you can find the caudal vena cava. Trace this vessel from the diaphragm cranially to where it drains into the right atrium. This large vein drains the lower part of the body.

11. Continue to trace the caudal vena cava caudally through the diaphragm and into the abdominal cavity, where it lies to the right of the aorta. To see this vein and its tributaries, dissect away the peritoneum because the caudal vena cava and the aorta are retroperitoneal. The tributaries usually accompany the arteries of the same name.

12. The **hepatic veins** drain blood from the liver into the caudal vena cava. To locate these veins, using a probe or a scalpel blade, gently scrape away tissue on the right cranial surface of the liver. Several hepatic veins may be located in this manner.

13. The **phrenicoabdominal veins** (formerly known as adrenolumbar veins) drain the adrenal glands and the body wall. The right vein drains into the caudal vena cava, and the left may either drain into the caudal vena cava or into the **renal veins.** These veins may be located by examining the dorsal muscle wall cranial to the kidney (see Figures 10.13 and 10.15).

14. Locate the *renal veins,* which carry blood from the kidneys into the caudal vena cava. Note that the right kidney, and thus the right renal vein, is more cranial than the left one. Do all the dissection on

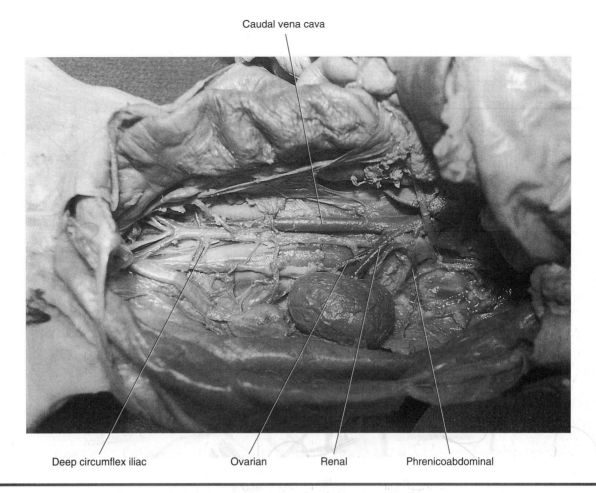

Caudal vena cava

Deep circumflex iliac   Ovarian   Renal   Phrenicoabdominal

Figure 10.15: Veins of the mid-abdominal dorsal wall of the cat.

the left kidney, leaving the right kidney intact under the peritoneum for later dissection. Also note the white **ureters** exiting from the kidney and traversing retroperitoneally toward the bladder. Find these, but do not cut them (see Figures 10.13 and 10.15).

15. Below the renal veins are the long, thin, paired **ovarian veins** or **testicular veins.** They are most easily located by tracing the vessels from the gonads back toward the caudal vena cava. If the cat is a female, locate the **ovaries,** small oval bodies near the cranial ends of the **uterus,** below the kidneys. The ovarian artery and vein should be visible. If the cat is a male, the **testicular vein, artery,** and **ductus deferens** can be seen passing through the **inguinal canal** (the opening in the caudal abdominal body wall). Trace these blood vessels cranially toward the aorta and caudal vena cava. The left testicular and ovarian veins enter the left renal vein in the cat; and on the right they enter the caudal vena cava (see Figures 10.13 and 10.15).

16. Several pairs of **lumbar veins** enter the dorsal surface of the caudal vena cava at intervals in the abdominal cavity. Spread the dorsal median musculature apart and lift the caudal vena cava gently to reveal these veins.

17. The **right** and **left deep circumflex iliac veins** (formerly known as iliolumbar veins) enter the caudal vena cava near its termination. These vessels drain the abdominal wall muscles (see Figures 10.13 and 10.15).

18. The **common iliac veins** join to form the caudal vena cava (see Figures 10.13 and 10.16). Trace the route of the common iliac veins toward the hind legs. On the left side, the first vessel joining this vein is the **internal iliac vein,** which drains the rectum, bladder, and internal reproductive organs. The common iliac vein continues distally from this point as the **external iliac vein.** When it passes

External iliac   Common iliac   Caudal vena cava

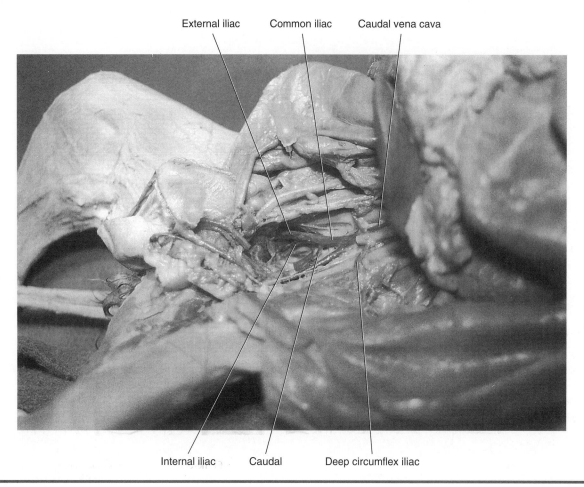

Internal iliac   Caudal   Deep circumflex iliac

Figure 10.16: Veins of the caudal abdomen of the cat.

out of the body wall, it is the **femoral vein.** There is a slight variation on the right side; the first vein to join the common iliac is the **caudal vein,** a short distance from the caudal vena cava (see Figures 10.13 and 10.16). Trace the femoral vein to its formation by the **medial saphenous vein** superficially and the **popliteal vein** from deep between the muscles (see Figures 10.13 and 10.17). If this vein were to be traced to the lateral side of the leg, you would note that it is formed by the *cranial tibial vein* and the *lateral saphenous vein.*

19. The *hepatic portal system* may or may not be injected; if it is, it will be injected with yellow latex (Figure 10.18). A portion of the **portal vein** was dissected in a previous chapter with the digestive system (see Figure 9.19 in chapter 9). As mentioned previously, a portal system is a vessel between two capillary beds—in this case between the capillaries of the digestive organs and those of the liver. Thus, the portal veins carry blood from the digestive organs to the liver. As you follow the portal vein caudally, note that it is formed by the union of the **splenic vein** (formerly known as the gastrosplenic vein), from the stomach and spleen, and the larger **cranial mesenteric vein** coming in caudally from the small intestine. There are two veins that join the cranial mesenteric just dorsal to the mid-pancreas area: the **pancreaticoduodenal** from the pancreas and duodenum and the **caudal mesenteric** from the colon.

PART B: DISSECTION OF THE ARTERIES

1. Locate the **pulmonary artery** emerging from the cranioventral surface of the heart. Because it is attached to the right ventricle, it will be filled with blue latex. Trace its path to the lungs, noting that it branches into the **right** and **left pulmonary arteries.** Figure 10.19 illustrates the important arteries identified during dissection.

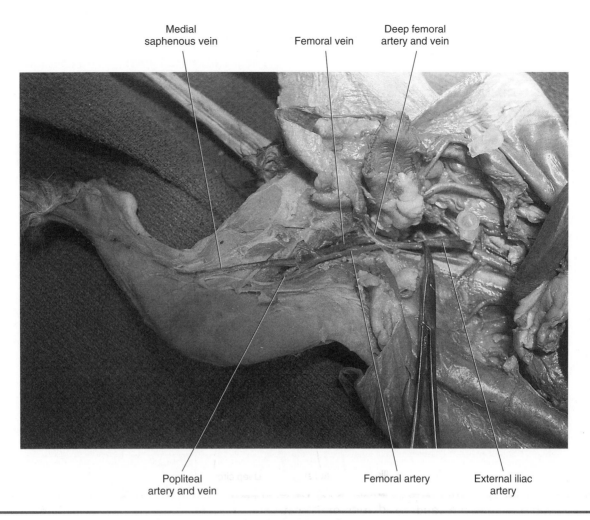

Medial saphenous vein

Femoral vein

Deep femoral artery and vein

Popliteal artery and vein

Femoral artery

External iliac artery

Figure 10.17: Medial view of the vessels of the hind leg of a cat.

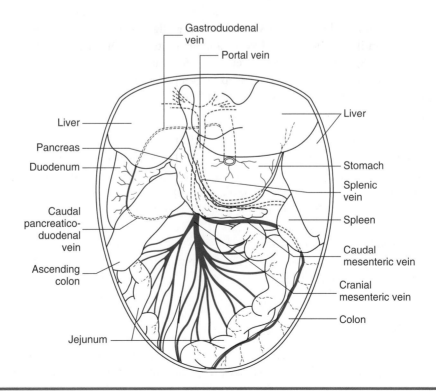

Gastroduodenal vein

Portal vein

Liver

Pancreas

Duodenum

Liver

Stomach

Splenic vein

Spleen

Caudal pancreatico-duodenal vein

Ascending colon

Caudal mesenteric vein

Cranial mesenteric vein

Colon

Jejunum

Figure 10.18: Hepatic-portal venous system in dogs and cats.

2. Move the *right auricle* to the right to see a large, white artery called the **aorta** emerging from the left ventricle and passing beneath the pulmonary artery; its path forms an arch that moves dorsally and to the left above the heart. The arch is known as the **aortic arch,** and when the aorta becomes attached to the dorsal thorax it is called the **thoracic aorta.** This is injected with red latex, but due to the thickness of the wall of the aorta, the color often cannot be seen (do not remove the wall of the aorta in your dissection).

3. The first major branch off the aortic arch is the **brachiocephalic trunk,** or **brachiocephalic artery.** Trace this vessel forward. The first branch off the brachiocephalic is the **left common carotid artery;** it then branches into the **right subclavian artery** and the **right common carotid artery** (Figures 10.19 and 10.20).

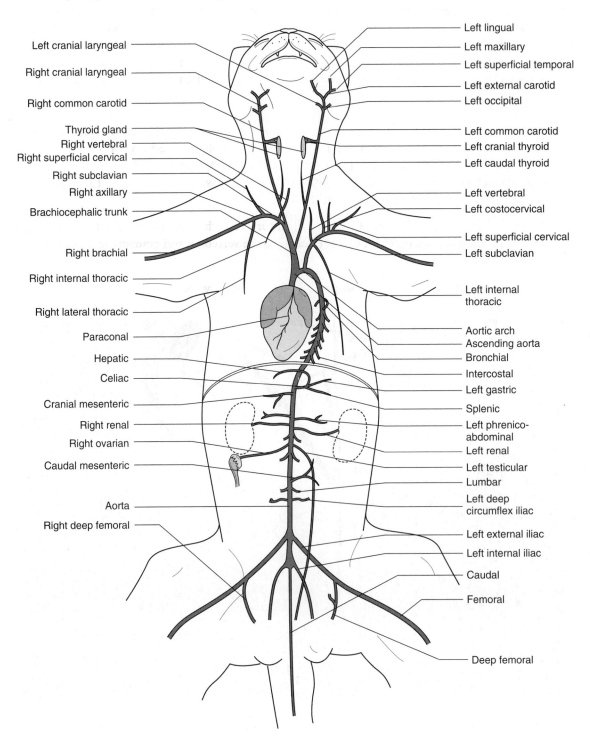

Figure 10.19: Arterial system of the cat.

4. Follow the path of the **left common carotid artery.** It ascends cranially in the neck, coursing between the small internal jugular vein and trachea. As you follow it cranially, you will note some small branches arising from this artery, such as the **caudal thyroid artery,** which is often hard to find and usually not injected (see Figures 10.19 and 10.20). The first major branch to locate is the **cranial thyroid artery,** which originates at the level of the junction between the trachea and cricoid cartilage. It supplies blood to the thyroid gland and muscles of the larynx. The next two branches off the common carotid are the **occipital artery,** which courses dorsally and supplies the back of the neck, and the **cranial laryngeal artery,** coursing ventrally to the larynx. The next branch is the **lingual artery,** which courses ventral and cranially. After this branch, the artery continues dorsocranially as the **external carotid artery.**

In the cat, the *internal carotid artery* is extremely small and need not be located. If you do find it, it emerges from the common carotid just cranial to the occipital artery and courses dorsally. The external carotid divides into the **superficial temporal artery,** coursing dorsally and supplying the parotid salivary gland and local musculature, and the **maxillary artery,** coursing ventrocranially and branching into numerous other arteries that deliver blood to the maxilla and mandible (see Figures 10.19 and 10.20).

5. Return to the aortic arch and locate the **left subclavian artery,** the next branch off the aortic arch just to the left of the brachiocephalic artery. This artery supplies the left side of the chest and left front leg. From the left subclavian, its first branches will be the **internal thoracic artery,** which goes to the sternum and supplies blood to the ventral intercostal muscles. Note that it joins its corresponding vein (see Figures 10.19 and 10.21).

6. The **left vertebral artery** arises from the dorsal surface of the left subclavian artery, nearly opposite the internal thoracic artery. This courses dorsally to the vertebrae and cranially to supply the brain (see Figures 10.19 and 10.21).

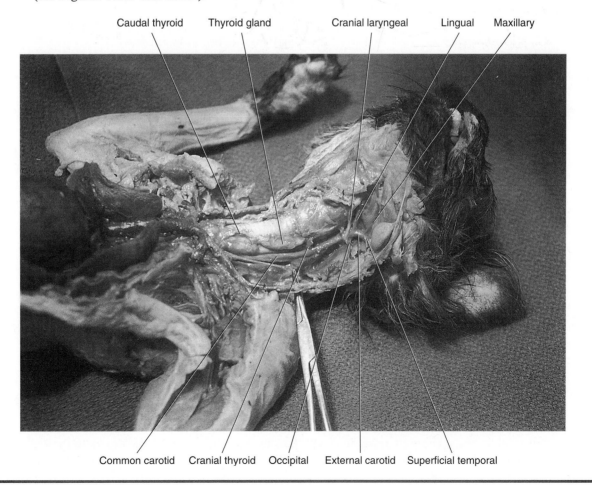

Caudal thyroid   Thyroid gland   Cranial laryngeal   Lingual   Maxillary

Common carotid   Cranial thyroid   Occipital   External carotid   Superficial temporal

Figure 10.20: Cranial arterial system of the cat.

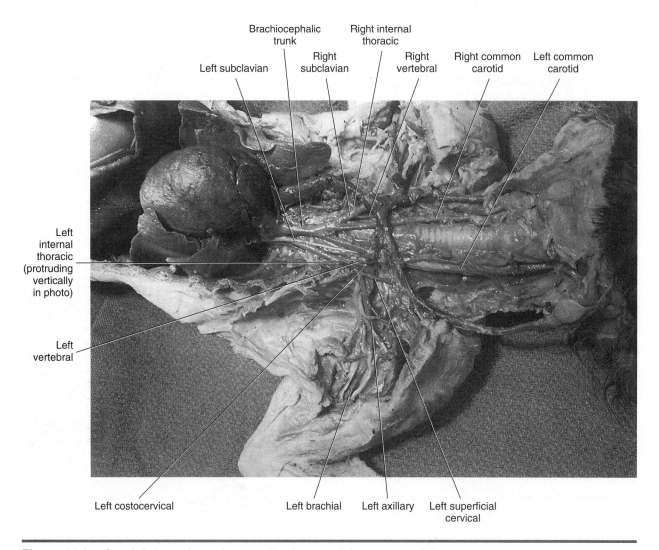

Left subclavian   Brachiocephalic trunk   Right subclavian   Right internal thoracic   Right vertebral   Right common carotid   Left common carotid

Left internal thoracic (protruding vertically in photo)

Left vertebral

Left costocervical   Left brachial   Left axillary   Left superficial cervical

Figure 10.21: Cranial thoracic and appendicular arterial systems of the cat.

7. The next two branches off the left subclavian artery are the **costocervical trunk** and then the **thyrocervical artery,** which supplies some of the neck, back, and dorsal intercostal muscles. Next locate the **superficial cervical artery,** which ascends for a short distance and then branches. It also supplies some of the neck and shoulder muscles (see Figures 10.19 and 10.21).

8. The left subclavian artery becomes the **axillary artery** as it emerges from the thoracic cavity and passes cranial to the first rib. When it becomes attached to the musculature of the front leg, it is known as the **brachial artery.** Follow the brachial artery until it branches to form the **superficial brachial artery** and the **median artery.** The superficial brachial gives off the **collateral ulnar artery** which courses toward the medial aspect of the olecranon process. Just distal to the elbow joint on the medial aspect of the leg, the median artery gives off the **ulnar artery** which courses caudally and medial to the ulna. Just proximal to the carpus, the median artery gives off the **radial artery,** which courses medially across the carpal bones and continues caudally and medially to the accessory carpal bone. All of these arteries together supply the forearm. On the right side, the **superficial cervical artery** is the first branch off the axillary as it enters the axilla. The **right vertebral artery** emerges from the right subclavian just distal to the point at which the right common carotid artery emerges (see Figures 10.19 and 10.21).

9. Returning to the aortic arch, trace the aorta caudally. Pull the viscera in the thorax to the right to expose the aorta in the thoracic cavity. As this vessel passes through the thorax it is called the *thoracic aorta.* Remove the pleura to expose this vessel in the thorax.

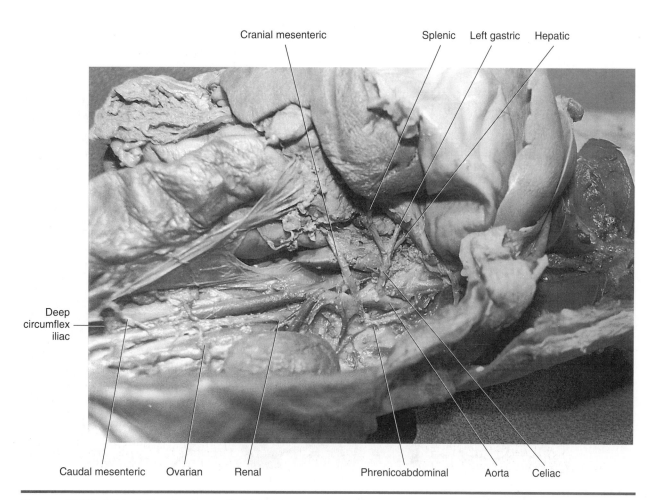

Cranial mesenteric   Splenic   Left gastric   Hepatic

Deep circumflex iliac

Caudal mesenteric   Ovarian   Renal   Phrenicoabdominal   Aorta   Celiac

Figure 10.22: Mid-abdominal arteries in a cat.

10. Note the **intercostal arteries** emerging from the thoracic aorta to supply the intercostal muscles. In addition, there are several **bronchial arteries,** which supply the lungs, and **esophageal arteries,** which supply the esophagus.

11. As the aorta emerges above the diaphragm, its first major branch from the **abdominal aorta** is the short **celiac artery.** Dissect away the fat and peritoneum that cover the cranial aspect of the abdominal aorta to locate this artery. The celiac artery divides into three branches: first, the **hepatic artery,** coursing above the stomach and to the liver; immediately distal to this is the **left gastric artery,** supplying the lesser curvature and cranial stomach; then the artery continues as the **splenic artery,** the largest of the branches, and goes to the spleen (see Figures 10.19 and 10.22). The hepatic branches into the **right gastric, gastroduodenal,** and **cranial pancreato-duodenal arteries** on its path to the liver. The hepatic artery was located previously in the lesser omentum, to the left of the portal vein, during the dissection of the digestive system (see chapter 9, Figure 9.19).

12. A short distance caudal from the celiac artery is the **cranial mesenteric artery** (see Figures 10.19 and 10.22). It supplies the small intestine and a portion of the large intestine.

13. The paired left and right **phrenicoabdominal arteries** (formerly known as the adrenolumbar arteries) course from the aorta to the dorsal body wall (see Figures 10.19 and 10.22). They supply the adrenal glands, diaphragm, and muscles of the body wall.

14. The next pair of arteries are the **renal arteries** that supply the kidneys (see Figures 10.19 and 10.22).

15. Just caudal to the kidneys, off the abdominal aorta, the right and left testicular or ovarian arteries emerge (see Figures 10.19 and 10.22). These are small, thread-like arteries emerging from the ventral surface of the aorta. If your cat is a male, follow the **testicular arteries** from the inguinal canal to the aorta. The testicular artery supplies each testis. If your cat is a female, trace the **ovarian arteries** to the ovaries.

16. Push the descending colon to one side to see the single **caudal mesenteric artery** (see Figures 10.19 and 10.22). This arises from the ventral surface of the aorta a short distance from the previous testicular or ovarian arteries. It supplies the descending colon.

17. Lift the aorta gently to observe the **lumbar arteries.** There are approximately seven pairs of lumbar arteries emerging at intervals along the abdominal aorta in the cat. These supply the abdominal wall.

18. Another pair of arteries called the **deep circumflex iliac arteries** (formerly known as the iliolumbar arteries) emerge next and course laterally to the dorsal caudal musculature (see Figures 10.19 and 10.23).

19. The aorta terminates in a series of branches, first the paired **external iliac arteries,** then the **internal iliac arteries** (see Figures 10.19 and 10.23). The external iliac becomes the **femoral artery** as it passes out through the body wall (see Figure 10.17). It courses next to its corresponding vein and branches with it; the nomenclature for these branches are the same as for the veins. The internal iliac artery supplies the gluteal muscles, rectum, and uterus.

20. The aorta terminates by branching into the small **caudal artery,** which courses down the median ventral surface of the sacrum and enters the tail (see Figures 10.19 and 10.23).

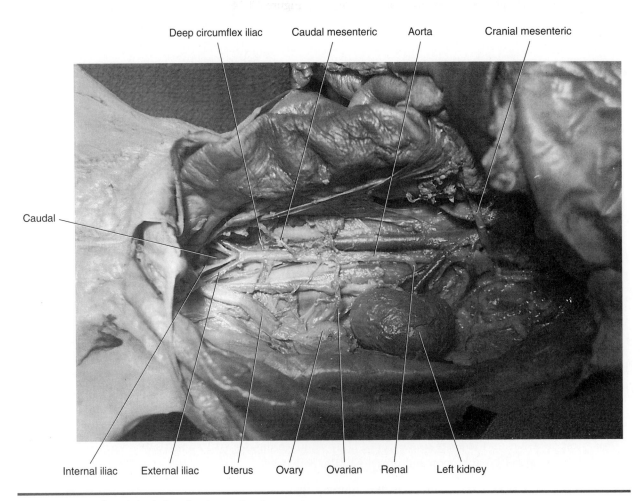

Figure 10.23: Caudal abdominal arteries in a cat.

The Electrocardiogram

An **electrocardiogram** is a recording of the electrochemical changes (depolarization waves) that travel through the heart immediately prior to each beat. The heart's ability to beat is intrinsic, and thus it does not depend on impulses from the nervous system to initiate its contraction. The heart will continue to contract rhythmically even if all nerve connections are severed. However, because it is connected to the autonomic nervous system, this system has a controlling effect on the heart and can either increase the heart rate or decrease it.

Within the heart is the **intrinsic conduction system,** or **nodal system.** This system consists of specialized noncontractile myocardial tissue, called **purkinje fibers** of the **purkinje system,** that conduct the wave of depolarization through the heart in an orderly, consistent, and sequential manner that enables it to beat as a coordinated unit. The definition of *depolarization* is the state of a neuron that occurs immediately after a sufficiently strong stimulus is applied and results in an influx of sodium ions. This changes the membrane potential from approximately -70 mV to $+30$ mV. *Repolarization* follows depolarization; during this, potassium ions rapidly diffuse out of the neuron. This causes the membrane to return to its resting potential (see chapter 15 for a more detailed explanation).

The **sinoatrial node** (the **SA node**), also called the **pacemaker,** has the highest rate of discharge and provides the stimulus to initiate the heartbeat. It also sets the rate of depolarization for the heart as a whole. The impulse then spreads across the both atria and is immediately followed by atrial contraction (atrial systole). The impulse then is picked up by the **atrioventricular node** (or **AV node**). At the AV node the impulse is momentarily delayed; in an animal with an average heart rate of 72 beats per minute, the delay is approximately 0.1 sec. It is shorter in animals with faster heart rates and longer in animals with slower heart rates. This allows the atria time to completely contract. From there the impulse passes through the **AV bundle** (or **Bundle of His**) and splits into the **right** and **left bundle branches** as it travels down the **interventricular septum.** Branching off the bundle branches are numerous purkinje fibers attached to the myocardium, called **terminal conducting fibers.** The impulse passes through these to initiate ventricular contraction (ventricular systole; Figure 10.24).

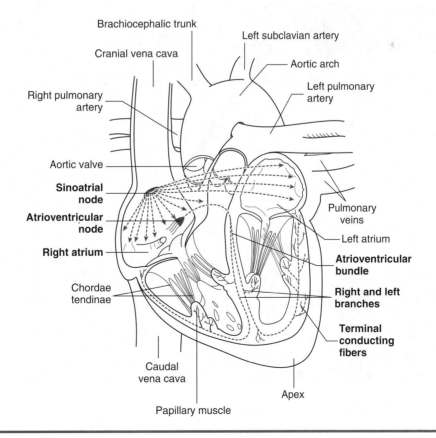

Figure 10.24: The intrinsic conduction system of the heart, all species.

The electrocardiograph records the electrical currents generated by the heart as they eventually spread through the body and are picked up by the machine's electrodes. The electrocardiogram (ECG or EKG) records three main recognizable waves, called **deflection waves.** The first is the **P wave,** representing depolarization of the atria, which occurs immediately prior to atrial systole. The next is the **QRS complex,** representing ventricular depolarization, and it is immediately followed by ventricular systole. The third wave is the **T wave,** representing ventricular repolarization. A small atrial repolarization wave occurs at the same time as ventricular depolarization, and thus, this wave is buried within the QRS complex (Figure 10.25).

The direction of the ECG deflection waves (up or down) depends on where the positives and negative electrodes are attached on the limbs of the patient. Traditionally, a dog lies in right lateral recumbency and has electrodes attached at the elbows and knees. The right knee is the ground electrode. Cats may be positioned like dogs or sternally. The electrocardiograph machine is able to change the polarity of the electrodes without having to move them. A typical ECG records three standard limb leads, three augmented limb leads, and four or more chest leads. The three standard limb leads—Leads I, II, and III—record two of the electrodes during a reading. Lead I is positive on the left front leg and negative on the right front leg; Lead II is positive on the left hind leg and negative on the right front leg; and Lead III is positive on the left hind leg and negative on the left front leg. The augmented or unipolar leads, known as aVR, aVL, and aVF, compares the positive electrode to each of the standard limb leads. The chest leads are placed at various locations on the external chest wall.

In the normal animal, the lead that usually gives the P wave and R wave in a positive deflection is Lead II because of the heart's position in the chest. As the wave of depolarization moves toward the positive electrode, it will record it as a positive deflection. If the wave moves directly toward the lead, it will be a strong positive deflection; if it is moving 45° obliquely toward the positive electrode, it will be a small positive deflection; if it is moving 90° to the positive electrode, the deflection will usually be minimal and be as much positive as negative. If the wave moves 45° away from the positive electrode and toward a negative electrode, it will be a small negative deflection; and finally, if the wave moves directly away from the positive electrode and directly toward the negative electrode, it will be strongly negative. If the QRS complex's wave deflections are equally pos-

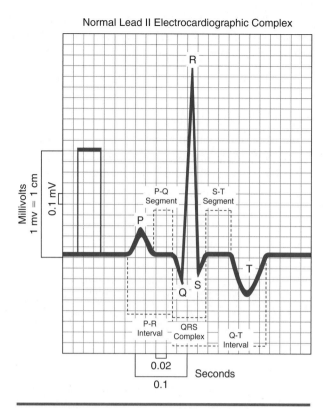

Figure 10.25: The normal lead II electrocardiogram pattern. The machine was standardized at 1 cm = 1 mV, and the tract was recorded at 50 mm/sec paper speed.

itive and negative, this lead is called the **isoelectric lead.** The size of the atria or ventricles also affects the size of the wave deflection, giving veterinarians an indication of whether there is heart enlargement. Therefore, measurements of the P wave and R wave are performed as follows (see also Figure 10.25). The units of the following are all in millivolts (mV).

P wave: Measure from the baseline to the top of the P wave.

R wave: Measure from the baseline to the top of the R wave (Note: it is not measured from the bottom of the Q or S waves to the top of the R wave, but from the baseline.)

Along the horizontal axis of the ECG, certain measurements are taken. Remember, *intervals* include the wave deflections; *segments* do not. The units of the following are in seconds (sec).

P-R interval: Measure from the start of the P wave to the start of the Q wave. This interval is misnamed and should be called the P-Q interval, but it is so named because the Q wave is often absent. This represents the time for atrial depolarization.

P-Q segment: Measure from the end of the P wave to the start of the Q wave. This represents the time the wave is within the AV node.

QRS complex: Measure from the start of the Q wave to the end of the R wave. This represents the time for ventricular depolarization.

S-T segment: Measure from the end of the S wave to the start of the T wave. This represents the time period between the end of depolarization of the ventricles and the initiation of repolarization.

Q-T interval: Measure from the start of the Q wave to the end of the T wave. This represents the length of time for ventricular contraction and repolarization.

Note that along the horizontal axis is a measure of time in seconds, and the vertical is measured in millivolts (mV). The standard measurements are taken at 50 mm/sec, and 1 mV = 1 cm on the vertical axis. This makes each small box equal to 0.02 sec horizontally and 0.1 mV vertically.

EXERCISE 10.3
RECORDING THE ELECTROCARDIOGRAM

Procedure

1. Place a dog in right lateral recumbency on a table with a non-metallic surface (otherwise you may get 60 cycle-per-second interference), and place the electrodes on the skin as indicated on the electrode leads. Standard to the industry is that the white electrode is right front (RF) leg, the black electrode is left front (LF) leg, red is left hind (LH) leg, and green is right hind (RH) leg (note that green is the ground wire). The brown electrode is the chest lead, but we will not use this electrode. Place a small amount of electrode gel or alcohol at the connection sites.

2. Turn on the ECG machine and push the standardize button to be sure it deflects 1 cm at 1 mV. (If needed, adjust it so 1 cm = 1 mV). Also make sure it is running at 50 mm/sec.

3. Record leads I, II, and III for a short distance on the paper. If you would like to, you may also record AVF, AVL, and AVR at this time.

4. Reduce the paper speed to 25 cm/sec. Note the printed marks at the top of the ECG paper. Each little box is 1 mm square, and there are 75 little boxes between each mark, which we will call one section. If you do the math you will see that at 25 mm/sec it will take 3 seconds to traverse 75 boxes (one section), and 6 seconds for 2 sections. Counting the number of R waves in two sections, then multiplying by 10 gives you the heart rate, or beats per minute for the animal. (There are 10 6-second intervals in each minute; therefore, you multiply the beats counted by 10).

5. Using the ECG, record the information in the following spaces and compare it to your instructor's computations. Remember to label your units.

P-R wave: _____   Q-T interval: _____

P-Q segment: _____   P wave: _____

QRS complex: _____   R wave: _____

S-T segment: _____   heart rate: _____

QUESTIONS
1. What would be the multiplier that is needed to determine the heart rate for two sections of ECG paper if the machine was run at 50 cm/sec?

2. If there were a delay or block in the impulse as it passed through the AV node, which two measurements would be increased?

Discussion

Regarding question 1, if the paper was running twice as fast as previously, there would be half the number of R waves recorded. Therefore, the multiplier would have to be doubled to 20.

To answer question 2, because the impulse is in the AV node during the P-Q segment, this measurement would be increased in length. Also, because the P-R interval includes the P-Q segment, this too would be increased in length.

EXERCISE 10.4
BLOOD PRESSURE MEASUREMENT USING A DOPPLER

With humans, blood pressure is measured using a stethoscope over a peripheral artery. A blood pressure cuff encircles the arm above the artery, and the pressure is read on a pressure manometer. The cuff and manometer are called a *sphygmomanometer*. Normally the blood flow through an artery is inaudible, but the inflated cuff causes turbulence of blood flow in the artery, which is audible. The cuff is inflated until the artery is completely compressed, with no flow of blood through it and thus no sound.

As pressure is slowly released from the cuff, the manometer needle drops. At the point where the arterial blood pressure exceeds the pressure of the cuff, blood will flow through the artery and will be audible. This point is called the **systolic blood pressure** and represents peak systolic pressure at maximum ventricular contraction. Continuing to release pressure drops from the cuff and the artery; the point at which the blood flow again becomes inaudible is the **diastolic blood pressure.** This is the pressure within the vascular system when the heart is in diastole.

Animals do not have peripheral vessels of sufficient size distal to where a cuff could be placed to hear an audible pulse. Therefore, devices such as dopplers must be used to hear the blood flow through the artery. Because the doppler records blood flow without turbulence, it is limited to systolic blood pressure measurement.

Procedure

1. Using an electric clipper and #40 blade, clip away the hair on a dog between the carpal pad and metacarpal pad (or if necessary, clip on the dorsum of the hock).

2. Place a human infant or newborn blood pressure cuff around the dog's foreleg, just below the elbow (or hind leg just below the knee). Do not inflate yet.

3. Put some ECG gel on the clipped area; in the area just medial to the midline of the ventral aspect of the leg, place the doppler head and listen for the audible heartbeat. When it is consistently audible, inflate the cuff until it become inaudible, then slowly release the pressure until it becomes audible once again. This point is the systolic blood pressure. Perform this three times and record your results below.

    first measurement _____

    second measurement _____

    third measurement _____

   QUESTION
   1. Were all three measurements different, or were they the same each time?

Discussion

If the measurements were done correctly and the animal's blood pressure was not affected by pumping the cuff, they should be the same each time. However, like humans, animals may get nervous and release epinephrine into their system, causing vasoconstriction, which increases peripheral resistance and increases blood pressure. Even though the act of measuring may hinder the accuracy of the measurement, the change is not so significant as to cause a misdiagnosis.

CLINICAL SIGNIFICANCE

Auscultation of the valves of the heart is an important skill for both the veterinarian and the veterinary technician to learn. During dissection, when you made your cuts through the pulmonary artery and aorta, they were made through the cranial surface of the heart on the left side. It was mentioned that these two valves, because of the way the heart is positioned in the thorax, are cranial to the mitral valve, or left AV valve. It is thus logical that we would auscultate for these valves on the left side of the chest.

If the diaphragm of the stethoscope is placed just above the sternum on the left side of the chest and slid forward until the heart sound is clearly audible, it will be over the mitral valve. This should be the fourth or fifth intercostal space at the costochondral junction. Next, slide the stethoscope head a little dorsally and cranially and listen for the heart sounds in this location; this will be the aortic valve and should be the fourth intercostal space. Then, slide the stethoscope head cranially and ventrally one intercostal space and listen to the pulmonary valve, which should be at the third intercostal space at the costochondral junction.

On the right side, the right AV valve is found at the fourth intercostal space at the costochondral junction, or wherever it is best heard. To remember the order in which these valves are found, if the valves were labeled on the left outside chest wall, reading from cranial to caudal they form the word P-A-M.

Veterinary Vignettes

*I*n veterinary school, the seniors used to delight in making junior veterinary students look stupid. When I became a senior, I personally never, ever took advantage of the poor lower forms of life.

The two senior students with whom I was assigned for my week of cardiology rotation had a lot of fun at my expense. The first game they played with me was "Find the Heart." On Monday morning they took me out to the calf pens and told me there was something wrong with the calf's heart and, if I had any brains at all, I should be able to diagnose the condition within a few minutes. I got out my stethoscope and proceeded to listen for the heart's sounds. I started on the left side, meticulously searching for the mitral valve. I searched and searched, listening intently, but I could not hear anything.

"Well, what do you hear?" Mutt (the tall one) asked me.

"Uh, nothing. I can't hear anything," I replied, beginning to show a little red on my face.

"Come here," said Jeff (the short one).

I walked over to him and, not being the brightest human on the planet at that mooment in time, did not remove the earpieces of my stethoscope from my ears.

Grabbing the head of my stethoscope and tapping on it, he yelled, "Does this thing work?"

I jumped back and pulled them from my ears.

"Well, that's not the problem. Listen to the other side of the chest; maybe the heart is over there," he told me.

Dumbly, I did as he suggested. All the while, these two were trying to keep straight faces and glaring at me like I was stupid. I wasn't stupid, just ignorant. As I listened, I was thinking, because the heart is in the middle of the chest, how could I hear it on the right side if I couldn't hear it on the left? This time I moved the stethoscope up and listened to the lung sounds. Now I could hear the muffled sounds of a heart.

I stood up and looked at them. "The heart's not in the chest is it?" I asked.

"What do you mean it's not in the chest?" Jeff replied.

"If it isn't in the chest, where would it be?" Mutt inquired.

I started listening to the cranial chest and palpated the neck. Deep in the neck, just in front of the thoracic inlet, I felt something beating and found a small heart.

"What's it called?" they asked.

"I don't know," I replied.

"An ectopic cordis," Mutt answered.

I had gotten the answer too quickly for them. Next I was guided to the small animal ward to play "Name the Murmur."

"Listen to this heart and tell us the type of murmur you hear and its cause," Jeff told me.

As I grasped the chest, I could feel the vibration of the heartbeat in my hand. Placing the stethoscope head against the chest, I immediately heard a *whoosh-whoosh* sound rather than the normal *lub-dub*. I had been taught that murmurs were abnormal heart sounds, often heard between the first and second heart sounds, but being only a rookie at this, my experience was lacking. It sounded more like a washing machine than a heartbeat.

"I have no idea what this is," I said.

"Well, is it systolic or diastolic?" Mutt asked me.

"I don't know. Both maybe?" I replied.

"That's an intelligent answer; you can do better than that, can't you?" Jeff remarked sarcastically.

"Describe it to us," Mutt demanded.

"Well, there's no first or second heart sound, just a loud whoosh, followed immediately by a second whoosh of a slightly lower pitch. Then it repeats itself continuously," I said, gesturing with my hands and shrugging my shoulders.

They had given me up as hopeless, I thought.

"It's a machinery murmur of a patent ductus arteriosus," Jeff told me.

"Better go read up on it before Dr. Frye asks you about it and you get it wrong," Mutt said, shaking his head.

They sent me off to the vet school library to read about it. This is what I found out: The ductus arteriosus is a vessel present in fetal life that shunts blood from the pulmonary artery to the aorta, thus by-passing the lungs (see Figure 10.1*B*). In dogs and cats, it normally closes by muscular contraction some time in the first eight days of life. When it does not close, the condition is known as a patent ductus arteriosus. Sometimes the vessel doesn't close completely and leaves a very small opening, and in those cases it usually causes no important hemodynamic changes. The female dog is most often affected, and there is a breed predilection in miniature poodles, German shepards, Pomeranians, Shetland sheepdogs, and other toy breeds. When the ductus stays patent, most dogs develop severe clinical signs in first six to eight weeks of life. Cats with this condition rarely survive past a few weeks of age.

The pathophysiology is complicated and depends on the size of the ductus and the direction of the blood flow through it. The result is left and/or right heart failure. Animals with the condition generally have a history of coughing and exercise intolerance. On physical exam, a veterinarian may find a persistent machinery murmur, a pounding femoral pulse, and a precordial thrill (which is the vibration I felt when I placed my hand on the dog's chest). Diagnosis is made by the clinical signs, radiography, ECG, ultrasonography of the heart, and angiography. If the blood flow is left to right, surgical ligation of the duct or implanting a plug (called an embolization coil) can help correct the problem. Because the condition may be hereditary, which has been confirmed in the miniature poodle, animals that survive should not be bred.

After reading up on the condition, I realized I had not done such a bad job in describing it. Later I found out I had secretly passed their test on both accounts, but they would never admit it to me.

SUMMARY

In this chapter we covered the anatomy and physiology of the heart and the blood-vascular system. You learned that there are three main circulatory systems of the blood vascular system: the carotid vascular system, the hepatic portal vascular system, and the systemic circulation. You will learn in Chapter 13 that there is a small portal system between the hypothalamus of the brain and the pituitary gland. In addition to learning the parts of the heart and its intrinsic vessels, you learned the names of the main vessels in the body of the cat. These names are similar in most species.

The vascular supply of the equine leg is beyond the scope of this book, but there are many excellent texts that cover this if you are interested. Some of the more important vessels and locations are the external maxillary artery as it crosses the mandible (to obtain a pulse rate), the jugular veins in the jugular furrow (for intravenous injections and drawing blood), the carotid arteries (so as to avoid them when working on the jugular vein), the common digital artery on the medial aspect of the leg (to obtain a pulse rate), and the medial and lateral posterior digital arteries just below the fetlock joint (may develop a prominent pulse in cases of laminitis).

Finally, learning how to take blood pressure, knowing what it means, and knowing the wave forms on an ECG are important skills every veterinary technician should learn.

CHAPTER 11

THE RESPIRATORY SYSTEM

OBJECTIVES:

- state the function of each component of the respiratory system
- describe the air flow from the nose to the alveoli of the lungs
- dissect the structures and organs of the respiratory system and identify them by name
- identify the parts of the larynx and its cartilages
- describe the anatomy of the lungs of different species
- understand the flow of air between the air sacs and lungs of the bird and their connection to the pneumatic bones
- understand the various methods of measuring lung volumes and capacities
- understand the concept of dead space

MATERIALS:

- cat cadaver, triple injected (order without skin attached)
- Mayo dissecting scissors
- probe
- 1 × 2 thumb forceps or Adson tissue forceps
- #4 scalpel handle with blade
- bone cutting forceps
- rubber gloves
- model of the larynx
- spirometer

Introduction

The **respiratory system** can be divided into two parts: the **upper respiratory system** and **lower respiratory system**. The structures of the *upper respiratory system* include the **external nares, nasal cavity, sinuses,** and **larynx.** The *lower respiratory system's* structures include the **trachea, bronchi, bronchioles,** and **alveoli.** The respiratory system performs **ventilation** and **respiration. Ventilation** is the process of breathing (also called **pulmonary ventilation**), which is the movement of gases in and out of the **lungs. Respiration** is a broader term that includes not only the act of ventilation but also the distribution of gases in the alveoli and the diffusion of oxygen into and of carbon dioxide out of the bloodstream. This exchange of gases in and out of the blood is called **external respiration. Internal respiration** is the exchange of gases between blood and tissues.

In the exercises in this chapter the important structures are listed in colored bold print. If a structure is mentioned prior to its dissection it will be italicized. Structures discussed prior to dissection may also be in bold print for special emphasis.

EXERCISE 11.1

THE UPPER RESPIRATORY SYSTEM

The functions of the upper respiratory system are as follows:

1. conducting air (in and out)

2. cleansing air before it is used by the body

3. humidifying air before it is used by the body

4. transferring heat

5. phonation (making sounds)

6. olfaction (smelling): The **olfactory** sensory receptors are **neuroepithelial cells,** which are chemical transducers. Contact induces a chemical phenomenon that excites the cells to depolarize. This message is carried by the first cranial nerve (the *olfactory nerve*) to the olfactory bulbs, up the olfactory tracts, into the *rhinencephalon* (the olfactory brain, part of the telencephalon of the cerebrum), and is interpreted there as an odor. The sensory receptors are located in the posteromedial and posterodorsal parts of the nasal cavity.

Procedure

1. Examine the **nose** of the cat. The air enters the **nasal cavity** through the **external nares** or **nostrils.** These openings vary in size and shape with each species. The horse's nostrils are large and easily dilated, whereas the pig's are small and rigid. The non-haired area of the nose is called the **planum nasale** (or **planum nasolabiale** in species with a large muzzle). The most rostral part of the nasal cavity is the **nasal vestibule,** which is the open chamber behind the nostrils that is lined with simple squamous epithelial cells.

   The two nasal cavities, left and right, are separated rostrally by a cartilaginous septum. Each cavity is divided further by the nasal bone in the center and by the **nasal turbinate bones** projecting from the lateral wall of the nasal passage. The bony turbinates are scroll-like in appearance and are covered by nasal mucosa to form the **nasal conchae.** These effectively divide the flowing air into channels: the **ventral meatus, middle meatus, dorsal meatus,** and **common meatus.** The mucosa is made up of *pseudostratified ciliated columnar epithelial cells, goblet cells,* and *nasal glands.* These channels increase air turbulence and thus warm, moisten, and filter air as it passes through.

   The main sinuses are the **maxillary** and **frontal sinuses,** which are located within the bones for which they are named. Large animals also have a **sphenoid** and **palatine sinus,** except the horse in which these two sinuses are fused into a **sphenopalatine sinus.** The cow and sheep have an additional sinus, the **lacrimal sinus.** These sinuses act as resonance chambers in phonation, and their mucosa, like that of the nasal cavity, warms and moistens the incoming air.

   From the nasal cavity the air passes through the **internal nares** and into the **nasopharynx.** On the dorsolateral wall of the nasopharynx are openings to the **auditory** or **eustachian tubes.** These act in pressure regulation with the middle ear, to which the tubes connect. (In the horse, each eustachian tube opens into a large ventral diverticulum within the nasopharynx known as the **guttural pouch,** which is located just lateral to the pharynx.) The nasopharynx is separated from the **oropharynx** by the **soft palate.** The position of the open epiglottis, either contacting the caudal edge of the soft palate or just beneath it, causes inhaled air to pass through the nasopharynx into the **laryngopharynx** (the area above the open epiglottis) and directly into the *larynx* (Figures 11.1 and 11.2).

2. Carefully remove the muscles from the ventral surface of the *larynx* in the neck to expose the cartilage (Figure 11.3).

   The **larynx** is a hard tube made of cartilages. The **thyroid cartilage** is the large, ventral cartilage visible after removal of the ventral neck muscles. The **cricoid cartilage** is located caudal to the thyroid cartilage. This cartilage is shaped like a signet ring, with the small band located ventrally and an expanded portion dorsally on both sides.

3. The **cricothyroid ligament** is the semi-transparent membrane in the space between the cricoid and thyroid cartilages. Make a median longitudinal incision through these two cartilages on the ventral ridge and through the **basihyoid bone** (the most ventral bone of the **hyoid apparatus**). Now find the two pyramid-shaped **arytenoid cartilages** on the dorsal surface of the larynx, cranial to the cricoid cartilage.

4. The **epiglottis** is the most cranial cartilage of the larynx and can be seen at the base of the tongue during the dissection of the mouth. This pointed, leaf-like cartilage is attached ventrally to the thyroid cartilage.

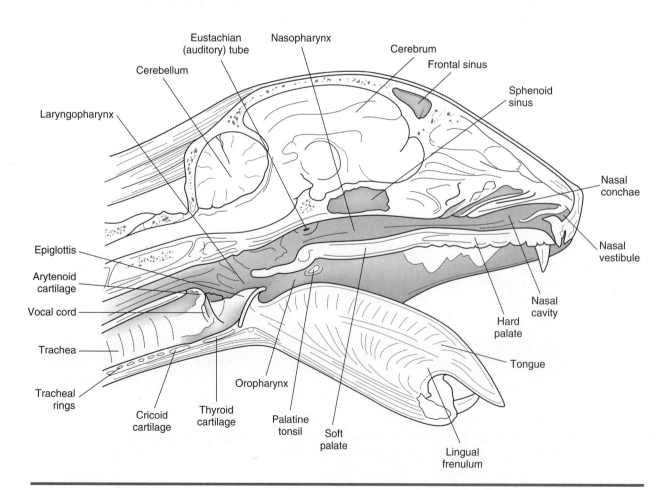

Figure 11.1: Sagittal view of the cat's head.

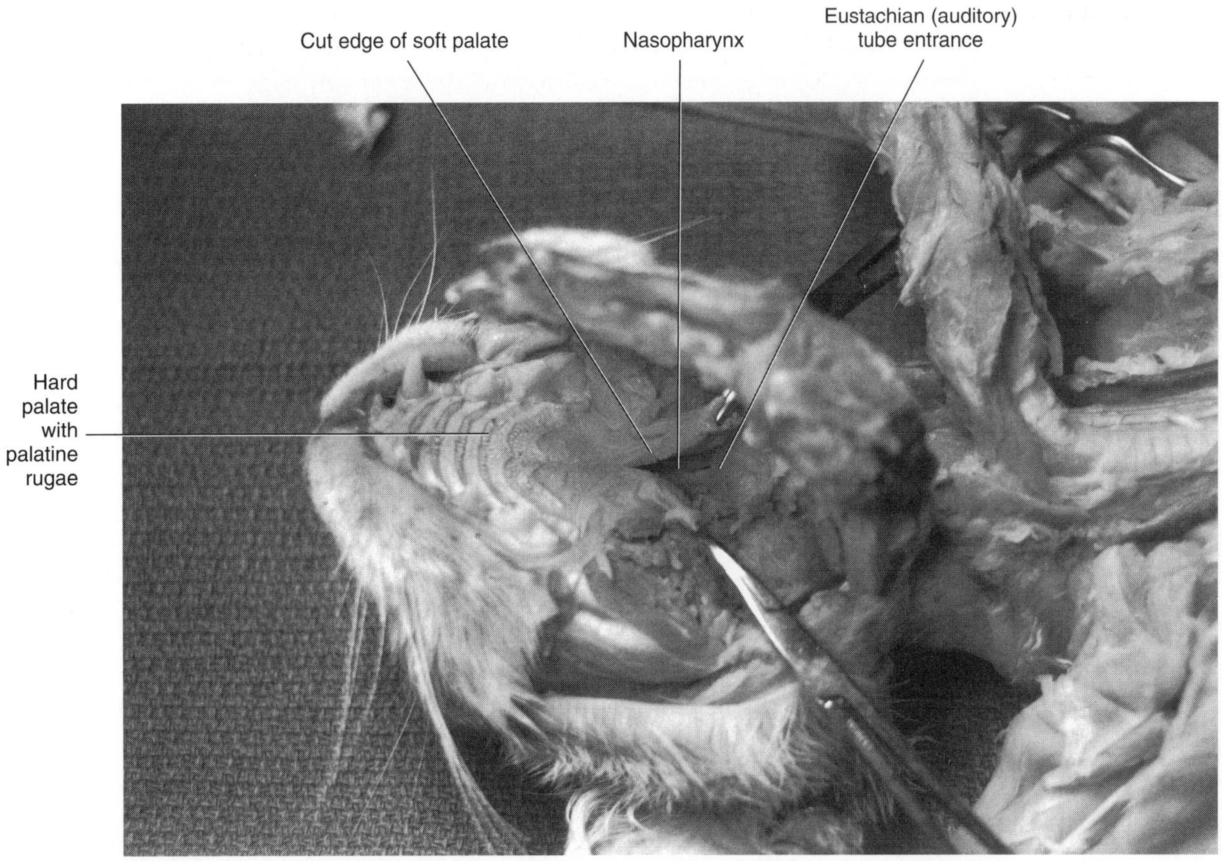

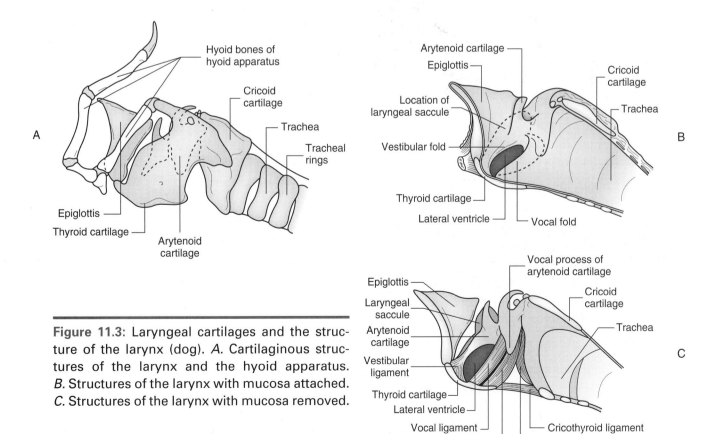

Figure 11.2: Nasopharynx and eustachian tube opening in a cat.

Figure 11.3: Laryngeal cartilages and the structure of the larynx (dog). *A.* Cartilaginous structures of the larynx and the hyoid apparatus. *B.* Structures of the larynx with mucosa attached. *C.* Structures of the larynx with mucosa removed.

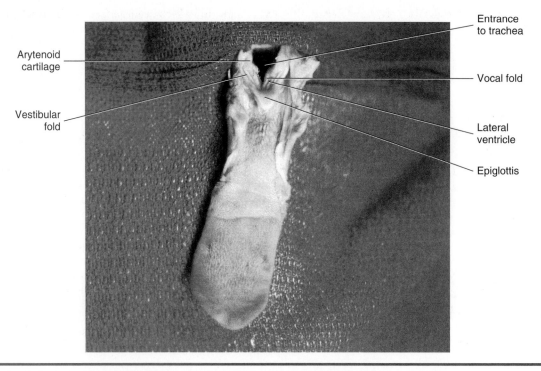

Figure 11.4: Isolated view of the tongue and laryngeal opening in a cat.

5. The most cranial and lateral folds of mucous membranes extending across the larynx are the vestibular folds; the caudal, large pair of folds are the **vocal folds** or **vocal cords.** In Figure 11.4 note that these structures form a V in the center ventrally. This is how you would view the entrance to the larynx prior to inserting an endotracheal tube. The median longitudinal cut you made through the ventrum of the larynx splits the apex of the V. Your view is upside down from the photo in Figure 11.4, but this does not alter the appearance of the structures studied. In fact, you can observe them more closely with the larynx opened.

The depression created between the vocal and vestibular folds is called the **lateral ventricle** (present in dogs, questionable appearance in cats). The *vocal cord* is actually a *vocal fold* of mucous membrane covering the internal **vocal ligament.** This ligament extends from the arytenoid cartilages to the thyroid cartilage. Just caudal to this ligament is the **vocalis** muscle (m. vocalis), the muscle that controls the tension on the ligament and thus the pitch of the sound that is produced. Sound production, of course, is the purpose of this structure. The **glottis** is the paired arythenoid cartilages and paired vocal folds that form the opening (the vestibulum) into the larynx.

EXERCISE 11.2

THE LOWER RESPIRATORY SYSTEM

The functions of the lower respiratory system are the conduction of air into the alveoli and the exchange of gases with the pulmonary blood.

Procedure

1. The **trachea** starts at the caudal end of the larynx and continues until it bifurcates just dorsal to the heart. Observe the large **right** and **left common carotid arteries** and small **internal jugular veins** on each side of the trachea. The **vagus nerve** is the white, flattened thread that courses adjacent to the common carotid artery. This nerve supplies many of the thoracic and abdominal organs.

2. Free the trachea laterally from the preceding blood vessels and nerves if this has not already been done. Find the **esophagus,** the muscular tube that runs dorsal and parallel to the trachea.

3. Locate the two small, dark-tan lobes of the **thyroid gland,** which are attached to the trachea on the lateral sides just caudal to the cricoid cartilage.

4. Follow the trachea through the **thoracic inlet** (the space formed by the vertebrae, first ribs, and manubrium sterni) and into the thoracic cavity. Observe the C-shaped rings of cartilage in the wall of the trachea. The rings are completed dorsally by muscle and connective tissue.

5. Examine the interior of the thoracic cavity. The thoracic cavity is divided by the **mediastinum** into two lateral **pleural cavities.** The **pericardial sac,** which contains the heart, is made up of the **visceral pericardium,** which is also the **epicardium** of the heart. The **parietal pericardium** is intimately attached to the **pericardial pleura** by a thin layer of areolar connective tissue forming the fibrous pericardium (Figure 11.5). Between the visceral and parietal pericardia is the sac in which

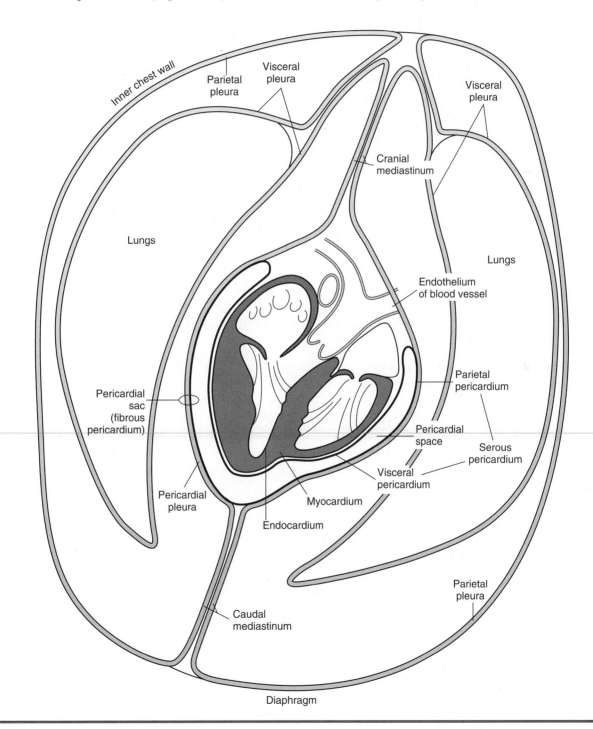

Figure 11.5: The pericardial sac and thoracic pleura.

the heart is located, and the fluid within is a lubricant that helps the heart beat easily. The *pericardial pleura* is an extension of the **mediastinal pleura.** The **parietal pleura** (like the visceral peritoneum) lines the inside of the body's wall. The **visceral pleura** covers the lungs and the space within is the pleura cavity. The *mediastinum,* which is composed of *mediastinal pleura,* is where the pleura meet along the midline and separate the two halves of the **thorax.**

6. Remove the heart to expose the bifurcation of the trachea; this is called the **carina.** Note the cuts you are making through the **pulmonary arteries** and **veins.** The pulmonary artery and its branches are blue because they are injected with blue latex, which also fills the right side of the heart. The *pulmonary veins* are red because the left side of the heart fills with red latex and backs up into the pulmonary veins. This color pattern is the opposite of what is found throughout the rest of the body, but it does accurately represent oxygenated blood (red latex) and deoxygenated blood (blue latex). Visualize the bifurcation of the trachea and the extrapulmonary **main bronchi** (Figure 11.6). Note how they divide and enter each lobe of the lungs.

7. Examine the **lungs** on either side of the **thorax.** Each lung is divided into lobes. There are four lobes on the right side: the **cranial, middle, caudal,** and **accessory lobes;** and there are two lobes on the left: cranial and caudal (Figures 11.7 and 11.8). Note in Figure 11.7B the appearance of the lung of a horse, which has a cranial and caudal lobe on each lung, and a smaller accessory lobe of the right lung.

These lobes were formerly known as the **apical** (cranial), **cardiac** (middle), **diaphragmatic** (caudal), and **intermediate** (accessory) lobes. Many veterinarians learned them by these names, and they are still used in older textbooks. See Table 11.1 for a comparison of the lungs in different species.

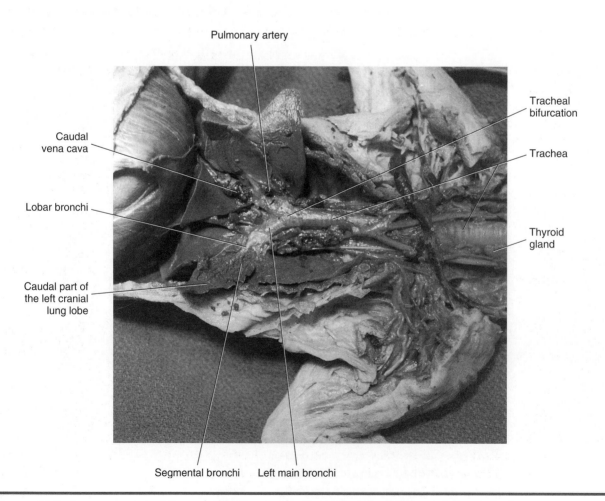

Figure 11.6: Arrangement of the mediastinum, pleura, and pericardial sac (all species).

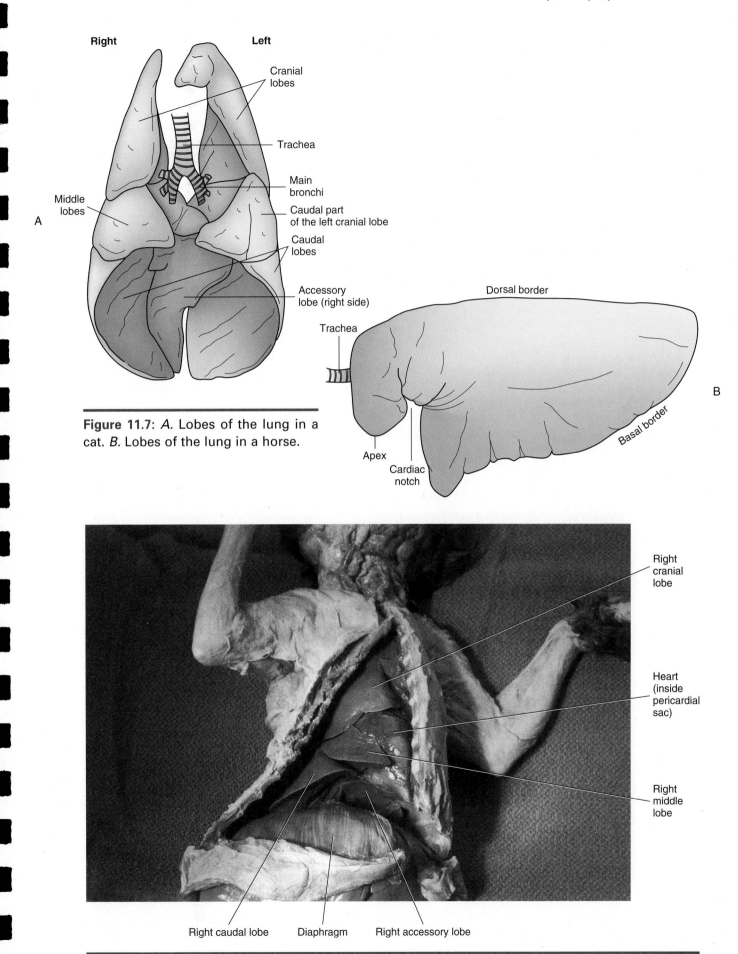

Right   **Left**

Cranial lobes

Trachea

Main bronchi

Caudal part of the left cranial lobe

Caudal lobes

Middle lobes

A

Accessory lobe (right side)

Dorsal border

Trachea

B

Apex

Cardiac notch

Basal border

Figure 11.7: *A.* Lobes of the lung in a cat. *B.* Lobes of the lung in a horse.

Right cranial lobe

Heart (inside pericardial sac)

Right middle lobe

Right caudal lobe   Diaphragm   Right accessory lobe

Figure 11.8: Lung lobes of the cat.

Table 11.1 Lungs of Various Species of Domestic Animals

	Cranial (apical)	Middle (cardiac)	Caudal (diaphragmatic)	Intermediate (accessory)
Horse	x		x	right
Ox	1 left 2 right	x	x	right
Sheep	x	1 right	x	right
Pig	x	1 right	x	right
Dog	x	1 right	x	right
Cat	x	1 right	x	right

Tease away the medial surface of the cranial or middle lobes to observe the branching of the bronchi and blood vessels. Observe the main lobar **bronchus** that runs the length of each lobe and the branches off of it.

8. Note that the lungs are attached to other structures in the thorax only by the root. The **root of the lung** is formed by the branching of the bronchi, the pulmonary artery and veins, the nerves, the lymphatic vessels and lymph nodes, all of which are encircled by pleura.

9. As the main *bronchus* enters the lungs, it becomes lobar **bronchus,** which is the tube that courses the entire distance down the lung lobe (Figure 11.9). The branches from this are the **segmental bronchi.** Segmental bronchi have less cartilage than the main and lobar bronchi. The branches off the segmental bronchi are the **terminal bronchioles,** which are too small to be dissected and appear as the tissue of

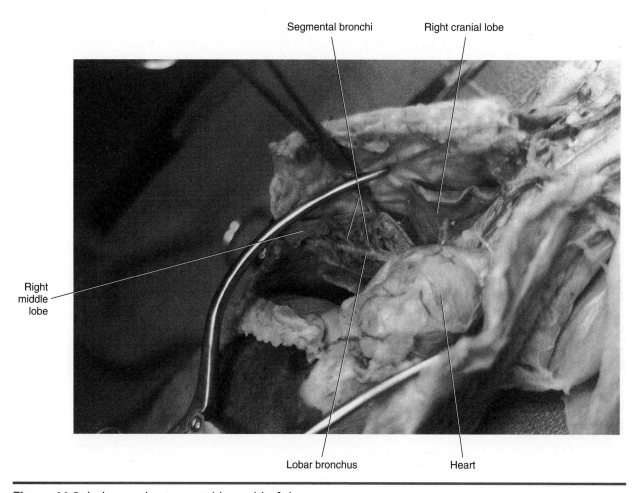

Figure 11.9: Lobar and segmental bronchi of the cat.

the lungs. The trachea and down the bronchial tree to the first ²/₃ of the terminal bronchioles, is lined with *pseudostratified ciliated columnar epithelium.* The last ¹/₃ is *simple ciliated columnar epithelium.* Branching off the terminal bronchioles are the **respiratory units,** each of which consists of a **respiratory bronchiole, alveolar duct, alveolar sac,** and **alveoli** (Figure 11.10). This is called a respiratory unit because all the structures contain alveoli, which are where external respiration occurs.

10. In the mid-cranial area of the thoracic cavity, where the **cranial mediastinum** is located, is a large, fluctuant-appearing gland called the **thymus gland** (Figure 11.11). This gland, divided into lobules, is large in young cats but smaller in older ones. Just dorsal to this gland, within the cranial mediastinum, are the **mediastinal lymph nodes.**

11. Examine the interior of the wall of the thoracic cavity and locate the **intercostal vein, artery,** and **nerve** on the caudal border of each rib.

12. The **crura of the diaphragm** (where it attaches to the dorsal wall) forms the caudodorsal area of the thoracic cavity. Find the **phrenic nerves:** white, thread-like structures directed caudally on each side of the pericardium to the cranial surface of the diaphragm (Figure 11.12). The phrenic nerves originate from the fifth and sixth cervical nerves.

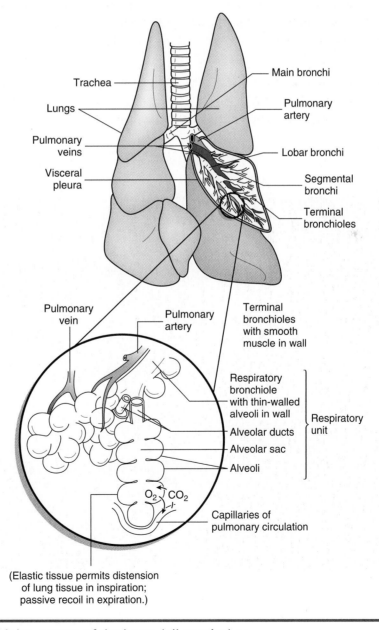

Figure 11.10: Bronchiolar system of the lungs (all species).

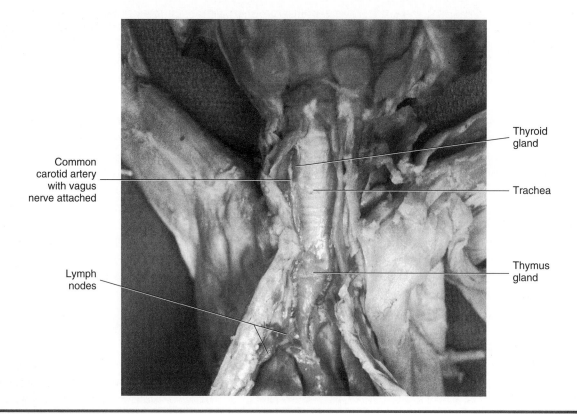

Figure 11.11: Thymus gland in the cranial thorax of a cat.

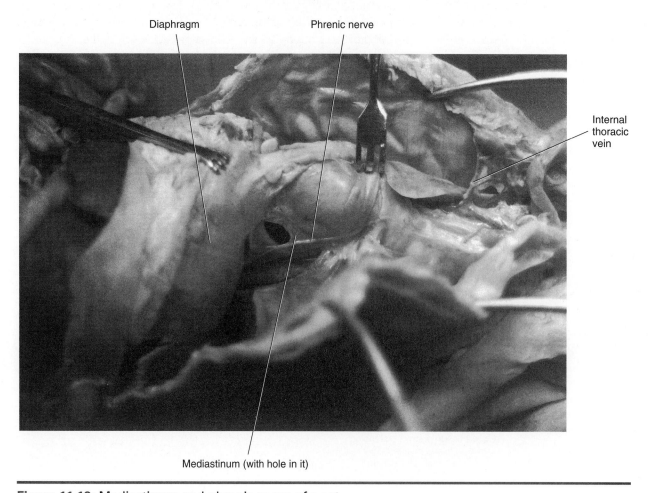

Figure 11.12: Mediastinum and phrenic nerve of a cat.

EXERCISE 11.3

THE LUNGS OF BIRDS

Procedure

Using Figure 11.13, find the structures of the respiratory system of the bird. Air enters the *external nares* through the *nasal cavity,* exits via the **choanal slit,** and continues through the *glottis* and *larynx* into the *trachea.* At the bifurcation of the trachea is the **syrinx,** or voice box. The trachea branches into bronchi, which pass through the ventral aspect of the each lung. As these bronchi enter the lungs, they lose their reinforcing cartilage and are called **mesobronchi.** These give rise to between four and six **ventrobronchi,** then branch into smaller **parabronchi,** which are connected to the **air capillaries.** The air capillaries are surrounded by blood capillaries, and gas exchange occurs between these two groups. The mesobronchi also continue down through the lungs and connect to the abdominal air sac. There are nine air sacs; four are pairs: the cranial thoracic, caudal thoracic, cervical, and abdominal air sacs. The unpaired air sac is the interclavicular air sac. In birds, the mesobronchi are considered primary bronchi, and the ventrobronchi are considered secondary bronchi. These secondary bronchi, in addition to giving off parabronchi, also connect to the cervical, cranial thoracic, caudal thoracic, and interclavicular air sacs.

Two breaths, or respiratory cycles, are required to move one pocket of air through the avian respiratory system (Figure 11.14). The first inhalation is made by expansion of the thoracoabdominal space (birds do not have a diaphragm), and most of the air moves directly into the abdominal air sacs. The first expiration pushes air into the lungs. The second inspiration moves the air into the cranial pair of air sacs, and the second expiration moves the air out through the trachea. More than any other animal, birds truly receive fresh air during breathing because airflow is pushed into the lungs, not pulled.

Bird's lungs are attached to the inside of the thoracic cavity, so holding a bird tightly can compromise air flow drastically. Air sacs also communicate with pneumatic bones. The interclavicular air sac extends to the humerus, sternum, syrinx, and pectoral girdle. The abdominal sac often extends to the legs and pelvic girdle.

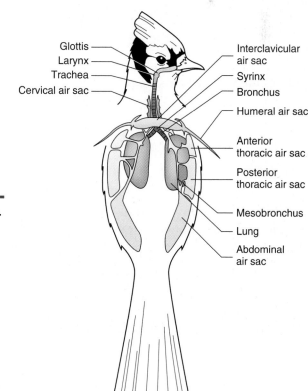

Figure 11.13: Avian lungs and air sacs.

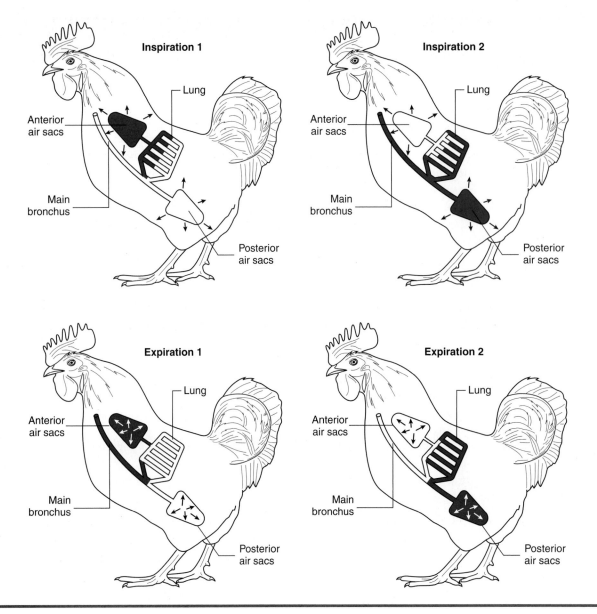

Figure 11.14: Air flow through the respiratory system of birds (white areas represents air flow).

Lung Volumes and Capacities

tidal volume (V_T): Volume of air moving in (or out) during normal breathing.

inspiratory reserve volume: Volume of air it is possible to inhale after a normal inspiration.

expiratory reserve volume: Volume of air it is possible to exhale after a normal expiration.

residual volume: Volume of air in the lungs, even after maximum expiration.

vital capacity: The total volume of air that can be exhaled after a maximum inspiration.

functional residual capacity: The sum of the expiratory reserve volume and the residual volume. Func-tional residual capacity equals the volume of gas left in the lungs at the end of a normal expiration, and this remaining gas will mix with the next tidal volume. Functional residual capacity also shows that the alveoli do not fill with fresh air on each inspiration.

dead space: The space occupied by air at the end of an expiration. Dead space is also the volume of the respiratory system that takes no active role in external respiration; in other words, it is the volume of air that occupies the space between the nose and the respiratory unit. Only the respiratory unit has alveoli in which external respiration can occur. If an endotracheal tube is inserted and extends through the mouth, as in gas anesthesia, this added volume is added dead space.

EXERCISE 11.4

MEASURING LUNG VOLUMES AND CAPACITIES

Procedure

Because dogs and cats are reluctant to breathe into a hose on command, this exercise will use a spirometer with you as the subject. In the spaces provided, record the volume of air for each measurement.

1. Without using the spirometer, count and record the your normal respiratory rate (RR).

   RR = _____respirations/minute

2. Insert the disposable cardboard mouthpiece in the open end of the valve assembly of the spirometer. Practice exhaling though the spirometer instead of through your nose so an accurate measurement can be obtained.

3. Conduct the test three times for each required measurement. Write down each measurement and the average of the three.

4. Measure tidal volume (V_T). Inhale a normal breath, and then exhale normally into the spirometer mouthpiece. Record the volume and repeat the test twice more. (Normal human value = 500 ml)

   trial 1 _____ml   trial 2 _____ml

   trial 3 _____ml   average _____ml

5. Compute the minute volume (V_E).

$$\text{minute volume } (V_E) = V_T \times RR$$

   minute volume (V_E) = _____ml × _____resp/min

   minute volume (V_E) = _____ml

6. Measure expiratory reserve volume (ERV). Inhale and exhale normally two or three times, then, after exhaling, insert the spirometer mouthpiece and exhale forcibly as much of the additional air in your lungs as you can. Record the result, and repeat the test twice again. (Normal human value = 1000 to 1200 ml)

   trial 1 _____ml   trial 2 _____ml

   trial 3 _____ml   average _____ml

7. Determine vital capacity (VC). Breathe in and out two or three times, then bend over and exhale all the air possible. Next, as you raise yourself upright, inhale as fully as possible (you must strain to inhale the maximum amount). Quickly insert the mouthpiece and forcibly exhale as much as you can. Record your result and repeat the test twice again. (Normal human value = 4500 ml or a range of 3600 to 4800 ml)

   trial 1 _____ml   trial 2 _____ml

   trial 3 _____ml   average _____ml

8. Calculate inspiratory reserve volume (IRV) with the following formula. (Normal human value = 2100 to 3100 ml.)

$$IRV = VC - (V_T + ERV)$$

   Record your average IRV = _____ml

QUESTION

1. How does your computed value compare with normal for each of the previous calculations?

CLINICAL SIGNIFICANCE

Knowledge of the anatomy of the lung lobes is critical when considering the areas to **auscultate.** *Most veterinarians start listening to the area of the chest wall just above the elbow joint, caudal to the scapula. This is where the trachea and carina are located. The sounds auscultated here are called* **bronchial sounds** *and are loudest at this point because of the air turbulence created at this bifurcation. Listen to the lower end of the trachea in the neck area to learn what this sounds like.*

Place the stethoscope head in the location discussed previously and listen for similar sounds. In large animals, this sound is heard only in a limited area of the chest because bronchial sounds are not conducted any distance in normal lung tissue. In the dog, and in particular in small breeds of dogs and in the cat, the bronchial sounds may be heard over a considerable part of the area of auscultation. Next, slide the stethoscope head ventrally to listen to the cranial lobe; then the middle and then the caudal lobes are auscultated in order. The normal sounds of the lung lobes are quieter and called **vesicular sounds.** *They are best described as the sound produced if the letter F is whispered softly.*

These sounds may change in various pathological situations. If the lung lobe, or part of a lung lobe, becomes consolidated with fibrin and fluid due to pneumonia, the sounds are called **referred bronchial sounds.** *A good auscultator can discern accurately where the line of consolidation occurs across the ventral aspects of the lung lobes. Abnormal lung sounds produced by pathological processes within the tracheo-bronchial tree are termed* **adventitious sounds.** *The point of maximum intensity is usually near the diseased area.*

Adventitious sounds may be either discontinuous (crackles) or continuous (wheezes). **Crackles** *are characterized as either* coarse *or* fine. *Coarse crackles are bubbling or gurgling sounds, whereas fine crackles are often characterized as Velcro or cellophane-type sounds. One mechanism for causing coarse crackles is the bursting of bubbles of secretions within an airway, such as in pulmonary edema or bronchopneumonia. Fine crackles may occur when an airway is closed and during inspiration suddenly opens, as in fibrosis, inflammation, or interstitial pulmonary edema.* **Wheezes** *are continuous musical or whistling sounds generated by air passage through a narrowed airway, as might be caused by stenosis, foreign bodies, bronchospasms, mucous plugging, or tumors.*

Veterinary Vignettes

Nike was a 12-lb. "terror" according to Mrs. Robertson. Corralling him long enough to administer his puppy shots was an impossible task for one person. I summoned my best technician, who, by the way, can hold anything still for 10 seconds. Fortunately, I can administer a DHLPP in less time than that and can even manage to get the needle into the animal rather than my thumb . . . usually. Nike was a perpetual-motion poodle, with every body part moving simultaneously: squirming, wiggling, running, jumping, and going every direction at the same time. But running was his specialty! He would have been about 9 months old the day he ran full speed through an 8-cm wide opening in a wooden fence. Unfortunately, Nike's chest was just slightly under 10 cm in diameter. It was not the best day of his life.

Mrs. Robertson had seen the whole event. The neighbor's cat made it through the opening with no problem with Nike in hot pursuit. With a howl of pain, he let his owner know he had made a huge error in judgment. Mrs. Robertson was frantic upon entering my office.

"He's having trouble breathing," she exclaimed.

I listened to his chest and noted that the lung sounds on the right side were normal, but on the left side, the area in which vesicular sounds could be heard was diminished. A quick x-ray revealed that his left lung was not inflating as much as it should, and there was fluid accumulating in the pleural cavity. The thoracocentesis showed the fluid was fresh blood. This condition is known as a hemothorax.

Fortunately, no ribs were fractured that I could see. The ribs have some give to them because of their ligamentous attachment at the vertebrae and the cartilaginous portion that attaches to the sternum. Nike's blood work-up revealed normal clotting times. After sedation, we drained the blood from his chest through a small catheter so as not to damage the lung tissues. Simultaneously, we treated him for shock.

It was necessary to drain his chest two more times that afternoon and evening. The next day we were still able to get blood from the chest, but the amount was diminishing. For that reason, a thoracotomy (surgically opening the thoracic cavity) was not performed. However, we placed a small-gauge chest tube in the thoracic cavity on his left side because we were concerned about the possibility of a large blood clot forming within the thoracic cavity. If this happened, a thoracotomy would be a necessity. We kept track of the total amount of blood drained from the chest and did repeated blood counts. We wanted to determine whether a blood transfusion would be necessary. By the next afternoon, we could no longer draw any blood through the chest tube.

The look of relief on Mrs. Robertson's face when she went home with Nike just made my day. Well, at least for 2 hours, until she came rushing back in with Nike breathing hard again!

"I just let him out to go to the bathroom. I didn't think he'd try to run through the fence again!" she cried.

I looked at the poor little beast and just shook my head; we were going to have to do it all over again.

SUMMARY

For a veterinarian or veterinary technician to perform a good physical exam on a patient's respiratory system, a detailed knowledge of the patient's anatomy is required. This includes the upper respiratory system as well as the lower respiratory system. Technicians, as part of their duties as veterinary anesthetists, routinely intubate patients using an endotracheal tube. To do this properly, knowledge of the internal anatomy of the larynx is needed.

In addition to understanding the anatomical structures of the respiratory system, we also covered some of the physiological parameters that can be measured or estimated. Because much of the anesthesia performed on animal patients is gas anesthesia, an understanding of respiratory physiology is needed to understand the principles of anesthesiology. Key concepts in respiratory physiology include the movement of gases through the respiratory tree to the alveoli, subsequent exchange of gases with the blood, and the concept of dead space. This chapter also introduced you to auscultation of the lungs and the normal and abnormal sounds that can be heard.

CHAPTER 12

THE URINARY SYSTEM

OBJECTIVES:

- name and describe the vascular supply to the kidneys
- name and identify the parts of the nephron using diagrams and models
- describe the function of the urinary system
- name and locate the major anatomical structures of the kidney using pig and cat kidneys
- understand blood pressure regulation by the kidneys
- understand the physiological principles and factors involved in the animal's ability to concentrate urine
- measure the specific gravity of urine and understand what it means when it is elevated, fixed, and abnormally low

MATERIALS:

- cat cadaver, triple injected (order without skin attached)
- pig kidney, injected with red and blue latex
- necropsy knife
- Mayo dissecting scissors
- probe
- 1 × 2 thumb forceps or Adson tissue forceps
- #4 scalpel handle with blade
- bone cutting forceps
- rubber gloves
- models of the nephron
- refractometers
- intravenous (IV) fluids and administration kit
- 2 live dogs

Introduction

The **urinary system** is one of the most important excretory pathways in the body. The other excretory pathways are: (1) the alimentary tract; (2) the biliary system, which eliminates waste into the digestive tract via the common bile duct; (3) the lungs, from which gases and chemicals are exhaled; and (4) the exocrine glands of the skin. Metabolism of nutrients produces waste products, such as carbon dioxide, nitrogenous wastes, ammonia, and detoxified organic and non-organic compounds that must be eliminated from the body if normal function is to continue. The urinary system is primarily concerned with removal of nitrogenous wastes from the body. In addition to their function as excretory organs, the **kidneys** help maintain the body's electrolyte, acid-base, and fluid balance; thus, they also act as major homeostatic organs.

To perform these functions, the kidneys filter the blood, taking fluid and chemicals into their functional tubular system, called the **nephron,** where the fluid and chemicals are then processed. The nephron selectively keeps certain chemicals in the fluid, such as toxins and nitrogenous by-products; it allows needed chemicals to be absorbed back into the blood, such as glucose, sodium, and chloride; and it allows excessive amounts of certain ions (such as hydrogen ions) to leave the blood via the fluid. After this process, the liquid to be excreted is **urine.**

The nephron also controls the amount of fluid absorbed back into the bloodstream and the amount eliminated from the body, thus helping maintain the correct amount of hydration in the system. Failure to perform these functions adequately (due to a malfunctioning kidney) will result in the animal's death.

Kidneys of various animals differ in structure and complexity. They can be monolobed, as in the pig,

dog, cat, and horse, or multilobed, as in ruminants and fowl. Some animals have complex duct systems, called the **calyxes,** leading from the **renal papilla** to the **renal pelvis;** others have no duct system and the urine flows directly into the renal pelvis. Figure 12.1 shows the kidneys of various domestic animals.

The Vascular Supply to the Kidneys

The **renal artery** enters the kidney and branches into **interlobar arteries,** coursing between the lobes of the kidneys. These lobes contain separate areas of the **medulla** (Figure 12.2). This is easily seen in complex kidneys of the pig and ruminants, but in the monolobed kidneys of the cat, dog, and horse, the division between the medullary areas is not distinct; the interlobar arteries course within the medulla rather than between the lobes. At the division between the medulla and the **cortex,** the interlobar arteries branch in an arcing pattern and are called **arcuate arteries.** The arcuate arteries send multiple smaller **interlobular arteries** linearly through the cortex toward the periphery of the kidney.

At multiple sites along the interlobular arteries, microscopic **afferent arterioles** branch. Each terminates in a tufted ball of capillaries called a **glomerulus** (Figure 12.3). From each glomerulus, an **efferent arteriole** emerges that quickly branches to surround the tubules of the nephron and forms the **peritubular capillary network.** The part of this network that sends vessels adjacent and parallel to the **loop of Henle** is called the **vasa rectae** (straight vessels) and is divided into **arteriae rectae,** which take the blood down next to the loop, and **venae rectae,** which return blood to the **interlobular veins.** The venous system has corresponding veins that course with the arteries. From the interlobular veins, the blood travels through the **arcuate veins** into the **interlobar veins,** which converge to form the **renal vein** and go out of the kidney.

The Nephron: Anatomy and Physiology

As mentioned previously, the **nephron** (tubular system of the kidney) selectively absorbs its contents into the blood or excretes them via the urine (see Figure 12.3). The nephron starts as a blind sac; its **visceral layer** virtually shrink-wraps around the *glomerus's* capillaries, and the **parietal layer** forms the outer wall of the sac, which is continuous with the wall of the **proximal convoluted tubule.** This blind sac is called **Bowman's capsule.** In chapter 5, Figure 5.3, the example of simple squamous epithelium

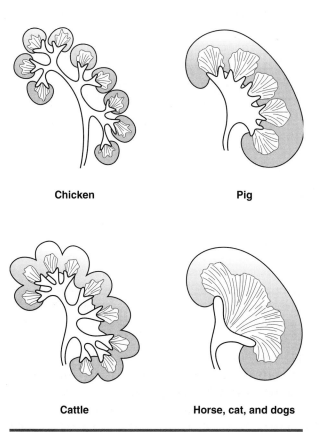

Chicken

Pig

Cattle

Horse, cat, and dogs

Figure 12.1: Structure of kidneys of various domestic animals.

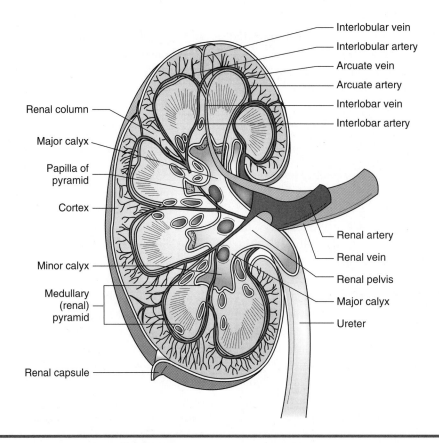

Figure 12.2: Frontal section of a pig kidney.

was the parietal layer of Bowman's capsule. Refer to this figure and identify the glomerulus, Bowman's capsule, and the lumen of the capsule.

Together, Bowman's capsule and the glomerulus are called the **renal corpuscle.** The fluid entering the lumen of Bowman's capsule is called **glomerular filtrate.** This fluid then proceeds into a convoluted tube proximal to the loop of Henle, called the **proximal convoluted tubule.** From there it goes down the **descending loop** of the **loop of Henle** and up the **ascending loop** to the **distal convoluted tubule.** Multiple distal convoluted tubules join in a common duct known as a **collecting duct.** The many collecting ducts within the kidney converge to produce larger **papillary ducts,** which terminate in a part of the kidney called the **renal papilla.** The fluid, which has become urine, flows out of the renal papilla either into a minor calyx or directly into the renal pelvis in simple kidneys like those of the cat.

Histologically, the cells of Bowman's capsule are simple squamous epithelium as mentioned previously. You saw in chapter 5, Figure 5.6, an example of simple cuboidal epithelium of the kidney. The simple cuboidal cells of the proximal convoluted tubule are unique in that they have microvilli, known as a *brush border* (Figure 12.4). The thin descending loop is simple squamous tissue again, but the thicker ascending loop is simple cuboidal or low columnar. The distal convoluted tubule is also simple cuboidal tissue, as are cells of the collecting ducts, most of which are called **principal cells,** except for a few **intercalated cells.** The intercalated cells also are cuboidal, but they have microvilli at the apical surface and a large number of mitochondria. The principal cells have receptors for both **antidiuretic hormone (ADH)** and *aldosterone,* the two hormones that regulate water and sodium resorption. Intercalated cells play a role in the homeostasis of blood pH.

The bottom of the loop of Henle is rotated such that the top of the ascending loop is adjacent to the afferent arteriole (see Figure 12.3). As the afferent arteriole approaches the glomerulus, it is surrounded by a special type of cells, called **myoepithelial cells** (see Figure 4.6 in chapter 4). These cells have some characteristics of smooth muscle and some of epithelial cells. Because they are juxtapositioned around the arteriole and glomerulus, they are called **juxtaglomerular cells.** A cluster of special cells that make up the wall of the ascending loop where it touches the afferent arteriole is called the **macula densa** (meaning dense spot). These are special *osmoreceptors* that communicate the *tubular osmolality* to the juxtaglomerular cells. The juxtaglomerular

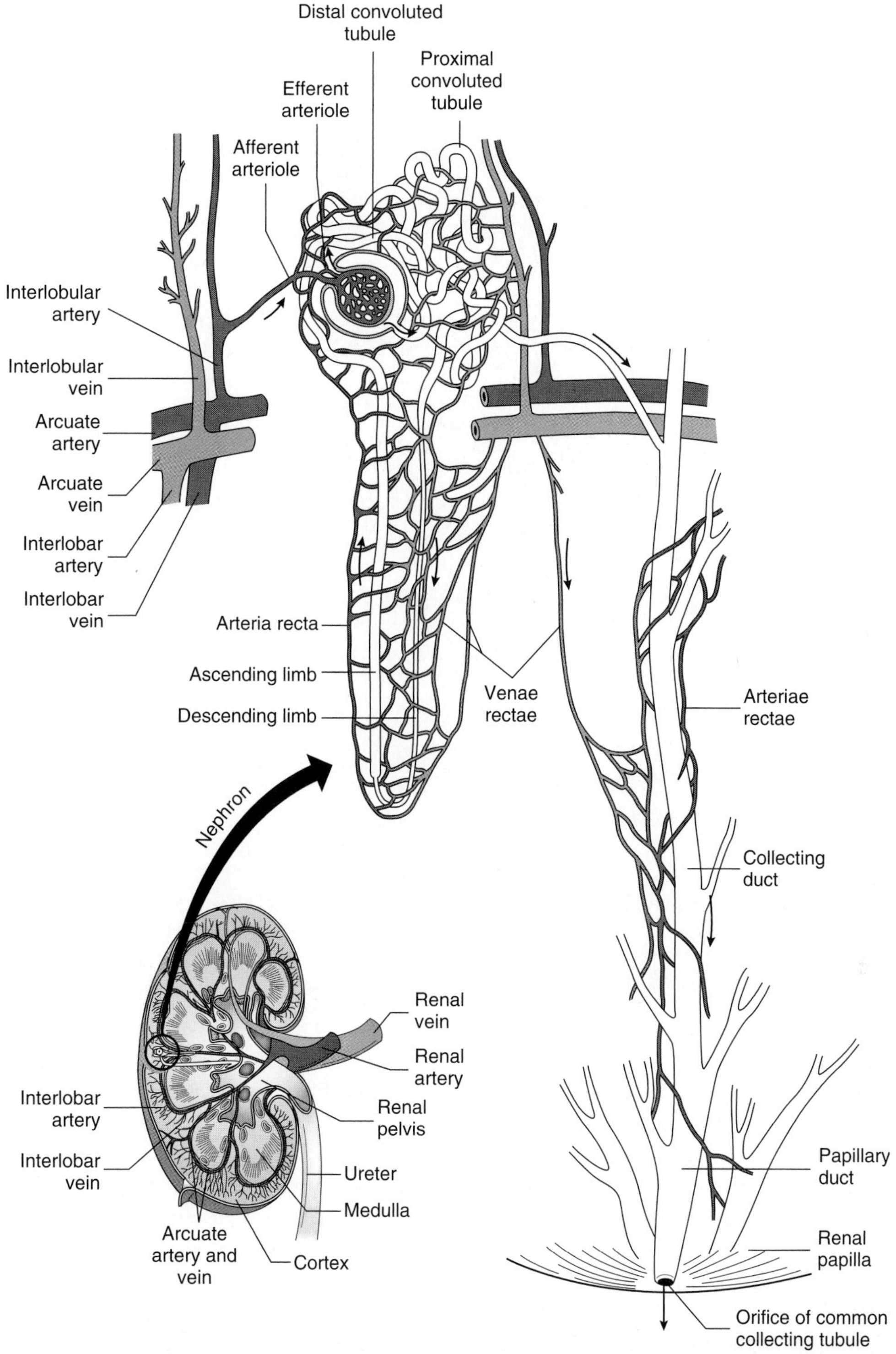

Figure 12.3: The nephron and its blood supply (all animals).

cells, combined with the macula densa, constitute the **juxtaglomerular apparatus.**

Relatively high blood pressure must be maintained to facilitate the filtering process within the glomerulus. Both the afferent and efferent arterioles are equipped with smooth muscle, so the amount of blood entering the glomerulus and pres-

sure within can be controlled by constricting either the afferent arteriole, the efferent arteriole, or both. The juxtaglomerular cells produce a chemical called **renin** (do not confuse with *rennin*), which is secreted into the blood under the following conditions: (1) when the blood pressure in the afferent arteriole falls; (2) when the sodium concentration

Histological Features of the Renal Tubule and Collecting Duct

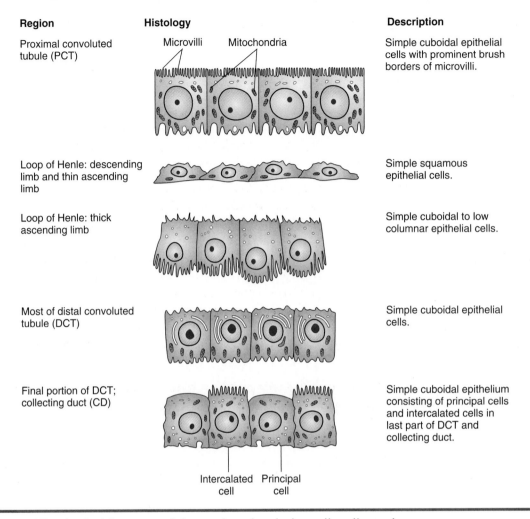

Region	Histology	Description
Proximal convoluted tubule (PCT)	*Microvilli* *Mitochondria*	Simple cuboidal epithelial cells with prominent brush borders of microvilli.
Loop of Henle: descending limb and thin ascending limb		Simple squamous epithelial cells.
Loop of Henle: thick ascending limb		Simple cuboidal to low columnar epithelial cells.
Most of distal convoluted tubule (DCT)		Simple cuboidal epithelial cells.
Final portion of DCT; collecting duct (CD)	*Intercalated cell* *Principal cell*	Simple cuboidal epithelium consisting of principal cells and intercalated cells in last part of DCT and collecting duct.

Figure 12.4: Histological features of the nephron's tubular cells, all species.

of the blood's plasma decreases; (3) when the distal tubular osmolality decreases; or (4) when sympathetic nerve fibers innervating the afferent arteriole are stimulated. In the blood, renin stimulates the production of **angiotensin,** which acts as a vasoconstrictor to increase blood pressure. Renin also acts on the adrenal gland to secrete **aldosterone,** a hormone that causes the kidney to conserve sodium ions.

In the exercises in this chapter the important structures are listed in colored bold print. If a structure is mentioned prior to its dissection it will be italicized. Structures discussed prior to dissection may also be in bold print for special emphasis.

EXERCISE 12.1

DISSECTION OF THE KIDNEY

In this exercise you will use both your cat cadaver and a pig kidney. Complete the following steps in the dissection procedure.

Procedure

1. Get out your cat cadaver. In your previous dissection of the arteries and veins you found the *renal arteries* coursing to the *kidneys*. The kidneys and ureters were found to be retroperitoneal and located along the dorsal wall. The right kidney should be relatively intact if your dissection has been careful

to this point. Each kidney is surrounded by adipose tissue, called **perirenal fat.** The right kidney is positioned higher than the left (Figure 12.5).

2. Remove the adipose tissue and the peritoneum that covers the ventral surface of the left and right kidneys.

3. Identify the **renal artery** and **renal vein,** which carry blood to and from the kidney (see Figures 12.5 and 12.6).

4. Find the **ureters** (the narrow, white tubes that drain the urine from each kidney) and trace them from the **hilus** (the indentation on the medial border of each kidney) to the **urinary bladder** (see Figures 12.5 and 12.6). The right and left ureter attach to the right and left dorsolateral aspect of the bladder, respectively, just cranial to where the urethra begins.

5. The oval **urinary bladder** is connected to the mid-ventral wall by a **median suspensory ligament** and to the lateral walls by **lateral ligaments,** which contain a large amount of adipose tissue. Make a longitudinal incision in the ventral bladder wall and locate where the ureters enter on the dorsal wall. The triangular area formed by the opening of the two ureters and the opening of the urethra is called the **trigonal area** of the bladder. The part of the bladder that narrows, leading to the urethra, is the **neck.** The **fundus** of the bladder is the expanded part. What type of epithelial cells line the bladder and urethra? If you remember your histology, you will know they are transitional epithelial cells. The bladder's smooth muscle layers are unique. Whereas most hollow or tubular structures contain an inner circular and outer longitudinal layer of smooth muscle, the bladder has a third, additional outer circular layer. This presumably aids in contraction of the organ to expel the urine (see Figures 12.6 and 12.7).

6. Locate the urethra, the duct that conducts urine from the caudal part of the bladder to the exterior. The urethra will be dissected with the reproductive system.

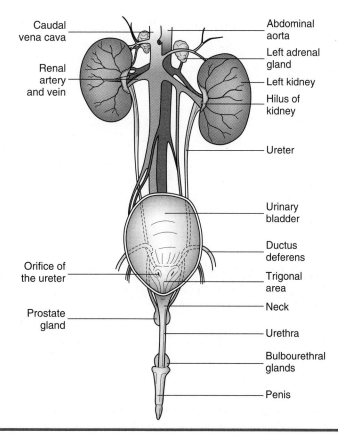

Figure 12.5: Urogenital system of the male cat.

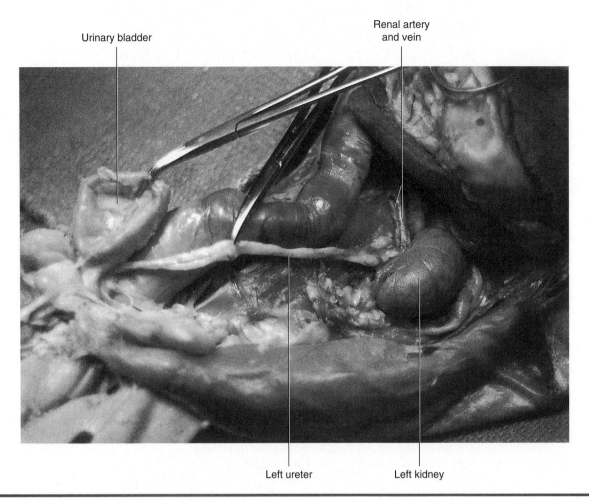

Urinary bladder

Renal artery and vein

Left ureter

Left kidney

Figure 12.6: Kidney, ureter, and bladder of a cat.

7. Remove the left kidney and make a longitudinal cut through it with a necropsy knife or scalpel blade, dividing it into two halves. Remove the **renal capsule,** a thin layer of connective tissue, from around the outside. The **renal cortex** is the outer, lighter-brown layer of the kidney immediately beneath the capsule. This layer contains renal corpuscles. The central layer of the kidney, the **renal medulla,** contains one large, dark **renal pyramid** (Figure 12.8). Identify the single **renal papilla,** called the renal crest in the cat. It is the rounded projection at the bottom of the renal pyramid of the medulla. It drains urine into the **renal pelvis,** which opens to the *ureters.*

8. Find the branches of the renal artery that lead into the kidney, the *interlobar arteries* within the medulla, the *arcuate arteries* at the division between the medulla and cortex, and the *interlobular arteries* within the cortex. The corresponding veins should be identified; they can be found coursing adjacent to the arteries.

9. Obtain a pig kidney, and as you did with the cat's kidney, make a longitudinal cut with a necropsy knife. Divide the kidney into two halves so it appears as the one shown in Figure 12.9.

10. Locate the **hilus** of the pig's kidney. All vessels enter and exit in this area. Find the **renal artery** and **vein** and the **ureter** exiting the **renal pelvis,** which is the funnel-shaped expansion of the ureter inside the **renal sinus.** The renal sinus is the hollow interior of the kidney; in other words, if the membrane that makes up the renal pelvis were removed, the space remaining within the kidney would be the renal sinus (see Figure 12.9).

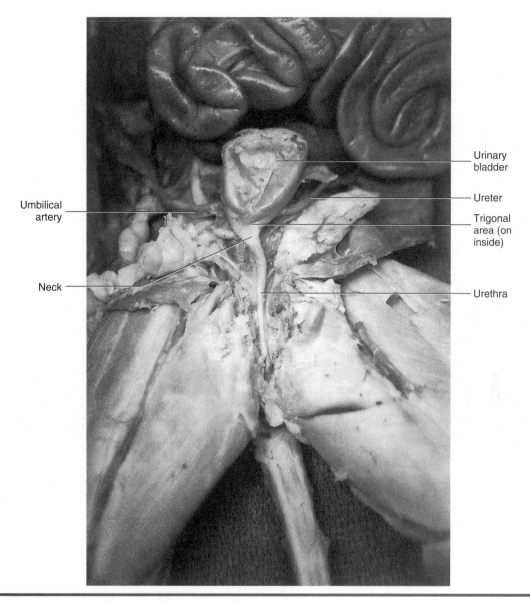

Figure 12.7: Urinary bladder, ureter, and urethra of a cat.

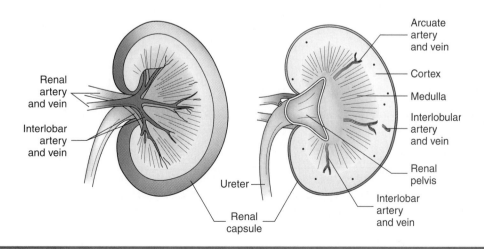

Figure 12.8: The cat kidney.

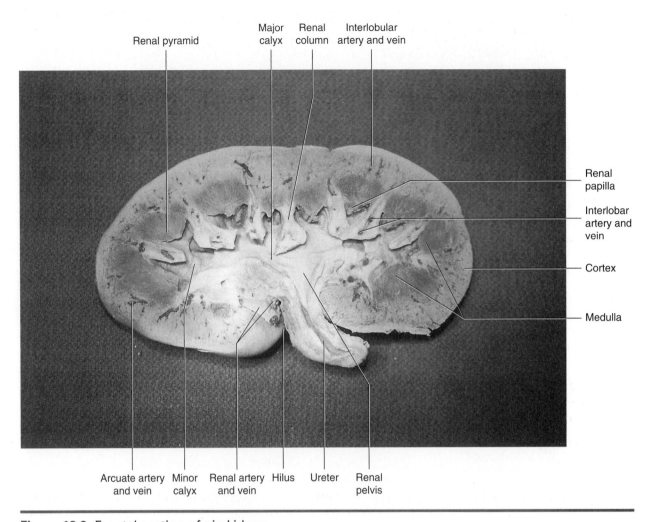

Renal pyramid | Major calyx | Renal column | Interlobular artery and vein

Renal papilla

Interlobar artery and vein

Cortex

Medulla

Arcuate artery and vein | Minor calyx | Renal artery and vein | Hilus | Ureter | Renal pelvis

Figure 12.9: Frontal section of pig kidney.

11. The renal pelvis is joined by two (or more) large tubules, each called a **major calyx.** These calyxes are formed by the joining of smaller **minor calyxes** (see Figure 12.9). These do not exist in the cat.

12. Notice the darker brown areas of the renal medulla; each of these is called a **renal pyramid** (three-dimensionally it looks like an upside-down pyramid within the kidney). The rounded apex, or point, of the pyramid that empties into a minor calyx is the **renal papilla.** It is a sieve-like region that drips urine from the *papillary ducts.* The striated appearance of the pyramids is caused by the numerous *collecting ducts* and papillary ducts, as well as the bottoms of some of the loops of Henle that extend down into the medulla. The areas between the pyramids are called the **renal columns.** These contain the **interlobar arteries** and **veins** (see Figure 12.9).

13. Follow the interlobar arteries and veins toward the periphery of the kidney where they branch. Note that they arc, branching along the division between the medulla and the cortex. These are the **arcuate arteries** and **veins,** which send branches into the cortex as the **interlobular arteries** and **veins** (see Figure 12.9). The small, latex-injected vessels within the cortex are interlobular vessels.

EXERCISE 12.2

PHYSIOLOGY OF THE URINARY SYSTEM

One of the most important functions of the urinary system is maintaining the correct fluid balance in the body. The kidney's ability to concentrate or dilute urine is dependent on its ability to resorb sodium into the **interstitium** and respond to the presence or absence of **antidiuretic hormone (ADH)** in the system. In the nephron, sodium is actively reabsorbed from the glomerular filtrate into the kidney's interstitium from the proximal convoluted tubules, the ascending loop of Henle, and the distal convoluted tubules. This makes the interstitium hyperosmotic in the area of the loop and collecting ducts (which are in the medulla). The reabsorption of water from the collecting ducts is dependent on the permeability of the cells lining these ducts. The permeability of these cells is influenced by the secretion of *ADH* from the posterior pituitary gland at the base of the brain. ADH is a hormone to stop *diuresis* (the production of urine), and therefore, the name is based on what it doesn't do. This can be confusing. It is better to think of ADH based on what it does do; thus, think of it as the water resorption hormone. Under the influence of ADH, the cells become more permeable and water is resorbed from the collecting ducts (and distal convoluted tubules) into the hyperosmotic interstitium. It is then picked up by the vasa rectae and returned to the blood-vascular system.

Secretion of ADH is based on the blood's plasma volume. If increased, the animal will have a decreased secretion of ADH; if plasma volume is decreased, there will be an increased secretion of ADH.

Procedure

1. Locate two live dogs and designate them A and B for this exercise. Weigh both dogs and record your results here:

   dog A: _____

   dog B: _____

2. Deprive the dogs of water for six hours (or overnight) and permit them to have only food with low moisture content. Let the dogs urinate, then cage them and wait another three hours. After three hours, collect urine by free catch or cystocentesis and record the time and specific gravity using a refractometer.

   time: _____ am/pm

   dog A: _____

   dog B: _____

3. Continue to deprive dog A of water until completion of the test, but let dog B drink as much as it wants and administer 200 ml of subcutaneous fluids.

4. Weigh both dogs and record your findings.

   dog A: _____

   dog B: _____

For the next three hours, collect urine and measure the specific gravity each hour.

time: _____am/pm	time: _____am/pm	time: _____am/pm
dog A: _____	dog A: _____	dog A: _____
dog B: _____	dog B: _____	dog B: _____

5. Graph the change in specific gravity with time on the x axis and specific gravity on the y axis.

QUESTIONS

1. At the end of the test, which of the two dogs was producing the most ADH?

2. Which dog's collecting ducts were very impermeable to water toward the end of the test?

Discussion

Withholding water and testing urine specific gravity is called the **water deprivation test** and is usually done by withholding water from an animal for 12 hours. Once the specific gravity rises above 1.030 the test is considered complete because this indicates the kidney's ability to concentrate urine. Dog A should have had a gradually increasing specific gravity, providing its kidneys have the ability to concentrate urine. Dog B's urine specific gravity should have risen while being deprived of water, then dropped again.

When a patient is deprived of fluids, the body's response is to conserve water by releasing ADH, thus resorbing water from the collecting ducts and distal convoluted tubules and causing the specific gravity to rise. Administering fluids and letting the animal drink reverses this need and stops ADH secretion. When an animal's kidneys fail and no longer respond to ADH, or when there are not enough nephrons surviving to do the job, the urine specific gravity falls into the fixed range of 1.008 to 1.012. In this state, the urine osmolality is essentially isotonic. Therefore, the answer to question 1 is dog A and to question 2 is dog B.

CLINICAL SIGNIFICANCE

Dogs and cats with certain diseases benefit from dietary restrictions that lessen their clinical signs, diminish the course of the pathology occurring, and lengthen their lives. Examples of these restrictions are sodium-restricted diets for heart patients and protein-restricted diets for renal patients. Patients with heart problems benefit from a diminution in plasma volume, because the heart does not have to pump as much fluid through the body. The body responds by releasing aldosterone and increasing the ADH production. However, if sodium is restricted, the body's ability to absorb sodium from the intestines and reabsorb it into the kidney's interstitium becomes diminished; thus, the animal is not able to increase its plasma volume.

Patients with renal pathology and a diminished number of functional nephrons cannot adequately excrete nitrogenous wastes. These wastes are breakdown products of protein metabolism; amino acids have a nitrogen-hydrogen (amino) group at each peptide bond that links them together to form a peptide chain and a protein molecule. Restricting the amount of protein and ensuring that the protein the animal does consume has a high biological value (so that virtually all is used and not excreted) helps relieve the kidney's tubules of the work of eliminating these wastes.

Veterinary Vignettes

Believe it or not, veterinarians are human, and they have patients they like and patients they dislike. I have scars from the ones I still dislike. Pointing to my hands, I can say, "This scar is from Killer, and this one is from Frodo," and so on. But some of our patients are very dear to us.

One day I heard a scratching and whining at the front door of my home. When I opened the door one of my Golden Retriever patients was sitting on my front porch. He was lost and stopped by because he couldn't find his way home. I called his owner, and he came over and picked him up. How the dog knew where I lived still baffles me, but I will always remember that dog.

Sometimes the patient I am treating belongs to me, and then, of course, there is no escaping the emotional involvement. I used to own a beautiful yellow Lab named Sandy. On a warm, sunny summer day, while I was out playing Frisbee with my wife and son, he decided to lie down in the tall grass under an apple tree and consume 2 lbs of apples. In the time it took us to walk home, his stomach was already bloating. I drove him to the local emergency animal hospital as quickly as I could. I could have taken him to the facilities at the college where I was employed, but I knew I didn't have time.

The veterinarian on duty and I tried tubing the dog to relieve the pressure, but the stomach had already twisted. Even I was shocked by the speed with which this had occurred; it had been no more than 15 minutes from the time I first noticed he was in distress. To relieve the pressure, we had to trocar the stomach by inserting a large bore hypodermic needle through the skin and into the stomach to let off some gas.

Personally, I don't like operating on my own animals; I've done it, but I don't like it. In this situation there was no alternative; the emergency clinic did not allow outside veterinarians to operate in their facility—not even on their own pets. It was very distressing to me because I lost control of the situation. Sandy survived his surgery to correct the rotated stomach, which is known as a volvulus.

Postoperatively, he started having numerous ventricular premature contractions (VPC), and I found a pneumothorax (air in the pleural cavity) to complicate his condition. How and why the spontaneous pneumothorax occurred was a mystery, both to me and to the clinician who did the surgery. I treated both; the VPCs diminished with medical treatment, and I had to insert a chest tube to correct the pneumothorax. It was almost 10 days before I could be certain the lung lobe had sealed.

Over the next month Sandy continued to improve, but he still was just not doing as well as he should have been. He started consuming large amounts of water. As soon as I noted this, I began testing his urine specific gravity. It had dropped to around 1.005, which is below the fixed level. For some reason he had developed a case of diabetes insipidus, a disease caused either by a lack of ADH from the posterior pituitary gland or by renal insensitivity to ADH. After consulting with the local internal medicine specialist, and because of Sandy's subsequent lack of response to treatment, I decided to put him down. He had been through enough, and after two months of the daily battle, we both gave up.

I have euthanized my own animals before, but for some reason, I couldn't do it this time. Fortunately for me, I have friends in the business that could come to my home and do it for me. A few times I had gone to client's homes to put their animals down for them, and after this experience I understood just why they had been so grateful. We learn about the practice of veterinary medicine in school, but we learn about the art of veterinary medicine in practice. I still have a painting of Sandy in my home office. He lived 12 grand years, and I still miss him 10 years later!

SUMMARY

In this chapter, you learned that the kidneys of domestic animals vary in both shape and size. Yet despite these variations in gross appearance, the functional part of the kidney, the nephron, is similar in all species. The anatomy and physiology of the nephron was studied in detail. Thus, you learned the mechanism by which the kidney is able to concentrate urine and maintain a stable hydration level within the body. By dissecting both the cat kidney and pig kidney, the comparative anatomy of the two kidneys was demonstrated. You also dissected the other parts of the urinary system and learned the parts of the bladder.

THE ENDOCRINE SYSTEM

- identify and name the major endocrine organs in the dissected cat and the brain of a sheep
- list the hormones produced by the hypothalamus, the pituitary gland, and the target organs
- understand the feedback mechanism that controls release of these hormones
- understand the basic functions of the hormones produced by the endocrine glands

- cat cadaver, triple injected (order without skin attached)
- sheep brain, with pituitary gland attached
- Mayo dissecting scissors
- probe
- 1 × 2 thumb forceps or Adson tissue forceps
- #4 scalpel handle with blade
- rubber gloves
- blood chemistry machine and serum glucose tests
- one live dog

Introduction

The **endocrine system** is system of glands that produce hormones, which are used by the body as chemical messengers. These hormones are released directly into the blood and transported throughout the body. Whereas the nervous system is able to effect rapid changes in the body (with the electrochemical impulses generated by *neurons*), hormones tend to produce a slower change. Although they are released into the bloodstream, specific hormones affect only the biochemical activity of a specific organ or organs. The organ that responds to a particular hormone is referred to as the hormone's **target organ.**

The target tissue's response seems to depend on the hormone molecule's ability to bind with specific receptors (proteins) occurring on the cells' plasma membranes or within the cells. Hormones are most often either steroid- or amino-acid-based molecules. When the hormone binds with the target organ's cells, it stimulates changes in the organ's metabolic activity. For example, thyroid-stimulating hormone causes an increase in the metabolic activity of thyroid cells, which increases production of the thyroid hormones. An increase in thyroid hormones affects many cells of the body by stimulating their metabolic activity.

Some endocrine glands produce only hormones; others produce both hormones and other cell types, such as reproductive cells. Examples of the former are the **anterior pituitary, thyroid, parathyroid,** and **adrenal glands.** Examples of the latter are the **testes** and **ovaries.** Both types of glands are derived from epithelial tissue in development, but the endocrine, or ductless, glands secrete their products into the blood or lymph. In chapter 5, the difference between exocrine and endocrine glands was discussed. You may remember that exocrine glands secrete their

their products onto an epithelial surface; whereas endocrine glands secrete directly into the bloodstream. The pancreas is a gland that has both endocrine and exocrine glandular tissue. Pancreatic islet cells contain endocrine tissue, which produces **insulin** and **glucagon;** and the exocrine glands produce trypsin, chymotrypsin, amylase, and lipase (see Exercise 9.4 in Chapter 9).

Anatomy and Physiology of the Hypothalamus and Pituitary Gland

The **pituitary gland,** or **hypophysis,** is attached to the **hypothalamus** at the base of the brain by a slender stalk called the **infundibulum.** The pituitary gland consists of two major functional areas: the **anterior pituitary,** or **adenohypophysis,** and the **posterior pituitary,** or **neurohypophysis.** The adenohypophysis consists of three portions: the **pars distalis, pars intermedia,** and **pars tuberalis.** The neurohypophysis contains the **pars nervosa** and is directly connected to the hypothalamus by the infundibulum. The shape of the pituitary gland and the degree of development of its relative parts varies in domestic animals (Figure 13.1).

The **hypothalamus** is an important regulatory center for the nervous system; it regulates body temperature, thirst, hunger, sexual behavior, and defensive reactions such as fear and rage. Despite its many functions, the hypothalamus is quite small. It consists of the **optic chiasm, infundibulum, tuber cinereum,** and **mammillary body** and hypophysis (Figures 13.2 and 13.3).

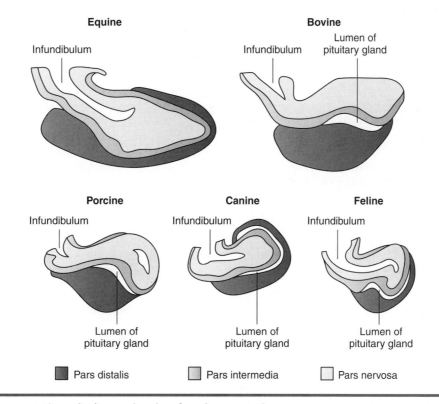

Figure 13.1: Comparative pituitary glands of various species.

Figure 13.2: The hypothalamus and pituitary gland.

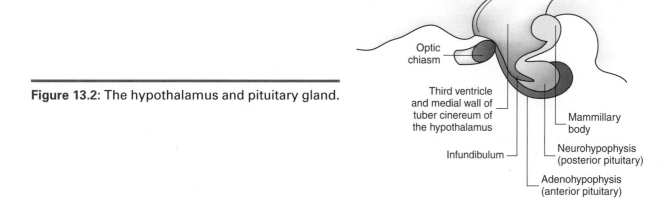

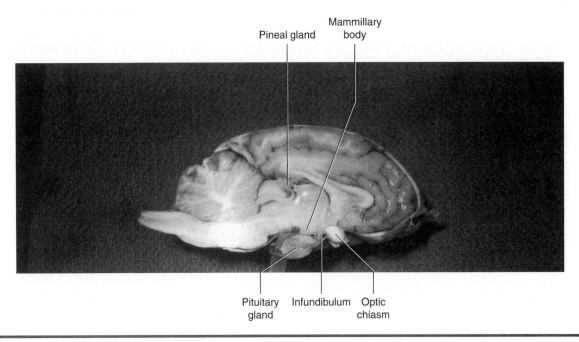

Pineal gland   Mammillary body

Pituitary gland   Infundibulum   Optic chiasm

Figure 13.3: Sagittal section of the sheep brain showing the endocrine structures.

Table 13.1   Hormones of the Hypothalamus and Adenohypophysis

Releasing and inhibiting hormones	Anterior pituitary hormones
Somatotropin releasing hormone (STH-RH)	Somatotropic hormone
Somatotropin release-inhibiting hormone (STH-RIH)	Somatotropic hormone
Thyrotropin releasing hormone (TRH)	Thyroid-stimulating hormone
Corticotropin releasing hormone (CRH)	Adrenocorticotropic hormone
Prolactin releasing hormone (PRH)	Prolactin
Prolactin inhibiting hormone (PIH)	Prolactin
Gonadotropin releasing hormone (GnRH)	Follicle stimulating hormone and luteinizing hormone

The hypothalamus is also a crucial regulator of the endocrine system and pituitary gland. The hormones produced by the *neurons* in the ventral hypothalamus are of two types: **releasing hormones,** which stimulate specific cells of the anterior pituitary to produce certain hormones, and **inhibitory hormones,** which have an inhibitory effect and suppress production. They are listed in Table 13.1; the hormones produced by the hypothalamus are on the left, and those hormones of the anterior pituitary that are affected are listed on the right.

Two other hormones also are produced by the hypothalamus: **antidiuretic hormone (ADH)** and **oxytocin.** ADH is produced primarily in the **supraoptic nuclei** and oxytocin primarily in the **paraventricular nuclei.** These are special neurosecretory neurons that allow their product to slide down an axon into the *posterior pituitary gland,* or *neurohypophysis,* for release into the blood (Figure 13.4).

The releasing and inhibitory hormones produced in the hypothalamus are delivered to the anterior pituitary via the **hypophyseal portal system.** The capillaries in the ventral hypothalamus pick up these hormones and transfer them via the **hypophyseal portal veins** to the secretory cells of the adenohypophysis (see Figure 13.4).

The *adenohypophysis* is the glandular part of the pituitary gland (*adeno* means gland) and produces the **tropic hormones.** The tropic hormones (see Table 13.1) stimulate target organs (which are also endocrine glands) to secrete their hormones. The hormones of the target organ then exert their influence on other body organs and tissues. There are seven known hormones produced by the anterior pituitary:

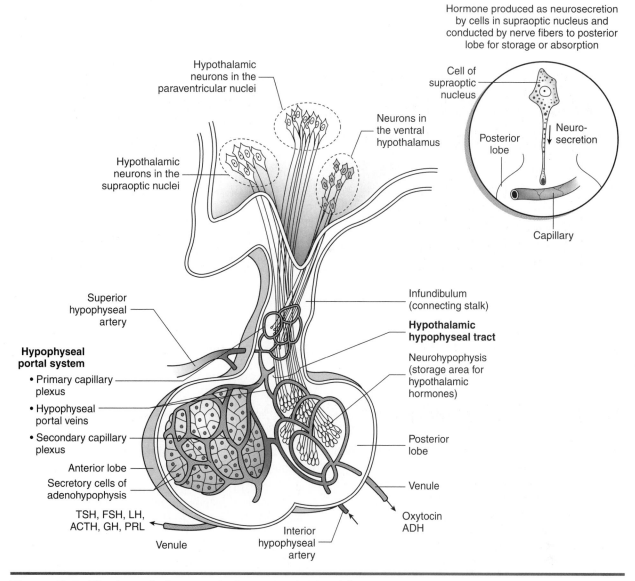

Figure 13.4: The hypothalamus, pituitary gland, blood flow, and neurosecretory pathways.

somatotropic hormone (STH): Also known as **growth hormone (GH),** STH is a general metabolic hormone that plays a role in determining body growth and size. However, it also affects many tissues of the body as it regulates metabolism of proteins, carbohydrates, and lipids. It acts as an anabolic hormone in its effect on metabolism of body proteins; in other words, it stimulates formation and build-up of tissues, and thus is an important hormone in repair and regeneration.

thyroid-stimulating hormone (TSH): Influences growth and activity of the thyroid gland. Increasing TSH increases production of the thyroid gland hormones.

adrenocorticotropic hormone (ACTH): Regulates the endocrine activity of the **adrenal cortex,** the outer part of the adrenal glands. Increasing ACTH increases production of the adrenal gland's glucocorticoid hormones.

follicle-stimulating hormone (FSH): Stimulates follicular development in the female and spermatogenesis in the male.

luteinizing hormone (LH): Stimulates follicular rupture and formation of the corpus luteum in the female and testosterone production in the male. In the male, it is called the **interstitial cell stimulating hormone (ICSH).**

prolactin: Stimulates milk production at parturition and replenishment of milk during lactation. It may also stimulate testosterone production in males.

melanocyte-stimulating hormone (MSH): Increases skin pigmentation in amphibians by stimulating dispersion of melanin granules in melanocytes. Its exact role in animals and humans is unknown.

The hormones secreted into the bloodstream from the neurohypophyis are as follows:

antidiuretic hormone (ADH): Studied in the last chapter, this can also be called the water resorption hormone.

oxytocin: Stimulates uterine contractions during birth and milk let-down in the lactating mother.

Physiology of the Glands of the Endocrine System

The hormones produced by target organs have a feedback effect on the hypothalamus and pituitary gland; as they are produced, their serum levels are read by the hypothalamus and secretion of the releasing factors is diminished. As levels drop, secretion of releasing factors increases. In this way, a constant level is maintained in the blood. To review, target organs regulated by the hormones of the adenohypophysis are the adrenal glands, thyroid glands, parathyroid glands, and reproductive organs.

Adrenal Glands

The *adrenal glands,* like many organs, have an outer area called the cortex and an inner, central area called the medulla. The *adrenal cortex* is divided into three zones:

1. The **zona glomerulosa** is the outer layer (under the capsule) that produces the **mineralocorticoids. Aldosterone** is 95% of the hormones produced by the zona glomerulosa and is responsible for sodium reabsorption by the kidneys.

2. The **zona fasciculata** is the middle layer and is the widest; it produces *glucocorticoids.*

3. The **zona reticularis** is the inner layer and produces a small amount of weak **androgens,** which are steroid hormones that have a masculinizing effect. The major

androgen secreted by the adrenal gland is dehydroepiandrosterone (DHEA). This hormone has a minimal effect in males, but in females it contributes to libido.

The **glucocorticoids** produced are steroid-based molecules. They are **cortisol (hydrocortisone), corticosterone,** and **cortisone.** Of these three, cortisol is the most abundant and is responsible for approximately 95% of the glucocorticoid activity. The metabolism of glucocorticoids is catabolic; that is, tissues are broken down to produce products needed for metabolism. The glucocorticoids' metabolic activities include:

1. increasing gluconeogenesis in liver

2. decreasing peripheral glucose use

3. increasing proteolysis in skeletal muscle

4. increasing circulating levels of amino acids

5. increasing mobilization of free fatty acids

6. producing an anti-inflammatory effect

7. producing an anti-allergic effect

8. decreasing the immune system's functionality

The **adrenal medulla** is not considered glandular tissue; instead it is *modified sympathetic nervous tissue.* Its products are **norepinephrine,** which is also present at post-ganglionic, sympathetic nerve endings, and **epinephrine (adrenalin),** which constitutes approximately 80% of the total secretion by this gland.

Both these hormones are sympathomimetic; that is, their effects mimic the sympathetic division of the autonomic nervous system. When released into the blood, the effects of these two chemicals on the body are as follows:

1. stimulate increase in heart rate and force of contraction

2. serve as a vasoconstrictors for increased peripheral resistance

3. increase blood pressure

4. counteract the depressing action of insulin on blood sugar, thus raising blood sugar levels

5. stimulate ACTH release

6. stimulate glucagon to initiate glycogenolysis in the liver

7. stimulate the breakdown of fats for energy

Thyroid Gland

The thyroid gland produces the hormones **triiodothyronine (T_3), tetraiodothyronine (T_4),** and **thyrocalcitonin (TCT).** Of T_3 and T_4, T_4 is the most prevalent in farm animals, dogs, and cats. In general, T_3 and T_4 are the hormones that regulate (1) basal metabolic rate and oxygen use, (2) cellular metabolism, and (3) normal growth and tissue differentiation. In addition, as production of T_3 and T_4 rises, so do the following metabolic activities:

1. glucose absorption and utilization

2. cholesterol synthesis (but under deficiency conditions, cholesterol levels rise because of decreased elimination via the bile)

3. potentiated effects of norepinephrine

The metabolism of thyrocalcitonin will be discussed with the parathyroid glands because its metabolic activity is associated with calcium and phosphorus balance in the body.

Parathyroid Glands

The parathyroid glands exist as small nodules within or near the thyroid glands, usually two on each side. One is located external to and the other internal to the thyroid gland. They produce **parathyroid hormone (PTH),** which controls calcium levels in the blood, and thus phosphorus levels. PTH acts to increase calcium mobilization from bone, calcium reabsorption from the intestines, and calcium reabsorption from the kidneys.

A decreased PTH level causes decreased blood calcium levels, blood phosphorus levels, and urine phosphorus levels. A pathologic decrease in blood calcium levels causes tremors, muscle twitches, tetany, and convulsions. This condition could be caused by loss of calcium from the blood due to increased demand from the mammary glands in a period of increased milk production. If sufficient PTH is not available for release, calcium absorption from the intestinal tract cannot keep up with the demand. This condition is called *eclampsia,* or *hypocalcemic tetany,* in dogs.

As previously mentioned, the hormone *thyrocalcitonin (TCT)* is produced by the thyroid gland. It keeps calcium in the bone by blocking reabsorption and moves calcium from blood back into the bone. TCT needs vitamin D as a coenzyme to work, and there is a relationship between PTH and TCT: If blood calcium levels are high, TCT is released; if blood calcium levels are low, PTH is released.

The Pancreas

The pancreas's endocrine activity is not controlled by pituitary hormones, and it has both endocrine and exocrine activity. The exocrine portion of the pancreas secretes digestive enzymes into the duodenum. Histologically, the endocrine portion of the pancreas is composed of numerous islands of cells scattered throughout its tissue, called the **pancreatic islet cells** (or **Islets of Langerhans**). The islets are comprised of three types of cells: **alpha cells, beta cells,** and **delta cells.** *Beta cells* produce *insulin; alpha cells* produce *glucagon;* and *delta cells* secrete somatostatin, gastrin, and a vasoactive intestinal peptide.

Insulin has the following anabolic effects.

1. facilitates transfer of glucose into cells (hypoglycemic effect)

2. once inside a cell, causes glucose to be used more rapidly

3. enhances glycogen formation

4. increases glucokinase (stimulates glucose to become glucose-6-phosphate)

5. increases amino acid uptake in skeletal muscle

The primary effect of insulin, as listed above, is to facilitate transfer of glucose into cells. However, during exercise cells do not need insulin for glucose to enter. This fact has a practical application when considering a dog with diabetes mellitus. A consistent amount of daily exercise should reduce the quantity of insulin required daily.

Glucagon has the following effects.

1. increases glyconeogenesis from glucose in the liver

2. increases gluconeogenesis from amino acids and lactic acid (via pyruvic acid in the liver)

3. increases insulin secretion

4. when given intravenously, relaxes smooth muscle

The hormones of the reproductive glands will be discussed in the next chapter. In the exercises in this chapter the important structures are listed in colored bold print. If a structure is mentioned prior to its dissection it is italicized. Structures discussed prior to dissection may also be in bold print for special emphasis.

EXERCISE 13.1

DISSECTION OF THE ENDOCRINE GLANDS

Most of these glands have been located at least once during previous dissections.

Procedure

1. Locate the paired **thyroid glands** on either side of the cranial end of the trachea, just below the cricoid cartilage (Figure 13.5).

2. Examine the surface of the thyroid glands for two tiny **parathyroid glands** on each side. The external parathyroid gland is located at the cranial pole of the thyroid gland and is external to the thyroid capsule. The internal parathyroid gland is located within the thyroid capsule (intracapsular) and lies within the parenchyma of the thyroid gland. Often these glands are not visible with the unaided eye because they are so small or previous dissection and the preservative have destroyed them (see Figure 13.5).

3. The **thymus gland** can be observed cranial to the heart, in the ventral medial area of the thorax (see Figure 13.5). This gland was probably destroyed during your dissection of the vascular system. The hormones produced by the thymus gland are thymosin, thymic humoral factor (THF), thymic factor (TF), and thymopoietin. They promote the proliferation and maturation of T cells (a type of white blood cell), which destroy microbes and foreign substances and may retard the aging process.

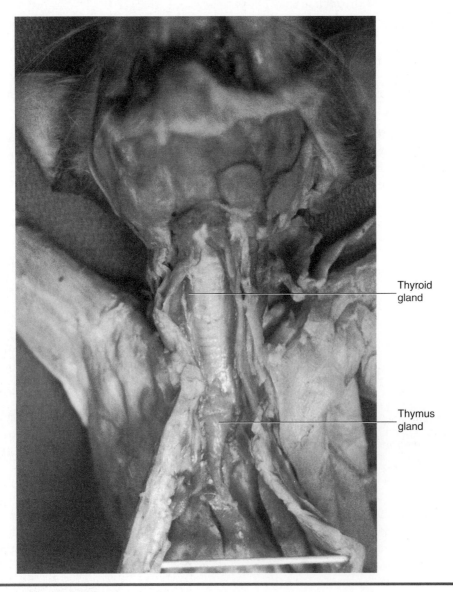

Thyroid gland

Thymus gland

Figure 13.5: Thyroid gland and thymus gland of a cat.

4. The **adrenal glands** are located cranial and slightly medial to each kidney (Figure 13.6). Each gland resembles a small lymph node and has the blue phrenicoabdominal vein traversing across it.

5. The **pituitary gland (hypophysis)** is attached to the **hypothalamus** by a stalk called the **infundibulum.** If you remove the brain of the cat, the pituitary can be seen on the floor of the cranial cavity in the hypophyseal fossa of the skull. In the sheep brain, with the meninges left intact the rounded bulb of the pituitary gland should be easily located (see Figures 13.3 and 13.7).

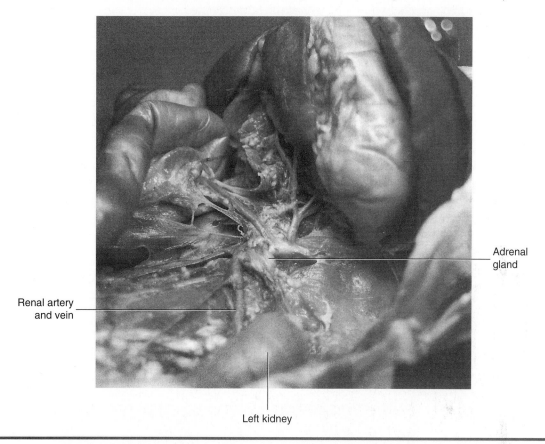

Adrenal gland

Renal artery and vein

Left kidney

Figure 13.6: Left adrenal gland of a cat.

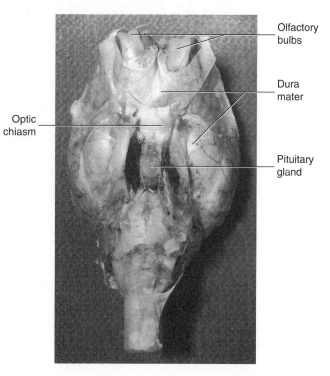

Olfactory bulbs

Dura mater

Optic chiasm

Pituitary gland

Figure 13.7: Ventral view of the sheep brain with dura mater and pituitary gland still attached.

6. The **pineal gland** may be found in either the cat or sheep brain above the corpora quadrigemina of the midbrain (see Figures 13.3 and 13.8).

7. The **pancreas** is located in the abdominal cavity, coursing adjacent to the duodenum and then curving within the mesentery medially underneath the greater omentum (Figure 13.9).

8. Locate the **ovaries** and **testes.** They will be described in more detail in the next chapter.

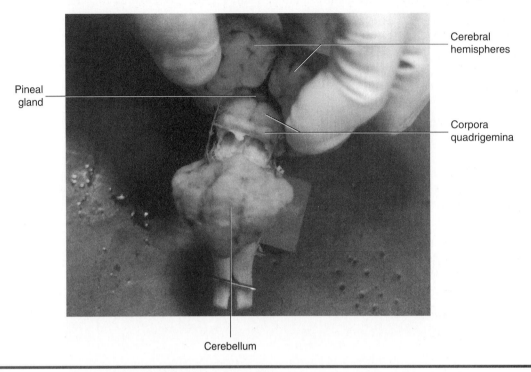

Figure 13.8: Caudal view of the pineal gland of a sheep.

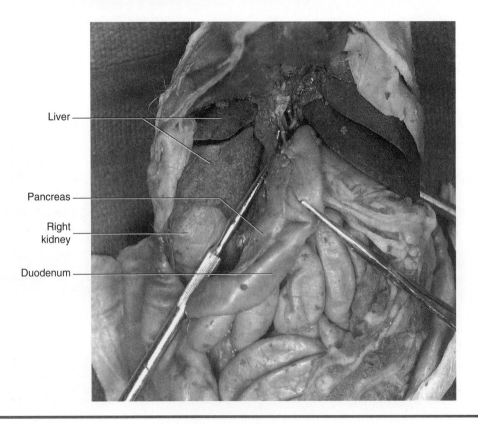

Figure 13.9: Pancreas of a cat.

EXERCISE 13.2
ORAL GLUCOSE TOLERANCE TEST

This test is used to confirm the diagnosis of a deviation from normal glucose metabolism. If glucose is administered orally to a normal monogastric animal, an alteration in glucose concentration is observed over time: In phase 1, glucose absorption into the circulation exceeds excretion. As the glucose level rises, hepatic glucose output is inhibited, and secretion of insulin is stimulated. In 30 min to 1 hour, the peak blood glucose level is reached and then begins to fall. Phase 2 is the decreasing phase of the blood glucose level, and the rate of removal exceeds that of entry. During phase 3, the glucose, after having returned to its original level, continues to fall to a minimal level before returning to the pre-test level.

Procedure

1. Select a live dog for the experiment and have the dog fast overnight. In the morning, take a fasting blood glucose serum level.

2. Administer glucose orally in the amount of 4 g per kilogram of body weight.

3. Take blood samples at 30- to 60-minute intervals for 3 to 4 hours.
   Blood: 30 minutes:_____

   1 hour:_____

   $1^{1}/_{2}$ hours: _____

   2 hours: _____

   $2^{1}/_{2}$ hours: _____

   3 hours: _____

   $3^{1}/_{2}$ hours: _____

   4 hours: _____

4. Make a graph of your findings using the time on the x axis and glucose on the y axis.

   ### QUESTIONS
   1. At what point in the test do you know the pancreas has begun to secrete insulin?

   2. Why do doctors use the term *hypoglycemic effect* when describing the action of insulin?

Discussion

As blood glucose rises, insulin is released. This causes the cells to start absorbing glucose. When this happens, blood glucose begins to drop as glucose is removed from the serum and put into the cells. Because glucose is removed from the blood, this process is described as having an *hypoglycemic effect*. A normal animal's fasting blood glucose is less than 120 mg/dL; this value should not exceed 160 mg/dL at the end of the first hour and should return to normal by the end of the second hour. In the diabetic animal (in whom insulin is not produced or the beta cells are not responsive), blood sugar levels during the test may rise to more than 180 mg/dL and will not return to normal during the testing period. In other words, much of the glucose stays in the blood and does not enter the cells.

CLINICAL SIGNIFICANCE

Hormonal problems can manifest themselves in many ways; some occur as pathology to the integumentary system. Two conditions in dogs cause bilaterally symmetrical alopecia, or hair loss equally on both sides of the body: hypothyroidism and hyperadrenocorticism. When testing for these diseases, you must either assay for the hormones produced by the gland or administer a stimulating hormone, then assay for the level of the hormones that should be produced.

For hypothyroidism, we assay for T_4—usually a free T_4—so results are not affected by changes in serum protein binding. Other tests that can be run are thyrotropin (TSH) stimulating test, thyrotropin-releasing hormone (TRH) stimulation test, and a TSH assay. For hyperadrenocorticism we can do an ACTH response test, a dexamethasone suppression test, or a plasma ACTH concentration. In these tests, failure to produce after a signal from the stimulating hormones indicates a suppressed gland, and overproduction indicates an overactive gland.

Veterinary Vignettes

The day Mr. Opecia and his dog Al entered my hospital was a dark day in the history of veterinary medicine, at least as far as I was concerned. On that day I realized that we may have entered a new era in which pharmaceuticals could be ordered from a catalog and delivered directly to the client's doorstep. In Al's case, my doctor-client/patient relationship had been abandoned and replaced by mail-order gerontology.

"Dr. Cochran, I found the greatest nutritional supplement to extend both human and animal life spans that has ever been developed. Al and I are both on it. It is a synthetic hormone that I order from overseas through a catalog," Mr. Opecia exclaimed, puffing out his chest and smiling broadly while reaching down to pet his dog on the head.

Al was a 10-year-old weimaraner with a pleasant disposition. He was not overly hyper and had a broad head and intelligent eyes. I wish I could have given the same compliment to Mr. Opecia. I stared in utter disbelief at this once proud dog; for Al was now almost completely hairless. I'm talking about having only 12 hairs in the middle of his back above the scapula. I counted them . . . 12. Well, 14, but two came out as I counted them.

Not only was there no hair, but there had to be 10,000 comedos, tiny little blackheads, all over his body. This was the most pathetic-looking dog I had ever seen. I didn't even want to touch him. Veterinarians don't get grossed out that easily, but this was definitely gross. Al looked up at me, and somehow I think he knew what I was thinking because his entire body turned red. I had embarrassed him! Now I really felt bad. Honestly, I have never intentionally or unintentionally embarrassed a dog before; it is not in my nature.

"Mr. Opecia, he has no hair!" I exclaimed.

"Yeah, but I bought him a doggy coat so he won't get cold."

"How do you know the supplement is working?" I asked him.

"Well, I'm on it, and I feel great," he explained.

I glanced quickly at the top of his head. Mr. Opecia was bald as a brick also.

"So why have you brought Al in today? What's wrong with him?" I asked.

"He's got some growths around his anus I thought you should look at" he replied.

There were three small, hard lumps in the tissue surrounding Al's anus. I did a fine-needle aspirate and found the typical large, foamy cells of a perianal adenoma.

"He's got a few benign tumors around his anus that will have to be removed," I told Mr. Opecia. I went on to explain that these tumors were stimulated by testosterone and castrating Al would prevent their recurrence. I also mentioned that the tumors might be due to the hormone pills Al was on.

"No way, Doc. There's no way I'm going to do that to Al. I want to use him as a stud dog; I want to breed him!"

Al looked up at his owner and stared in apparent disbelief. "Who would have me?" he seemed to be wondering.

My concerns about the pills were correct. They must have had some potent hormones in them, because when Al stopped taking them, he regrew most of his hair. He did regrow a few more tumors, but after those were excised, they stopped returning. Eventually, Al went on to become a doggie daddy. All's well that ends well, I suppose. Oh yes, and he lived for another 8 years.

Mr. Opecia also stopped taking the pills. He is still alive and well, and he currently works as a pharmaceutical salesman. He sells injectable growth hormone, if I remember correctly.

SUMMARY

Much of this chapter was devoted to explaining the anatomy and physiology of the hypothalamus, pituitary gland, and the target organs of pituitary hormone production. You learned that the pituitary gland varies in size and shape in the various species of domestic animals. The hormones produced by the hypothalamus were described, and the mechanisms by which they are transported to the pituitary gland were covered. You found out that the releasing hormones of the hypothalamus are transported to the anterior pituitary gland by a small portal blood system; whereas oxytocin and antidiuretic hormone flow down the neurosecretory axons to the posterior pituitary. These pituitary hormones are then carried to their target organs via the bloodstream.

Each hormone causes a specific physiological effect on the target organ. You learned that the production of hormones by the target organ is controlled by two mechanisms: a feedback mechanism to the hypothalamus, which controls output of releasing factors, or neurosecretion. You also learned that two hormones can have opposite physiological effects and homeostasis is maintained by a balance between the two hormones. For example, insulin acts to lower blood glucose, and glucagon acts to raise it; normal blood glucose levels are maintained by balanced production of these two hormones. Finally, you performed an oral glucose tolerance test to demonstrate insulin's action and its effect on serum glucose levels.

CHAPTER 14

THE REPRODUCTIVE SYSTEM

OBJECTIVES:

- recognize the types of uteri in domestic animals
- understand the relationship between the endocrine system and the reproductive system, and how the endocrine hormones affect the production of reproductive hormones and cells
- understand the estrous cycle, and what factors influence its stages
- name and locate the anatomical structures of the reproductive system using diagrams and the dissected cat
- understand the production of semen, its composition, and the organs that contribute to its formation
- name and describe the layers of the uterus
- understand the types of placentation in domestic animals
- know the signs of impending parturition
- identify the various stages of the estrous cycle from vaginal cytology in the dog

MATERIALS:

- cat cadaver, triple injected (order without skin attached)
- Mayo dissecting scissors
- probe
- 1 × 2 thumb forceps or Adson tissue forceps
- #4 scalpel handle with blade
- bone cutting forceps
- rubber gloves
- compound microscope
- stained slides of the various stages of the estrous cycle of the dog
- progesterone and luteinizing hormone assay test kits

Introduction

The purpose of the **reproductive system** is to perpetuate the species. Although it is considered a separate system, the endocrine system's hormones play a vital role in the function of the reproductive system, and the *target organs* include the essential organs of reproduction, or the **gonads,** which are the **testes** and **ovaries.** These produce the **germ cells: spermatozoa** in the male and **ova** in the female. During the period of female receptivity, which is hormone controlled, **copulation,** or mating, occurs; a **haploid sperm** fertilizes the **haploid ova (egg)** and a **diploid zygote** is produced. This union occurs in the **oviducts.**

Cell division in the zygote commences quickly to prepare it for implantation in the uterine lining. It implants and becomes an **embryo.** The embryo continues to develop, and when all of the body parts are recognizable, it is called a **fetus.** The length of **gestation,** or the **gestation period** (the time from fertilization to birth), varies with the species. Generally, the bigger the animal, the longer it takes. The act of giving birth is called **parturition.**

Anatomy of the Female Reproductive Tract

The ovaries are both *endocrine* (hormone producing) and *cytogenic* (cell producing). The **medulla** is the vascular center of the ovary. The **cortex** of the ovary is protected by a layer of dense irregular connective tissue called the **tunica albuginea.** This is surrounded by a layer of **germinal epithelium.** In animals that are seasonally polyestrus (having repetitious heat cycles), in the region of the cortex various stages in the life of the **follicle** may be found, such as developing and atrophying follicles, including the corpus lutea. In the development of the primary follicle, the germinal epithelium invades the *stroma* as a mass of cells; one cell of the mass becomes an **oocyte,** and the others become the follicular cells, which surround the oocyte. As the follicle matures, the oocyte enlarges and the follicular cells multiply. During follicular enlargement, a fluid-filled vesicle forms in the center. At this stage it is called a **vesicular follicle,** or **Graafian follicle** (Figure 14.1). At a certain stage of development the follicle ruptures, releasing the ovum. The follicular cells

multiply to fill in the vesicle, and the structure becomes a **corpus luteum (CL).** It appears as a large, round, yellow-looking structure, hence the common name *"yellow body."* The CL will persist if pregnancy occurs, but degenerate into a **corpus albicans** if pregnancy does not occur. A corpus albicans has the appearance of a white, scar-like area.

The open end of the uterine tube or **oviduct** *(Fallopian tube)* that receives the ejected ova is the **infundibulum,** which is a funnel-like structure adjacent to each ovary. The funnel has a fringed margin called the *fimbria.* The two oviducts, one for each ovary, are convoluted and travel from the ovary to the uterus. Their lining is highly folded and covered by ciliated simple columnar epithelial cells. They contain smooth muscle. Both the cilia and smooth muscle help move the ova and possibly the spermatozoa. As mentioned previously, the oviduct is the site of fertilization.

The **uterus** consists of three parts in domestic animals: the **neck** (where the cervix is located), the **body,** and the **horns.** The uterus is a hollow structure, lined by a mucous membrane called the **endometrium,** and is glandular. The endometrium's thickness and vascularity vary with the stages of the estrous cycle. In the horse and dog, the uterus is lined by simple columnar epithelium, and in the pig and ruminants by stratified columnar epithelium. The uterus has intermediate smooth muscle layers called the **myometrium,** which consist of thick,

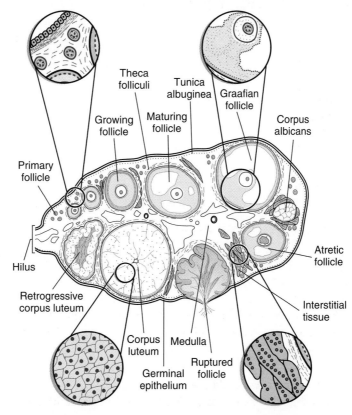

Figure 14.1: Ovary showing all stages of ova development and regression (all species).

inner, circular and thin, outer, longitudinal muscle layers. The outer serous layer, called the **perimetrium,** is visceral peritoneum and has a thin layer of connective tissue to attach it to the *myometrium.* The **cervix** is a heavy, smooth muscle sphincter, and it is tightly closed except during estrus and parturition. There are four types of uteri in domestic animals (Figure 14.2)

1. **bicornuate** uterus: in the pig (sow), dog (bitch), and cat (queen)

2. **bipartite** uterus: in the ox (cow), sheep (ewe), and horse (mare)

3. **simplex** uterus: in the primate

4. **duplex** uterus: in the rabbit (doe)

The **vagina** is located within the *pelvic canal* and extends from the cervix cranially to the vulva caudally. Its mucosa is glandless, and most of the mucous secretions come from the mucous cells of the cervix. It is lined with stratified squamous epithelium. The muscle layer is mainly circular smooth muscle. The *serosa* is only present in the small cranial part that enters the abdominal cavity; the caudal part in the pelvic canal is covered by connective tissue. The **fornix** is an area of depression around the cervix. (The fornix is absent in the pig.)

The **vulva** is that portion of female genitalia extending from the **external urethral orifice** to the exterior opening. It consists of the **vestibule of the vagina** and the **labia.** The vestibule of the vagina is the tubular portion of the vulva from the *external urethral orifice* internally to the labia externally, which structurally form the visible opening. The labia of the vulva consist of well-developed **major labia,** but the *minor labia* are poorly developed or absent.

At the *ventral commissure,* just deep to the external opening, is the **clitoris,** which is the homologue to the male *penis* and consists of similar parts (minus the urethra and muscles). The shaft is called a *corpora clitoridis,* and its size varies with the species. It ranges from 5 cm long in the mare to about 3 cm in the bitch, and it is considerably smaller in the queen. The *glans clitoridis* is the rounded and enlarged free end of the organ that occupies the *fossa clitoridis* at the ventral commissure of the vulva in female animals, and it has the same embryonic origin as the penis in the male.

The *mesentery* supporting the uterus is called the **broad ligament.** The broad ligament has three indistinct divisions:

1. **mesovarium:** supports the ovaries

2. **mesosalpinx:** supports the oviducts (In the bitch this forms a *bursa* around the ovary, called the **ovarian bursa.**)

3. **mesometrium:** supports to the uterus (see Exercise 14.1, #3)

There are two ligaments found within the *broad ligament.* The most cranial is the **suspensory ligament,** which is incorporated into the *mesovarium* and *mesosalpinx* and is attached to the body wall just caudal

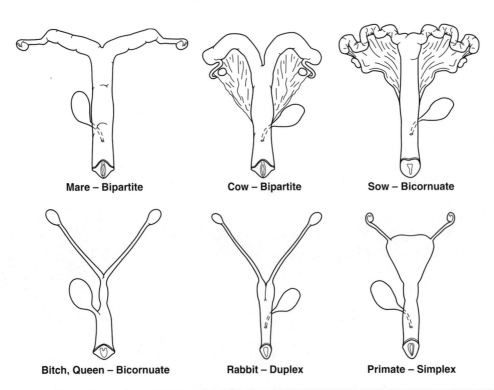

Mare – Bipartite   **Cow – Bipartite**   **Sow – Bicornuate**

Bitch, Queen – Bicornuate   **Rabbit – Duplex**   **Primate – Simplex**

Figure 14.2: Shapes and types of uteri of domestic animals.

to the kidneys. The second is the **round ligament** of the uterus. This is the lateral edge of a sheet of mesentery that projects lateral and perpendicular to the main part of the *mesometrium* (see Exercise 14.1, #3).

Anatomy of the Male Reproductive Tract

Like the ovaries, the **testes** (also known as **testis** [singular] or **testicles**) are both endocrine (hormone producing) and cytogenic (cell producing). They contain a mass of **seminiferous tubules** surrounded by a heavy connective tissue capsule called the **tunica albuginea.** The tubules are separated by fibrous septa called **trabeculae,** which form the support, or *stroma,* for the tubules. The *seminiferous tubules* combine to form the larger **rete testis,** which

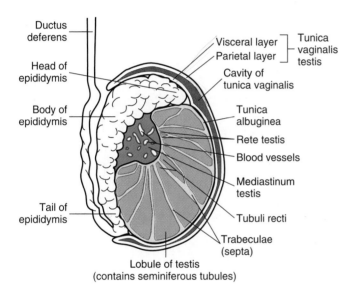

Labels: Ductus deferens; Head of epididymis; Body of epididymis; Tail of epididymis; Lobule of testis (contains seminiferous tubules); Visceral layer; Parietal layer; Tunica vaginalis testis; Cavity of tunica vaginalis; Tunica albuginea; Rete testis; Blood vessels; Mediastinum testis; Tubuli recti; Trabeculae (septa)

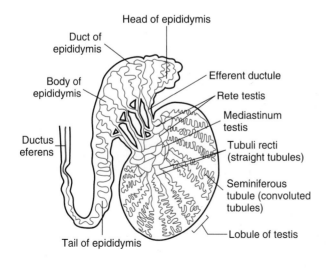

Labels: Head of epididymis; Duct of epididymis; Body of epididymis; Ductus eferens; Tail of epididymis; Efferent ductule; Rete testis; Mediastinum testis; Tubuli recti (straight tubules); Seminiferous tubule (convoluted tubules); Lobule of testis

Figure 14.3: The testis and epididymis (all species).

flow into the **epididymis** (via the **ductuli efferentes,** or *efferent ducts*). The **interstitial cells (Cells of Leydig)** are located in the connective tissue between the seminiferous tubules (Figure 14.3).

The *epididymis* contains three parts (see Figure 14.3): The **head** is found on the dorsal surface of the testicle; the **body** is the middle part; and the **tail** can be located on the ventral surface of the testicle (see Figure 14.3). The tubule inside is the *duct of the epididymis,* which is connected to the **ductus deferens** *(vas deferens).* This is a muscular tube that transports the spermatozoa and aids in their propulsion. The ductus deferens exits the tail of the epididymis, traveling within the spermatic cord, passes through the *inguinal canal,* and enters the urethra just caudal to the neck of the bladder and immediately cranial to the **prostate gland** (if the species has this organ).

The **spermatic cord** consists of the following.

1. **ductus deferens**

2. **testicular artery, vein, nerve,** and **lymphatics**

3. **mesoductus deferens** and **mesochium**

4. **tunica vaginalis visceralis** (visceral vaginal tunic)

The **tunica vaginalis parietalis** (parietal vaginal tunic) and cremaster muscle covers the spermatic cord (Figure 14.4)

The **scrotum** is the sac containing the *testes.* It is attached to the bottom of the testicle by the **scrotal ligament,** which is a remnant of the *gubernaculum* (a ligament that pulled the testis from the abdomen, through the inguinal canal, and into the scrotum during development).

The following are the **accessory sex glands** of male animals. The presence of these glands varies according to species.

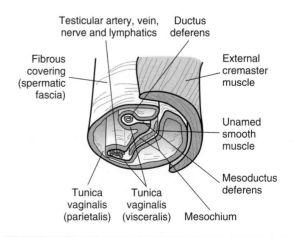

Labels: Testicular artery, vein, nerve and lymphatics; Ductus deferens; Fibrous covering (spermatic fascia); External cremaster muscle; Unamed smooth muscle; Mesoductus deferens; Tunica vaginalis (parietalis); Tunica vaginalis (visceralis); Mesochium

Figure 14.4: The spermatic cord (all species).

prostate: Surrounds the pelvic urethra, just caudal to bladder. The prostate is round in the dog, cat, and horse (these species have two-lobed prostates—a right and left) and more diffuse in other species. The prostate alkalinizes the semen.

vesicular glands (seminal vesicles): These paired glands lie just caudal to the neck of the bladder on both sides. Vesicular glands are hollow, pear-shaped sacs in the stallion and large, lobulated glands in the bull, ram, and boar. These glands are absent in the dog and cat.

ampullae: Glandular enlargements at the terminal end of the *ductus deferens*. They are well developed in the stallion, bull, and ram; small in the dog; and absent in boar and cat. Ampullae contribute to the **seminal fluid.**

bulbourethral glands (Cowper's glands): Small, paired glands on either side of the urethra just cranial to the bony ischial arch but caudal to other glands. Bulbourethral glands are found in all domestic animals, except the dog, and are very large in the boar (Figure 14.5).

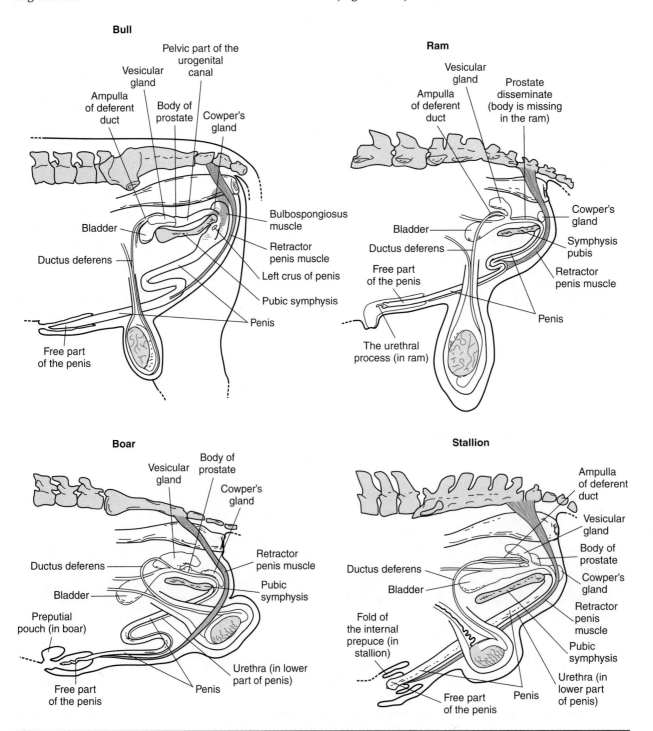

Figure 14.5: Comparative male reproductive systems in large animals.

The **penis** contains the **glans penis,** which is the head of the penis; the **body of the penis,** which is the main portion; and the **crura** (or roots), which anchor it to the ischium. It is very vascular, consisting of the erectile tissue. The two parts of this erectile tissue are the **corpus spongiosum penis** (which surrounds the urethra) and the **corpus cavernosum penis** (which makes up the bulk of the penis' erectile tissue, is located dorsal to the urethra, and may or may not be paired in structure). The **urethral process** extends the urethra out from the glans and is present in the horse, sheep, and goat.

The **bulbus glandis** is the caudal part of the penis in dogs. It swells to lock the male into the female during copulation by becoming entrapped (or tied up) within the pelvic canal. The bull, ram, and boar have a *sigmoid* (S-shaped) flexure of the penis (see Figure 14.5); the penis straightens during the erection process. The **os penis** is a grooved bone in the dog's penis. A **retractor penis muscle** (made of smooth muscle) returns the penis to normal position following the erection.

Physiology of the Female Reproductive System

Follicle stimulating hormone initiates development of the follicle. The *adenohypophysis* starts to liberate **luteinizing hormone** toward the end of follicular development, which causes the follicle to finish developing and rupture, expelling the ovum.

This process is called **ovulation** and is associated with **estrus** (heat) in most mammals. There are two types of ovulation, depending on the species. Animals that are **spontaneous ovulators** do not require copulation for ovulation to occur. Most species fall into this category, including the horse, ox, sheep, pig, goat, dog, primate, mouse, rat, and guinea pig. **Induced ovulators** do require copulation for ovulation to occur. These include the cat, rabbit, ferret, and mink.

After ovulation, the opened follicle becomes a **corpus hemorrhagicum** when the ruptured follicle forms a blood clot in the *antrum* area. It next develops into the **corpus luteum (CL)** and is influenced by *luteinizing hormone* to develop. The follicular cells lining the vesicle multiply to fill this open cavity. The CL persists if pregnancy occurs. Maintenance of the corpus luteum is caused by the hormones produced by the placental membranes, specifically *chorionic gonadotropin,* and *prolactin,* which is produced by the adenohypophysis. The CL persists throughout pregnancy in the cow, doe (goat), sow, bitch, and queen; whereas it terminates in late pregnancy in the mare and ewe. If the animal is not pregnant, but the CL remains, this is considered a pathological condition. The terminal stage of development is the **corpus albicans.** The CL degenerates to form this structure, which is a small amount of scar tissue remaining in the ovary.

There are five stages in the estrous cycle. Figure 14.6 shows the relationship between the stages of estrus and the development of the follicle and corpus luteum.

1. **proestrus:** The anterior pituitary gland (adenohypophysis) produces *follicle stimulating hormone* (FSH) and some *luteinizing*

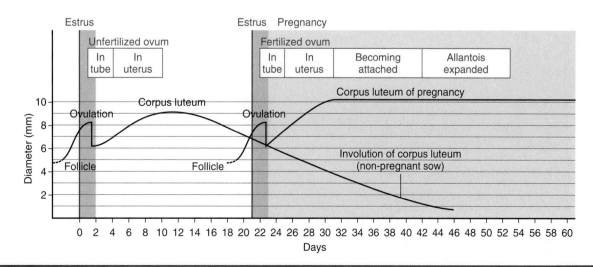

Figure 14.6: The estrous cycle of the sow, showing the relationship between stages of estrus and the period of follicular and corpus luteum (CL) development.

hormone (LH), which cause development of the follicle. The follicle starts producing **estrogen,** which stimulates the build-up of the walls of the vagina and uterus to prepare for copulation and pregnancy.

2. **estrus:** This is the period of female receptivity. Ovulation usually occurs during the end of this period in spontaneous ovulators. FSH levels are decreasing and LH levels are increasing.

3. **metestrus:** In this post-ovulatory phase, the CL starts secreting **progesterone,** which prevents further development of other follicles. The CL also is responsible for maintaining the uterine lining to support the fetus during pregnancy.

4. **diestrus:** This is the short period of quiescence in seasonally polyestrus animals. First the CL degenerates, then one or more follicles develops (depending on whether it is a litter-bearing animal) and estrus repeats.

5. **anestrus:** The CL degenerates during this period of quiescence between breeding seasons.

Oxytocin, as mentioned in the previous chapter, is produced by the posterior pituitary gland *(neurohypophysis).* It stimulates milk let-down in the mother, and in the presence of estrogen, oxytocin stimulates uterine contractions during *parturition.* It also stimulates oviducts to help move spermatozoa.

Prolactin from the anterior pituitary gland helps maintain the CL during pregnancy, stimulates mammary glands to fill up with milk at parturition, and stimulates replenishment of milk (production of prolactin is stimulated via suckling).

The **placenta** is the membranous structure that obtains nutrients and oxygen from the mother and delivers them to the fetus. The **chorion** is the outermost membrane and is in contact with the maternal uterus. The **amnion** is the innermost membrane closest to the fetus. The **amnionic sac** is the sac in which the fetus is located, and it is also called the *second water bag.* The **allantoic sac** is the space between the amnion and the chorion, which is lined by the **allantois** (or **allantoic membranes**) and called the *first water bag* (Figure 14.7).

The attachment of the *placenta* to the lining of the uterus varies according to species (Figure 14.8). The following are the types of **placentation.**

1. **diffuse:** *epithelial;* found in the pig and horse

2. **cotyledonary:** *epitheliochorial;* found in the sheep, goat, and cow

3. **zonary:** *endotheliochorial;* found in the dog and cat

4. **discoid:** *hemochorial;* found in primates

Figure 14.8 illustrated the types of placentation of various species, in addition the structure of the lining of the areas of placentation also differ. Figure 14.9 illustrates the types of tissues present of both the

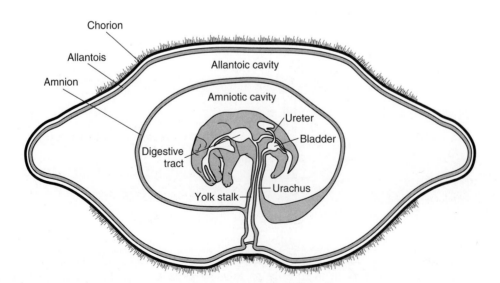

Figure 14.7: The fetal membranes in the dog.

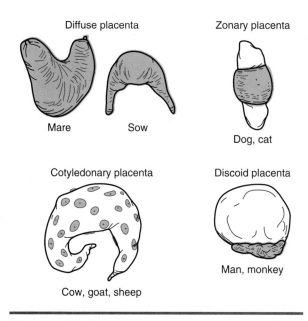

Figure 14.8: Types of placentation in various species.

maternal and fetal tissue of the placenta. The information above correlates to Figure 14.9.

Signs that *parturition* is approaching include an enlarged abdomen, enlarged mammary glands, mammary glands that fill up with colostrum, and clear mucus from the vulva. In large animals there will be a relaxed abdominal wall, sinking in of the flanks, dropping of the belly, and sinking of the rump on both sides of the tail head. In dogs, the first stage noted is the nesting stage; the dog is restless, tears up bedding, scratches, and has no appetite. The dog's temperature will decrease at least 2°F within 24 hours of the onset of labor. In dogs, this is called the *first definitive sign* of impending parturition.

Labor in dogs is characterized by intermittent straining that becomes more pronounced and continuous as delivery approaches. The fetal membranes appear as a sac filled with fluid, and the bitch will usually rupture the sac with her teeth (sometimes the sac is not seen). The puppy generally arrives within 30 minutes after the fetal sac appears. The interval between pups may vary greatly; however, puppies usually arrive in pairs, with the mother resting 1 to 2 hours between deliveries. The bitch normally licks the puppy as soon as it is born, and an experienced mother will sever the umbilical cord with her teeth. The placental membranes are delivered shortly after each puppy is born.

Physiology of the Male Reproductive System

The main hormone produced by the testes, and specifically by the interstital cells (Cells of Leydig), is **testosterone,** which is responsible for secondary sex characteristics and sex drive. Testosterone is an *androgen,* or *anabolic steroid.* Its production is stimulated by luteinizing hormone (from the anterior pituitary gland); however, in the male, the hormone is

Placental Microscopic Structure

Type	Maternal tissue	Fetal tissue	Gross form	Species
Epithelial			Diffuse	Pig, horse
Epitheliochorial			Cotyledonary or multiplex	Sheep, goat, cow
Endotheliochorial			Zonary	Dog, cat
Hemochorial			Discoid	Primates

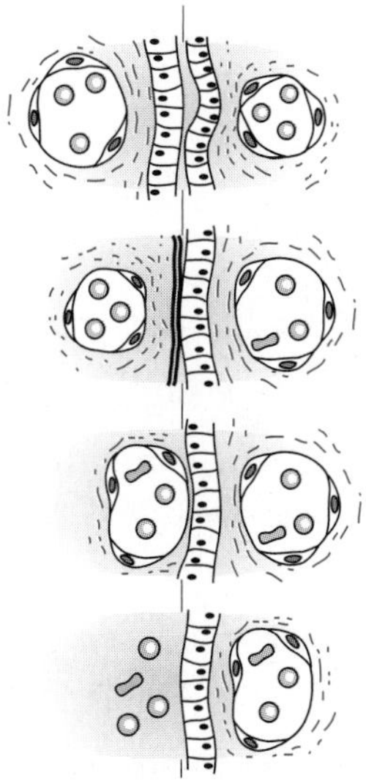

Figure 14.9: Layers of the placentation in various species.

called *interstitial cell stimulating hormone (ICSH)*. In males, *follicle stimulating hormone* is the anterior pituitary hormone that stimulates **spermatogenesis,** or the creation of spermatozoa.

The sperm has the following parts.

1. **head:** This oval structure contains the nucleus where the *haploid* number of chromosomes are located (half of those needed for an animal of the species). The head also has a cap called the **acrosome,** which contains enzymes to permit penetration into the ovum.

2. **midpiece:** This is the power plant of the sperm. Within the midpiece are numerous mitochondria that carry out the metabolism that provides adenosine triphosphate (ATP) for the sperm's locomotion.

3. **tail:** This section consists of a flagellum for propulsion (Figure 14.10).

The **seminal fluid** is produced by the *accessory sex organs,* and it is the medium for survival and activation of the sperm. The *prostatic secretion* has a specific action (in addition to those mentioned previously): It also alkalinizes the vaginal environment, preventing spermatozoa death.

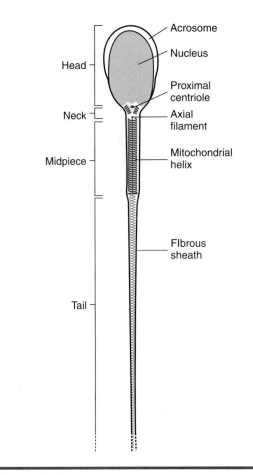

Figure 14.10: Parts of a spermatozoa.

DISSECTION OF THE REPRODUCTIVE SYSTEM

Although you may dissect the reproductive system of only one sex, you will need to know the parts of both sexes. After dissection, trade specimens with others in the lab to study the reproductive structures in a cadaver of the opposite sex. Follow the steps in the appropriate section below, depending on the sex of your cat cadaver.

In the exercises in this chapter the important structures are listed in colored bold print. If a structure is mentioned prior to its dissection it will be italicized. Structures discussed prior to dissection may also be in bold print for special emphasis.

Procedure

The Female Reproductive System

1. Locate the **ovaries** (Figure 14.11). These are small, oval structures just caudal to the caudal pole of the kidneys. They are enveloped within the **mesosalpinx,** which forms a bursa around the each ovary called the **ovarian bursa.**

2. The **uterine tubes** or **oviducts** (Fallopian tubes) are small, highly convoluted tubes within the mesosalpinx. The dilated end is called the **infundibulum** of the oviduct, but it is difficult to see. The oviduct's opening of is called the **ostium** of the tube and is surrounded by the **fimbria.**

3. Trace the oviducts to their termination at the **uterine horns,** or **cornua.** The descriptive name for this type of uterus is **bicornuate,** or two-horned. The ova are passed through the oviducts to the

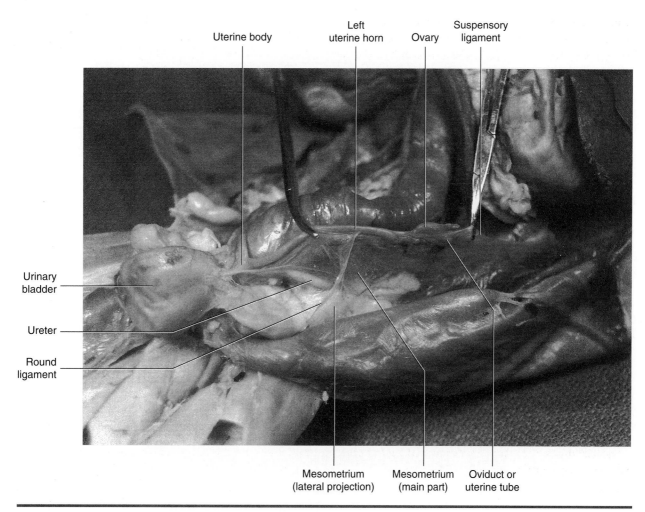

Uterine body | Left uterine horn | Ovary | Suspensory ligament

Urinary bladder

Ureter

Round ligament

Mesometrium (lateral projection) | Mesometrium (main part) | Oviduct or uterine tube

Figure 14.11: The broad ligament in the uterus of a cat.

uterine horns where they implant and develop. The fetuses tend to be equally spaced throughout the two horns. The **broad ligament** supports the uterine horns and **body of the uterus.** The broad ligament consists of three parts: the **mesovarium, mesosalpinx,** and **mesometrium,** but these parts are indistinguishable. The broad ligament is T-shaped; its lateral portion is a sheet of mesentery that is perpendicular to the main part of the mesometrium and anchors it to the dorsolateral wall (see Figure 14.11). The thickened, lateral edge of this lateral portion is called the **round ligament** of the uterus. The **suspensory ligament** is the thickened part of the mesovarium and mesosalpinx that connects each ovary to the dorsal body wall, just caudal to each kidney. The suspensory ligament must be broken by the veterinarian during an ovariohysterectomy so the ovary and ovarian artery can be elevated to the exterior and ligated.

4. The two horns come together in the middle to form the **body of the uterus,** which lies dorsal to the bladder and urethra (Figure 14.12).

5. To dissect the remainder of the female reproductive system, it is necessary to cut through the pubic bone to expose the pelvic cavity. Using bone cutters, cut through the pelvic muscles and along the *pubic symphysis* in the ventral midline. Be careful not to cut the *urethra,* which lies immediately beneath the pubic bone. Open the area further by abducting the thighs to clear the pelvic canal.

6. Locate the **urethra,** the tube carrying urine from the urinary bladder. Dorsal to the urethra, find the **vagina,** the tube leading from the body of the uterus. Note that about ²/₃ of the length of the vagina, toward the uterus, is slightly thicker than the rest. This is the **cervix** (see Figure 14.12).

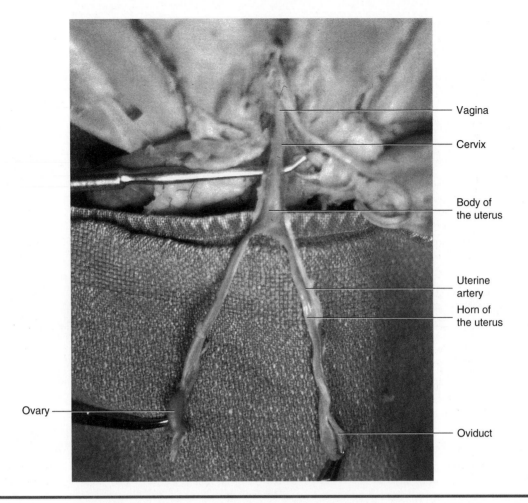

Figure 14.12: Parts of the female reproductive tract. The uterus has been everted and laid out on a surgical towel.

7. The vagina and urethra open into a common passage, called the **vestibule of the vagina,** which is the space from the external urethral orifice to the external or **urogenital opening** or (genital opening).

8. Examine the external part of the vulva, which is the predominant structure of the external genitalia. The major **labia majora** are the lips surrounding the urogenital opening. On the ventral wall of the vestibule, just deep to the opening, is the *fossa clitoridis*, which is difficult to identify.

9. Locate the **rectum,** the continuation of the descending colon, which should be found dorsal to the vagina.

The Male Reproductive System

1. Locate the **scrotum,** the sac ventral to the anus in cats. Early in fetal development, the testes are located caudal to the kidneys; they migrate before birth through the inguinal canal, on either side, and into the scrotum either just before or soon after birth. They are guided by a fibrous cord, which is called the *gubernaculum*. The remnant of this cord is the connective tissue attaching the testis to the inside of the scrotum and is called the **scrotal ligament.** The scrotum is covered with skin on the outside and lined with peritoneum on the inside. It is divided into two compartments by a *median septum*.

2. Open the scrotum to view and examine the testes. They are covered with **tunica vaginalis parietalis** (parietal vaginal tunic), an extension of the parietal peritoneum. Cut open this covering to view the testis

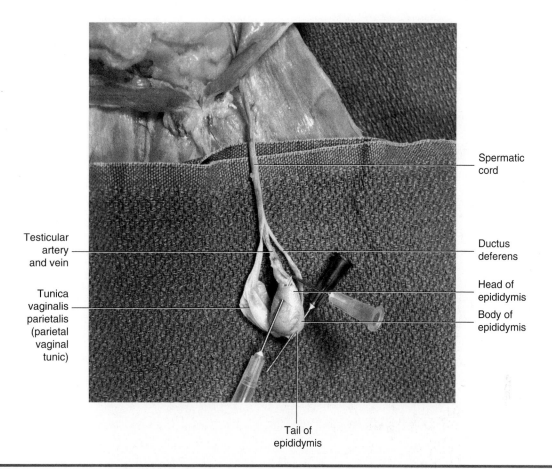

Spermatic cord

Testicular artery and vein

Ductus deferens

Tunica vaginalis parietalis (parietal vaginal tunic)

Head of epididymis

Body of epididymis

Tail of epididymis

Figure 14.13: Testis and component parts of a cat. The lower needle is under the head of the epididymis; the upper needle (dark hubbed) is under the tail.

within. The shiny surface covering the testis, epididymis, and the vessels is the **tunica vaginalis visceralis** (visceral vaginal tunic), an extension of the visceral peritoneum (Figures 14.13 and 14.14).

3. The **epididymis** is found tightly attached to the lateral aspect of each testis. The **head** of the epididymis is found at the cranial or dorsal aspect of the testis, the **body** on the side, and the **tail** of the epididymis at the caudal or ventral end of the testis. The natural position of the testicle varies with species; the testicle is oriented horizontally in the dog, cat, boar, and stallion and vertically in the bull and ram (see Figure 14.5). The epididymis is a long, coiled tube that receives sperm from the testis, stores it, and delivers it to the *ductus deferens* during ejaculation.

4. The **ductus deferens** carries the sperm from the epididymis through the **inguinal canal** and empties it into the **urethra.** As the ductus deferens leaves the testis it travels under the skin cranial to the scrotum (one on each side), through the **inguinal canal** and into the abdominal cavity. It is accompanied by the **testicular artery, vein,** and **nerve** as it courses to and through the inguinal canal. Collectively, these structures, covered with the tunica vaginalis visceralis and their serous membranous attachments, the mesoductus deferens and mesorchium, are called the **spermatic cord.** Trace the path of the ductus deferens from the testis through the inguinal canal; then follow it as it curves (leaving the artery and vein) and passes into the pelvic canal where it will join to the urethra (see Figure 14.14).

5. Locate the **penis,** ventral to the scrotum. The **glans penis** is the enlargement at the distal end of the organ. In adult, intact male cats, it possesses numerous rows of proximally directed cornified papillae known as *penile spines.* After copulation, these spines scrape the walls of the queen's vagina upon withdrawal of the penis, an act which is in part responsible for inducing ovulation. The penile spines regress almost completely after castration. The penis' opening to the outside is called the **external urethral orifice.** The **prepuce** is the skin that covers the glans penis.

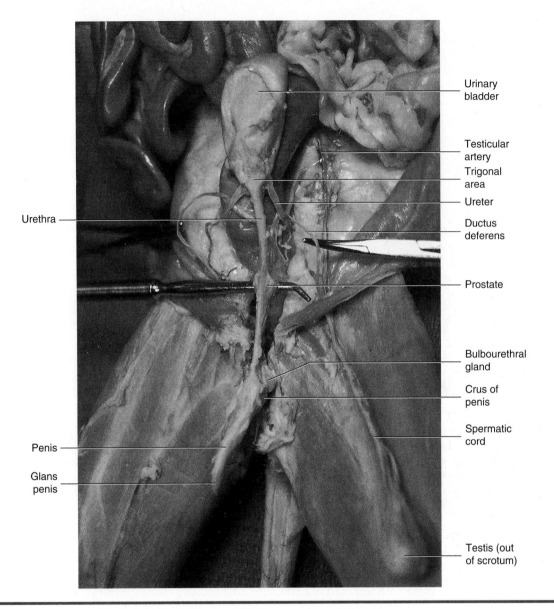

Urinary
bladder

Testicular
artery

Trigonal
area

Ureter

Ductus
deferens

Prostate

Bulbourethral
gland

Crus of
penis

Spermatic
cord

Testis (out
of scrotum)

Urethra

Penis

Glans
penis

Figure 14.14: Ventral view of the male reproductive system of the cat.

6. To dissect the rest of the male reproductive system, the pelvic cavity must be exposed. Cut through the pelvic muscles and the pubic symphysis at the ventral midline. Do not cut the urethra, which lies immediately beneath the pubis. Open the area further by abducting the thighs to expose this pelvic canal.

7. It is now possible to view the entire urethral pathway from the urinary bladder caudally. Trace this to the penis. From the distal penis working cranially, free the penis from the connective tissue and cut the **crura** of the penis where it attaches to the *ischium* on both sides. Using your finger, work cranially and free the rest of the urethra from the underlying colon. Any remaining connective tissue can be cut away, but make sure it is only connective tissue. The penis and urethra should now be free and can be rotated so their dorsal aspects can be viewed.

Immediately cranial to the crura of the penis are the paired **bulbourethral glands** (Figure 14.15). Carefully dissect the *ductus deferens* until it reaches their point of attachment to the dorsolateral surface of the urethra. Locate the **prostate gland** just caudal to the attachment of the ductus deferens. The urethra can be divided into three parts: the *prostatic urethra*, surrounded by the prostate gland; the *membranous urethra*, between the prostate gland and the penis; and the *penile urethra*, passing through the penis.

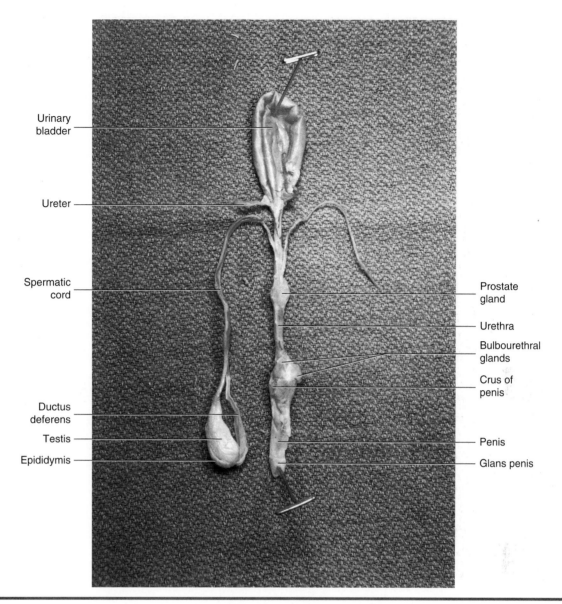

Urinary bladder

Ureter

Spermatic cord

Ductus deferens

Testis

Epididymis

Prostate gland

Urethra

Bulbourethral glands

Crus of penis

Penis

Glans penis

Figure 14.15: Dorsal view of the male reproductive tract of the cat.

EXERCISE 14.2

DIAGNOSIS OF BREEDING TIME IN THE FEMALE DOG

The estrous cycle of the bitch typically lasts for several weeks; however, the fertile period is short: only 48 to 72 hours. A female dog may exhibit receptive behavior, which is controlled primarily by changes in levels of the hormone estrogen, but this does not always coincide exactly with ovulation. Therefore, a combination of tests can be used to more accurately predict the best time to breed or inseminate a female dog.

The vaginal epithelium of the **anestrus** bitch is made up of noncornified stratified squamous epithelial cells. Under the influence of estrogen, the lining cells begin to cornify (take on keratin) in preparation for copulation. **Proestrus** is the time when the uterine lining, the *endometrium,* is developing for implantation of the embryo. The bitch will typically bleed during this stage because of red blood cell (RBC) *diapedesis,* or migration, into the uterine lumen. Vaginal cytology will show many uncornified squamous epithelial cells and many RBCs. Proestrus can last anywhere from four to 20 days. During proestrus, the epithelial cells begin to cornify and gradually change from a rounded shape to a more polygonal shape with rough, angular borders, and they sometimes appear folded (Figure 14.16). The cells also are progressively more densely stained. As the cell becomes more cornified, the nucleus begins to shrink, becoming *pyknotic*.

When there are no more RBCs visible (occasionally RBCs will persist into estrus) and all the cells are cornified, **estrus** has commenced. Estrus can last anywhere from four to 13 days. Ideal time to breed is when the cells' nuclei are completely pyknotic, but this is a subjective determination. **Metestrus** is characterized by the presence of white blood cells (WBCs) in the smear and new non-cornified squamous epithelial cells. Many of the old cornified cells also are still present. Ovulation generally occurs 24 to 36 hours before the appearance of

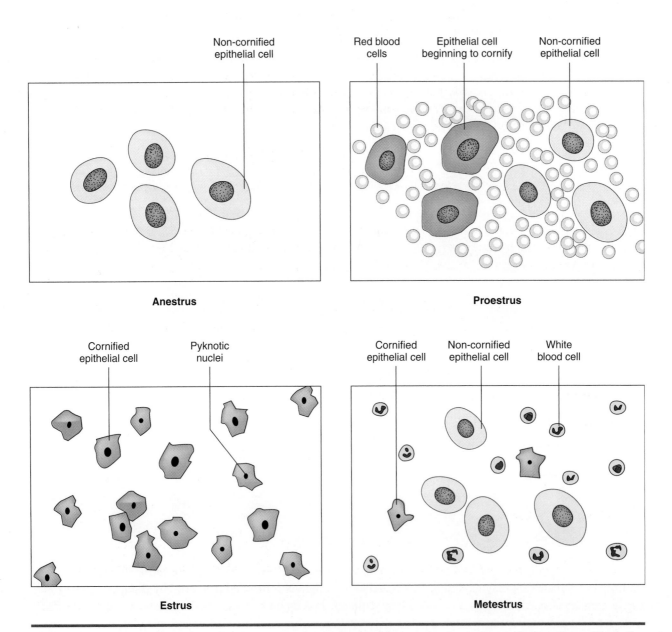

Figure 14.16: The stages of canine estrus as demonstrated by vaginal cytology. *A,* Anestrus: Note the large uncornified epithelial cells predominate. (Diestrus would look similar, but a mixture of non-cornified and early cornified epithelial cells would be present.) *B,* Proestrus: Note the appearance of many RBCs and epithelial cells beginning cornification. *C,* Estrus: All cells are cornified, and nuclei are becoming increasingly pyknotic. RBCs are no longer present. *D,* Metestrus: Note the reemergence of non-cornified epithelial cells. Some cornified epithelial cells are still present. White blood cells are numerous.

WBCs in the vaginal smear. Metestrus represents the *luteal phase* and begins when the bitch no longer accepts the male. If pregnancy does not occur, metestrus may last up to two weeks before anestrus or **diestrus** occurs. Diestrus is a relatively short phase of quiescence between estrus cycles in seasonally polyestrus animals.

Because of the wide variation in the length of the stages of the estrous cycle and the difficulty in predicting the exact time for ovulation, we can use serological testing to enhance the accuracy of our prediction. The following is one method of determining optimal breeding time.

1. On the first day of either noticeable blood spotting or vulvar swelling, perform vaginal cytology. Draw blood to perform a baseline progesterone assay. Continue performing vaginal cytology every other day until the vaginal epithelium is 50% cornified.

2. At this stage, start drawing blood daily to test the progesterone level. Prior to LH release, the level will remain low, generally between 0 and 1.0 ng/ml.

3. When progesterone rises to a range of 1.5 to 2.0 ng/ml, initiate daily testing of serum LH (ICG Status-LHTM Synbiotic Corporation, San Diego, CA). The progesterone will continue to rise, but it is the initial rise in progesterone that signals the ending LH surge. However, there is a variation from dog to dog in the time between the first rise in progesterone and the LH surge. Therefore, continue to draw blood and test for LH rise on a daily basis.

4. Ovulation will occur 2 days after the LH surge. The ova then require an additional 2 to 3 days to mature enough for implantation in the endometrium. Thus, the fertile period will fall between 4 and 7 days after the LH surge, with the most-fertile days being 5 and 6 (the day of the LH surge is counted as day 0).

5. The traditional method of counting the gestation period is 63 days from breeding. Using the LH surge offers a second method. Research has demonstrated that gestation is 65 days in length ($+/-1$ day) from the LH surge. This is true regardless of the day the bitch is bred.

Procedure

1. Using prepared slides, identify all four stages of the estrous cycle: anestrus, proestrus, estrus, and metestrus.

2. Draw blood on an anestrus dog; perform a progesterone assay.

3. Draw blood on a dog in heat (if available); perform an LH assay.

4. Record the test results.

   Progesterone: _____

   LH: _____

QUESTIONS

1. Will doing these tests one time enable you to predict the optimum time for breeding?

2. If a dog becomes pregnant and progesterone is continually assayed, what will happen to the level over time?

Discussion

Because we are looking for progesterone and LH rise, it usually takes two or more samples to observe this. However, it is possible to draw a sample on the day of LH rise and see an increased LH compared to the predicted normal value in dogs. We could tentatively predict ovulation time by this rise in the LH level.

If a dog becomes pregnant, the corpus luteum will be maintained throughout pregnancy, and progesterone will be secreted throughout gestation. Therefore, the level of progesterone assayed will rise and level off until near the time of parturition.

CLINICAL SIGNIFICANCE

Cryptorchidism is a condition in which one or both testes do not descend into the scrotal pouch during development. It is an inherited trait and is thought to be due to a single autosomal recessive gene, although recent research indicates more than one gene may be involved. Usually, testicle descent into the scrotum is expected to be complete by two months of age. Absence of palpable testes at two months is presumptive evidence of cryptorchidism. However, individual dogs have been known to have delayed descent of up to four months. Cryptorchidism is reported in all breeds, but the toy poodle, Pomeranian, Yorkshire terrier, and other toy and miniature breeds have a higher incidence. Unilateral cryptorchidism is more common than bilateral (75% of cases are unilateral and 25% bilateral). The right testis is twice as often retained as the left. The prevalence is the general dog population is 1.2% and 1.7% in cats.

The cryptorchid testes are usually retained just caudal to the inguinal ring, within the inguinal ring, or inside the abdominal cavity. If the testis is caudal to or within the inguinal ring, it often can be palpated. Ultrasound is the other method of determining the position of the retained testis.

The treatment is castration of both testes. Orchiopexy, the surgical placement of a retained testis into the scrotum, is considered unethical because it masks an inherited condition. Castration is strongly recommended to owners, because of the increased incidence of tumor formation in the retained testis. Owners are encouraged to castrate before the animal is four years old. Fifty-three percent of sertoli cell tumors and 36% of seminomas are found in cryptorchid testes. The risk of testicular neoplasia is approximately ten times greater in the cryptorchid testis than in normal testes.

Veterinary Vignettes

When the owner of a practice considers himself a horse doctor, and you are the newest vet on the block, or in this case in the hospital, you get all the cow, sheep, pig, and goat work. I pulled so many calves and did so many Cesarian sections on cows that I became known as C-Section Man. In a moment of sheer lunacy I even considered sewing a big, red C on my scrub shirt. A significant proportion of my work was done on just one farm whose owner had decided that maximum weight at weaning was better for the bottom line. Of course, any extra profit he made was being eaten up in veterinary bills. But being a pseudo-scientist and following a strict protocol for breeding, he was determined to make this work. He was doing a breeding program called a *three breed reciprocal cross:* He was trying to maximize weight at birth, milk production, and weaning weight.

I studied this during my Master's degree work in genetics using Hereford, Angus, and milking Shorthorn cattle. He was using Simmental, Hereford, and Angus. It was the Simmental genes he was counting on to give the increase in size, but the small Angus cows and young Hereford heifers did not have the pelvic size to expel a 110-lb calf. The results were disastrous.

"Dr. C, Mr. Barkley is on the phone; another one of his cows is having trouble giving birth he says." I was putting in the last suture on a spayed animal as the receptionist told me this.

"Ask him to wait a few minutes and I'll talk to him," I told her.

Within 10 minutes, I was in my truck heading for Barkley Breeding Farm with all possible haste. I have always thought there should be a special green flashing light veterinarians can put on their trucks in an emergency situation. This would not allow the veterinarian to drive in excess of the speed limit, but it would alert people to a doctor on his way to an emergency. I guess we could let people doctors use it also.

When I arrived, they had the cow in a large stall in the barn (at least at this place I didn't have to play cowboy and rope it first). Sure enough, it was a small Hereford cow, bred to a Simmental bull. I expected to see the front legs protruding from the vulva; then I could reach inside, find the head, and pull it out. In this case, there was nothing presenting at all.

"It was coming hind legs first, so we pushed it back in and tried to turn it," Mr. Barkley told me while looking toward the ceiling of the barn.

Well, that explained why nothing was showing, I thought. I put on a plastic sleeve, lubricated it well, and pushed my arm through the vagina and up into the uterus. Being tall, thin, and lanky, I had the perfect build for this type of work. First, I located the calf's hind legs and tail, then worked my way forward along the back until I found the neck and head. Then I felt it! What the heck?, I thought (or words to that effect). Why is the cow's spleen next to the calf's head? I quickly realized that they had pushed the calf so hard trying to turn it that it had ripped through the uterus and was now partly in the abdominal cavity. I couldn't tell if it was dead or alive.

"I'm going to have to do a Cesarian section. The uterine wall is torn open, and the calf's head and neck are in the cow's abdomen. I can't tell if it is still alive." I told Mr. Barkley. The look on his face told me he knew he had screwed up.

If a calf were dead, I would use a fetatome to cut it up and take it out in pieces. But in this case, it wouldn't matter. I was still going to have to open up the cow to repair the uterine wall. When I finally got the cow opened, I saw that the entire side of the uterus was torn, from its cranial tip into the pelvic canal. Most of the calf was in the abdomen and it was dead—it had been for some time. I had to struggle to remove it because of its size and muscular stiffness. I set it on the ground behind me.

"Push on its chest and see if you can get it breathing," Mr. Barkley shouted.

"Mind you, this is only opinion," I said. "But it's been my experience that when rigor mortis begins to set in, we rarely are able to bring them back. Sorry, Mr. Barkley, this one's dead."

I double-sutured the uterus and inverted the second layer, the serosa, inward to ensure a good seal and prevent adhesions.

"If I were you, Mr. Barkley, because of the long tear in the uterus, I would cull the cow," I told him. "Even though I did my best in suturing the uterus closed, I don't know if it would support another pregnancy."

I left, thinking he would abide by my advice. A year later, I got a phone call from none other than Mr. Barkley. "Hey Doc, you know that cow you told me to cull? Well, you were wrong. It just had the nicest looking calf you'll ever see! Thanks, Doc!"

Sometimes it's nice to be wrong.

SUMMARY

This chapter covered the anatomy of the female and male reproductive systems through examination of diagrams and by dissection of a cat. You learned the structural anatomy of both the testes and the ovaries. The process of ovum and sperm development was included. Emphasis was placed on the comparative anatomy of the genital systems of the various male domestic animals. You learned that there were different types of uteri, each with a specific type of placentation and vascular attachment to the chorion. The act of parturition was discussed in detail in the dog. In the discussion and exercise you learned how to identify the various stages of the estrous cycle through the use of vaginal cytology and a serum chemistry method to obtain more accurate results with regard to time of breeding. The anatomy of the male and its different accessory sex organs were also included.

CHAPTER 15

THE NERVOUS SYSTEM

OBJECTIVES:

- describe the anatomy of a neuron and a nerve using diagrams or a prepared slide
- understand the reflex arc from the sensory afferent nerve fibers, through the spinal cord, to the efferent nerve fibers
- understand the mechanism of a nerve impulse
- locate and name the major anatomical structures of the sheep's brain and meninges
- explain the purpose and function of each part of the brain
- locate and name the major nerves of the peripheral nervous system
- understand the transmission of an impulse across a synapse
- understand and describe the flow of cerebrospinal fluid from its origin
- name the cranial nerves and describe their function

MATERIALS:

- compound microscope
- prepared slide of a giant multipolar neuron and cross-section of mammal spinal cord
- cat cadaver, triple injected (order without skin attached)
- sheep brains
- Mayo dissecting scissors
- probe
- 1 × 2 thumb forceps or Adson tissue forceps
- #4 scalpel handle with blade
- bone cutting forceps
- rubber gloves
- percussion hammer
- live dog or cat

Introduction

The **nervous system** is the master integrating and coordinating unit of the body. It is continuously monitoring sensory input from internal systems and from the external environment, then processing this information. All thoughts, actions, and perceived sensations are a reflection of the nervous system's activity.

The nervous system can be divided into two parts: the **central nervous system (CNS)** and the **peripheral nervous system (PNS).** The CNS consists of the **brain** and **spinal cord,** whereas the PNS consists of the

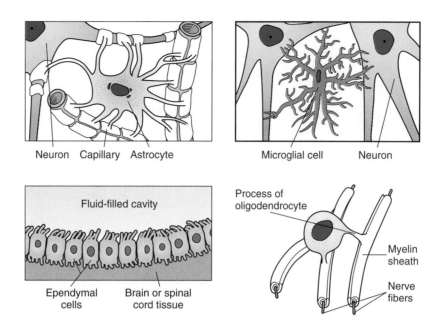

Figure 15.1: Supporting cells of the nervous system (all species).

spinal nerves from the spinal cord and the **cranial nerves** from the brain. The PNS can also be divided functionally into the **autonomic nervous system,** which acts automatically, and the **somatic nervous system,** which involves controlled muscular action.

Nervous tissue has just two different classes of cells: the **neurons** and their **supporting cells.** The supporting cells of the CNS are usually referred to as **neuroglia,** or **glial cells.** These cells include the **astrocytes, oligodendrocytes, microglia,** and **ependymal cells** (Figure 15.1). They hold the neurons and their processes in place and have been described as nerve glue, which is the meaning of the word *neuroglia.* Supporting cells of the PNS are **Schwann cells** and **satellite cells.** They serve the neuron by acting as phagocytes and by bracing,

protecting, and myelinating the tiny, delicate neuron's fibers.

In addition, these support cells play a role in the exchange between local capillaries and a neuron to control the surrounding chemical environment. Although the neuroglia resemble the neurons (because of their fibrous cellular extensions), they cannot generate or transmit nerve impulses. Their pathological importance in veterinary medicine is that, on rare occasions, they may be the source of tumor development.

In the exercises in this chapter the important structures are listed in colored bold print. If a structure is mentioned prior to its dissection it is italicized. Structures discussed before dissection may also be in bold print for special emphasis.

EXERCISE 15.1

THE NEURON

The **neuron** is the basic structural unit of nervous tissue (Figure 15.2). Neurons transmit messages as nerve impulses from one part of the body to another. Although neurons in different parts of the body differ structurally, they have a number of commonly identifiable features, which are discussed in the following text.

The **cell body** is where the nucleus is located, and the slender *processes,* or fibers, extend from it. The cell bodies can be found within the CNS or outside of it. **Ganglia** (sing. ganglion) are small masses of nervous tissue, containing primarily cell bodies of a neuron, that are located outside the brain and spinal cord. Collections of nerve cell bodies within the CNS represent the gray matter.

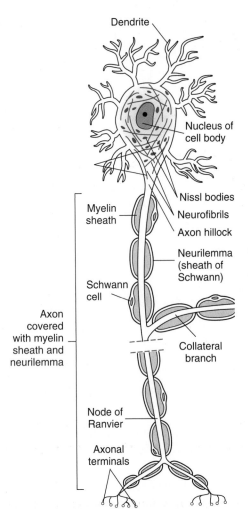

Dendrite

Nucleus of
cell body

Nissl bodies

Neurofibrils

Axon hillock

Myelin
sheath

Neurilemma
(sheath of
Schwann)

Schwann
cell

Axon
covered
with myelin
sheath and
neurilemma

Collateral
branch

Node of
Ranvier

Axonal
terminals

Figure 15.2: The structure of a neuron (all species).

Within the cell body of a neuron, **neurofibrils** can be found in the cytoplasm. These are the cytoskeletal elements of the neuron that help with support and intracellular transport. Also visible are **Nissl (chromatophilic) bodies,** which are elaborate types of endoplasmic reticulum involved in the metabolic activities of the cell.

Extending from the cell body are the neuron's processes, or fibers, which can take a variety of forms. **Dendrites** are the *receptive region* of the neuron, as they bear receptors for neurotransmitter substances released from adjoining axons and conduct the nerve impulse *toward* the cell body. Neurons have many dendrites.

Axons are another type of process. Generally axons only carry impulses in one direction: *away* from the cell body. However, we now know that some axons transmit impulses both to and from the cell body. Therefore, axons are now defined as *nerve impulse generators and transmitters.*

Neurons are classified into **unipolar, bipolar,** and **multipolar** (Figure 15.3). *Unipolar neurons* are sensory neurons that originate in the embryo as *bipolar neurons.* During development, the axon and dendrite fuse to form a single process that divides into two branches a short distance from the cell body. Both branches function together as an axon. The branch that extends to the periphery at its distal tip has unmyelinated dendrites attached. The other branch extends into the CNS and synapses with other neurons.

Bipolar neurons have one main dendrite and one axon. They are found in the retina of the eye, in the inner ear, and in the olfactory area of the brain. *Multipolar neurons* usually have several dendrites and one axon. Most neurons of the brain and spinal cord are of this type. Even though multipolar neurons have only one axon, they may branch into *collaterals.* Note that the term **nerve fiber** is a synonym for axon and is thus quite specific.

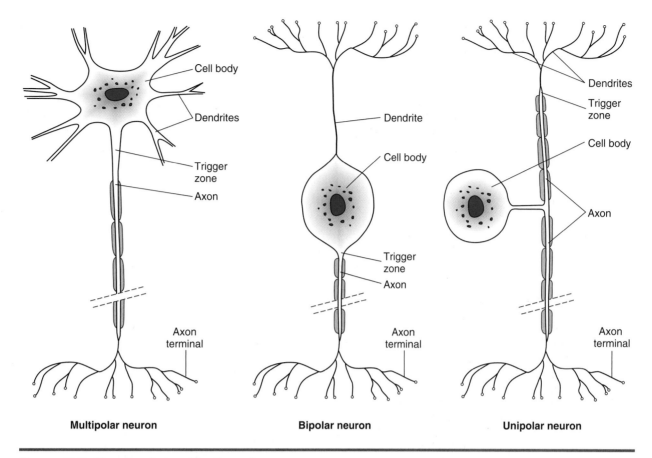

Figure 15.3: Structural classification of neurons.

The neurons communicate with one another or with other tissues (such as muscle) by transmitting impulses from the terminal end of the axon (**axonal terminal**) to a dendrite or cell (Figure 15.4). The terminal end of an axon meets a dendrite at a **synapse.** These axon terminals have **synaptic vesicles,** containing the **neurotransmitter substance,** and numerous **mitochondria** for energy. The membrane of the axonal terminal is called the **presynaptic membrane.** When the nerve impulse reaches the axonal terminals, some of the vesicles release the neurotransmitter substance to diffuse across the **synaptic cleft** (the tiny gap between the *presynaptic* and *postsynaptic membrane* of the dendrite). This substance then stimulates the dendrite, and thus, the impulse is picked up and carried to the receiving nerve's cell body. It was long thought that each type of axon produced only one neurotransmitter, but we now know that they may produce two or three different types of neurotransmitter substances, each with its own specific synaptic vesicle.

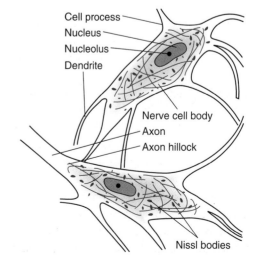

Figure 15.4: Neuron cell bodies and their processes.

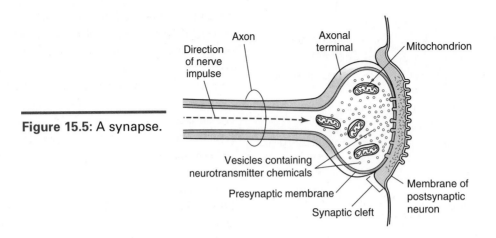

Figure 15.5: A synapse.

Procedure

1. First, examine the giant multipolar neuron smear slide using the 10× objective lens. Locate a dark-blue, large motor neuron. Center the neuron in your field of vision and change to the 40× objective lens.

2. Note that at the cell boundary there is a distinct separation between the blue and the pink background. This is the **plasma membrane.** The details of this membrane are not visible using the light microscope. These cells have numerous cytoplasmic extensions called **cell processes.** Some of these processes were cut off during the sectioning of the tissue. Each cell, if sliced perfectly down the middle, would have a long process called an **axon,** and numerous shorter ones called **dendrites.**

   If you are using a slide stained specifically for RNA, the **nucleus** (with mostly DNA) has a halo appearance around the RNA-rich **nucleolus** (the dark dot in the center). The **nuclear membrane** will not be visible on an RNA slide (Figure 15.5) unless it is stained with Hematoxylin and Eosin stain. Observe the granular structures within the cytoplasm; these are the **Nissl bodies.** Now look for **neurofibrils,** which are linear, darkly stained structures often found at the widened area where the axon leaves the cell body (the **axon hillock**).

The Nerve Impulse

Neurons have two major physiological properties: They are excitable (can convert stimuli into a nerve impulse); and they are conductive (can transmit the impulse). A number of characteristics of the nerve impulse differentiate it from an electric impulse that runs through a wire.

1. Nerve impulses are based on ion movement rather than electron movement.

2. Nerve impulses are slower than electric impulses.

3. Nerve impulses are active and self-propagating.

4. Nerve impulses require energy in the form of ATP.

5. Nerve impulses move at a constant amplitude and velocity.

The nerve impulse is described as an **action potential** (Figure 15.6). The difference in electrical charge on two sides of a membrane results in a voltage known as the **resting membrane potential,** and a neuron in this state is **polarized.** This resting potential, measured across the axonal cell membrane, is approximately -70 millivolts (mV). The value is negative because the inside of the nerve cell membrane is negatively charged relative to the exterior because of an excess of cations in the extracellular fluid (ECF), especially sodium (Na^+).

At rest, the plasma membrane is virtually impermeable to sodium. A balance is maintained (by an active sodium-potassium pump) in which there is more potassium ion (K^+) inside the cell and more sodium ion (Na^+) outside the cell. When an axon receives a stimulus, the plasma membrane briefly becomes more permeable, and sodium rushes into the cell, resulting in more positive ions inside the cell (the chloride ions [Cl^-] in the extracellular fluid stay

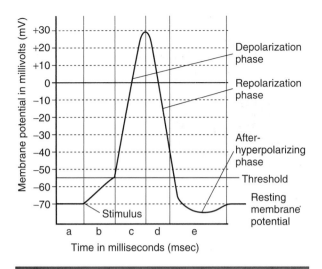

Figure 15.6: The action potential, or nerve impulse, and its phases. *A.* During resting membrane potential the voltage-gated Na⁺ channels are resting and voltage-gated K⁺ channels are closed. *B.* Rise to threshold potential: Sufficient stimulus causes depolarization to the threshold potential. *C.* Depolarization: The voltage-gated Na⁺ channels' activation gates are open. *D.* Repolarization: The voltage-gated K⁺ channels are open; Na⁺ channels are inactivating. Parts *C* represents the absolute refractory period. *E.* After hyper-polarization: voltage-gated K⁺ channels are still open; Na⁺ channels are in a resting state. Parts *D* and *E* represent the relative refractory period.

outside the cell). Thus, the interior of the cell starts to become less negative, and the outside less positive. This process is called **depolarization.**

If the stimulus is strong enough to depolarize the membrane to a critical level, called a **threshold stimulus** (approximately -55 mV), voltage-gated Na⁺ channels (special gates that open and close to allow ions to pass) rapidly start to open. Both the electrical and chemical gradients favor inward diffusion of Na⁺; this initiates the **depolarizing phase** of the action potential. The inflow of Na⁺ becomes so large that the membrane potential passes 0 mV and rises to $+30$ mV. The action potential is always the same size (amplitude) and is based on the **all-or-none principle:** Different neurons may have different thresholds for the generation of action potentials, but the threshold in any one neuron is usually constant.

When the voltage-gated Na⁺ channels close, the voltage-gated K⁺ channels are opening. This produces the **repolarizing phase** of the action potential and a process called **repolarization.** With the slowing of Na⁺ influx and the acceleration of K⁺ outflow, the membrane changes from $+30$ mV to -70 mV. However, the voltage-gated K⁺ channels continue to allow K⁺ to flow out; this overshoot is called **after-hyperpolarization** of the action potential. As the voltage-gated K⁺ channels close, the membrane potential returns to -70 mV. The entire action potential lasts about 1 msec, or 0.001 seconds, in a typical neuron.

When the Na⁺ is rushing in, the neuron is totally insensitive to additional stimuli and is said to be in an **absolute refractory period.** During the period of repolarization, it is nearly insensitive to further stimuli; however, a very strong stimulus may reactivate it. This period is called the **relative refractory period.** The sodium-potassium pump reestablishes the ionic balance soon after the action potential is completed, and because only minute amounts of sodium and potassium ions change place, once repolarization is completed, the neuron can quickly respond again to a stimulus.

Once generated, the action potential is **self-propagating:** It spreads its phenomenon along the entire length of the nerve fiber. It is never partially transmitted but is an all-or-none response (see previous discussion of this concept). The *nerve impulse* is the propagation of the action potential in which the disruption of the membrane permeability in one area of the axon causes disruption of the membrane permeability of the distal adjacent axonal area. This initiates another action potential in that axonal area, which stimulates the next axonal area, and so on down the axon. Because these axonal areas are small, propagation, and thus nerve impulse conduction, under these circumstances is slow.

As an animal develops and grows, either *in utero* or early in life, the nerves become **myelinated.** This results in faster nerve transmission and increased coordination. Animals that need to run immediately after birth to survive are born with fully myelinated nerves; others, such as kittens, do not complete myelination of motor neurons until 4 to 6 weeks of age. Figure 15.7 shows a **myelin sheath** around an axon. **Myelin** is a fatty substance within **Schwann cells** that segmentally envelopes the axon for its entire length. The area between the Schwann cells are called **nodes of Ranvier** (or **neurofibril nodes**). In myelinated nerves, these nodes act as the adjacent axonal area in which propagation occurs. In other words, the impulse jumps from node to node, progressing down the axon at a very fast rate, especially when compared with an unmyelinated nerve fiber.

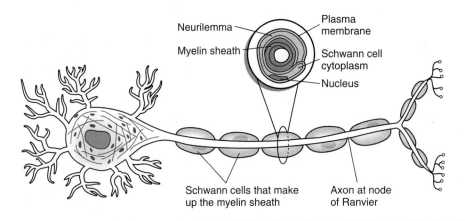

Figure 15.7: Axon with Schwann cells.

DISSECTION OF THE PERIPHERAL NERVES

There are approximately 38 pairs of **spinal nerves** in the cat. There are usually eight **cervical,** 13 **thoracic,** seven **lumbar,** three **sacral,** and seven **coccygeal nerves.** The first pair of cervical spinal nerves exits the atlas through the lateral vertebral foramen; the second pair exits between the atlas (C1) and the axis (C2). Each spinal cord segment produces a pair of spinal nerves; for example, cord segment C6 produces spinal nerve C6. Within the seven cervical vertebrae there are eight cord segments producing eight pairs of spinal nerves. From T1 to L2 in the dog, and to L3 in the cat, the cord segment is located entirely within its associated vertebrae; in other words, inside the seventh thoracic vertebrae is cord segment T7 that produces spinal nerve T7.

From T1 caudally, the spinal nerve emerges from the **intervertebral foramen** between its associated vertebrae and the next one or, in other words, just caudal to its associated vertebrae. Continuing with the previous example, T7 exits the vertebral column just caudal to the seventh thoracic vertebrae (between the seventh and eighth thoracic vertebrae). Caudal to L2, more than one cord segment can be found within the vertebral bodies (Table 15.1).

Even though the cord segments lie within a more cranial vertebra, their spinal nerves still exit caudal to the associated vertebral body. This means the nerves travel down the vertebral canal to reach the proper intervertebral foramen. Collectively, these spinal nerves come off the end of the spinal cord to form the **cauda equina,** so named for its resemblance to a horse's tail (Figure 15.8).

Table 15.1: Relationship between Vertebral Body and Cord Segments		
	Vertebral Body	**Cord Segment**
Dog	L1	L1
	L2	L2
	L3	L3, L4*
	L4	L5,* L6,*·† L7†
	L5	S1,†·‡ S2,‡ S3‡
	L6	Cy++(all coccygeal)
Cat	Cord segments S1, S2, S3, are located either within L6 or at the L5, L6 junction.	

*Patellar reflexes
†Pedal reflexes
‡Perineal reflexes

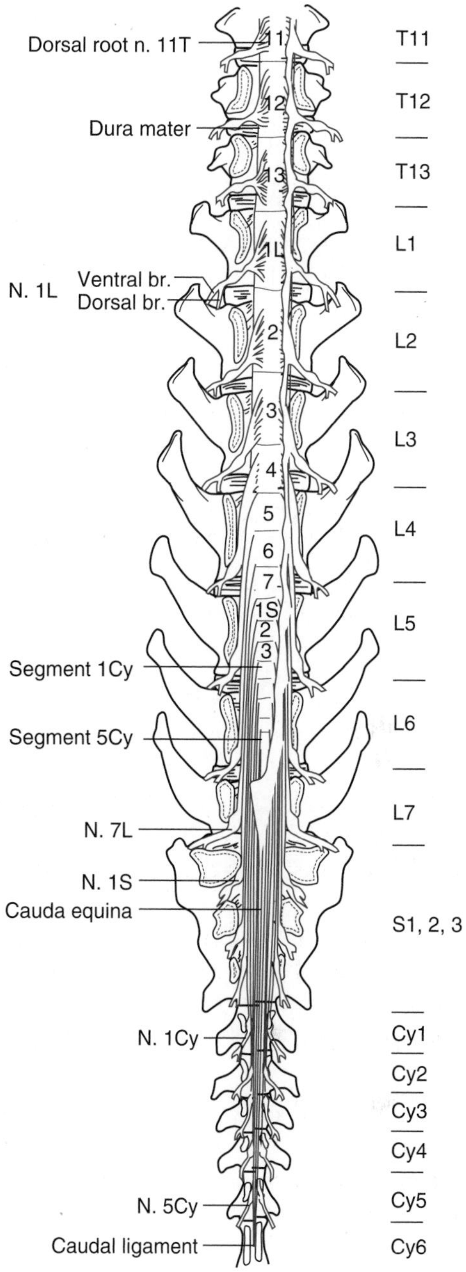

Figure 15.8: Spinal cord segment's placement within the spinal vertebrae and the cauda equina.

Labels on figure: Dorsal root n. 11T, Dura mater, N. 1L, Ventral br., Dorsal br., Segment 1Cy, Segment 5Cy, N. 7L, N. 1S, Cauda equina, N. 1Cy, N. 5Cy, Caudal ligament. Right-side markings: T11, T12, T13, L1, L2, L3, L4, L5, L6, L7, S1, 2, 3, Cy1, Cy2, Cy3, Cy4, Cy5, Cy6.

Procedure

1. As the spinal nerves emerge from the intervertebral foramen, they divide to form a **dorsal ramus** and a **ventral ramus.** The ventral rami form the major **plexuses,** the **brachial** and **lumbosacral,** which supply nerves to the front legs and hind legs as well as the skin and musculature of the ventral trunk. The dorsal rami are smaller and supply only the skin and musculature of the back.

2. The first four cervical nerves supply the lateral neck musculature. It is not necessary to locate these nerves.

3. The **phrenic nerve** was identified during the dissection of the respiratory system. The two phrenic nerves are formed by the fifth and sixth cervical nerves; they pass lateral to the heart on their way to the diaphragm.

4. The sixth through eighth cervical nerves and first thoracic nerve form the **brachial plexus,** which supplies nerves to the muscles of the front leg. This plexus was exposed during the dissection of the chest and arm muscles. In your previous dissection, we tried to preserve the nerves on the right front

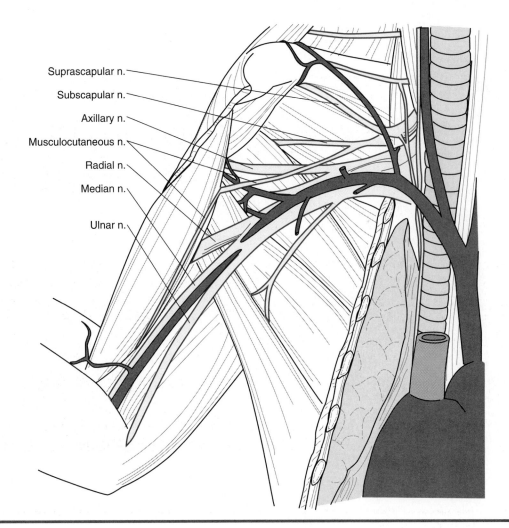

Suprascapular n.
Subscapular n.
Axillary n.
Musculocutaneous n.
Radial n.
Median n.

Ulnar n.

Figure 15.9: Diagram of the nerves of the brachial plexus of a cat.

leg by only dissecting the vessels on the left. Therefore, carefully cut through the pectoralis muscle group on the right side of the cat's chest until these nerves are exposed.

5. The most cranial of the nerves of the brachial plexus we will dissect is the **musculocutaneous nerve** (Figures 15.9 and 15.10). This small nerve may be identified as it passes along the medial edge of the biceps brachii muscle. This nerve supplies the biceps, coracobrachialis, and brachialis muscles.

6. The deepest of the nerves is dorsal and cranial to the brachial artery and passes between the triceps and the humerus: the **radial nerve.** It is the largest nerve of the plexus. Follow this nerve as it passes to the lateral surface of the upper leg and trace it distally. It divides into a superficial branch and a deep-muscular branch near the elbow. This nerve supplies the triceps, the supinator muscles, and the extensor muscles of the lower forelimb. The superficial branch innervates the skin of the cranial foreleg.

7. Coursing adjacent and cranial to the brachial artery is the **median nerve.** Follow it to the elbow region, through the supracondyloid foramen in the humerus, and to the forearm. It supplies the pronators and flexors of the lower forelimb, except for the flexor carpi ulnaris.

8. The **ulnar nerve** is the most caudal nerve of the brachial plexus that we will dissect. It emerges from its course adjacent to the median nerve and runs caudally from the brachial artery to the carpus. It passes between the medial epicondyle of the humerus and the olecranon process of the ulna at the elbow. This nerve supples the flexor carpi ulnaris and the ulnar head of the flexor digitorum profundus.

9. The next 12 thoracic nerves pass between the ribs, each with an artery and vein. They are known as the **intercostal nerves,** and they supply the intercostal muscles.

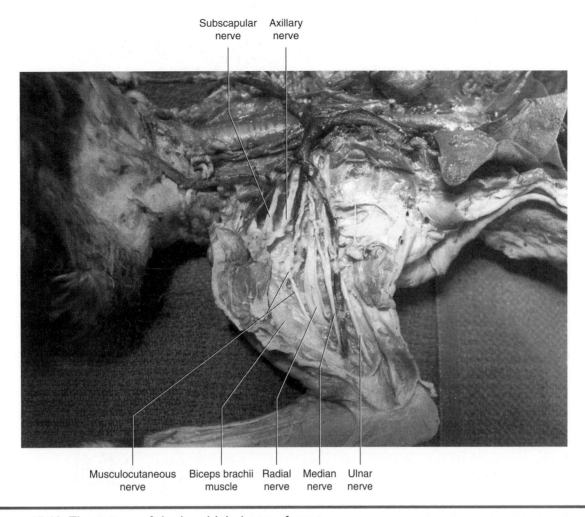

Subscapular nerve   Axillary nerve

Musculocutaneous nerve   Biceps brachii muscle   Radial nerve   Median nerve   Ulnar nerve

Figure 15.10: The nerves of the brachial plexus of a cat.

10. The **sciatic nerve** is the major nerve emerging from the **lumbosacral plexus** (Figures 15.11 and 15.12). This plexus is made up of the last four lumbar nerves and the three sacral nerves. Locate the sciatic nerve beneath the biceps femoris muscle. Trace it to the popliteal fossa where it divides into two branches, which supply the leg. This nerve supplies the flexor muscles of the stifle, extensors of the hock, and the digital flexors.

11. The **femoral nerve** also emerges from the lumbosacral plexus. This nerve can be located on the ventral surface of the thigh, coursing with the femoral artery between the gracilis and sartorius muscles (Figures 15.13 and 15.14). It supplies the sartorius muscles and the quadriceps femoris group (the extensors of the stifle). It continues as the *saphenous nerve* to supply the skin on the medial aspect of the leg. Trace the femoral nerve cranially to where it enters the ventral back muscles. Separate these muscles and continue the dissection to the femoral nerve's origin with other lumbar nerves. Follow these nerves caudally, finding the separation between the lumbosacral cord and the **obturator nerve** (see Figures 15.13 and 15.14). The obturator nerve supplies the adductors, pectineus, and gracilis.

12. The autonomic nervous system of the cat will not be dissected. Without dissection, you can observe (in the thoracic cavity) the two **sympathetic trunks,** or **chains,** located on each side of the vertebral column. Push the left lung to the right and locate the sympathetic trunks on the dorsal wall of the thoracic cavity. The sympathetic trunks are beneath the parietal pleura, parallel to the vertebral column. They resemble white threads.

13. The cranial nerves also will not be dissected. The **vagus nerve,** or **cranial nerve X,** has been noted previously, as it courses adjacent and attached to the common carotid arteries on both sides.

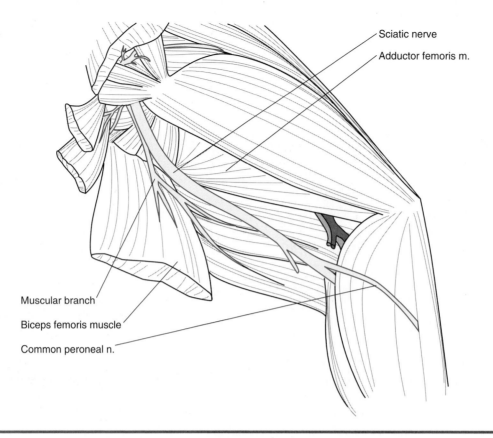

Figure 15.11: Diagram of the sciatic nerve of the lumbosacral plexus of a cat.

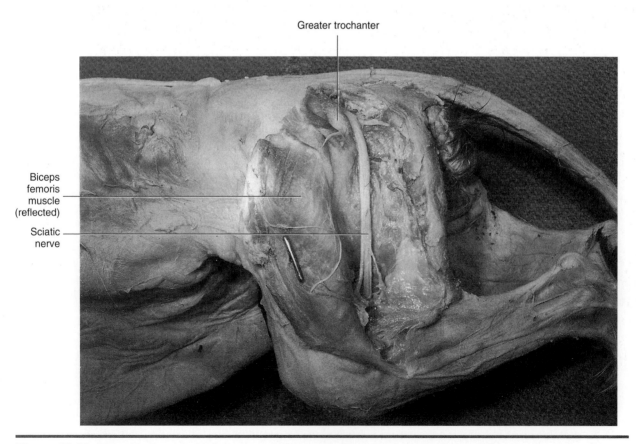

Figure 15.12: The sciatic nerve of a cat.

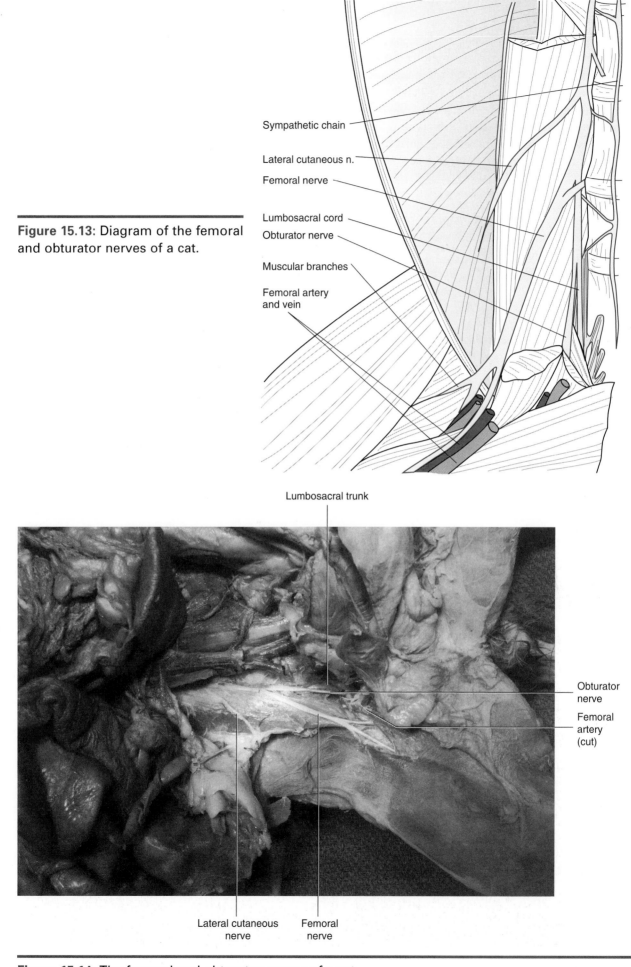

Sympathetic chain

Lateral cutaneous n.

Femoral nerve

Lumbosacral cord

Obturator nerve

Muscular branches

Femoral artery
and vein

Figure 15.13: Diagram of the femoral and obturator nerves of a cat.

Lumbosacral trunk

Obturator
nerve

Femoral
artery
(cut)

Lateral cutaneous
nerve

Femoral
nerve

Figure 15.14: The femoral and obturator nerves of a cat.

308 ■ Chapter 15

The Cranial Nerves

The **cranial nerves** are the nerves of the peripheral nervous system that come from the brain and the brain stem. Each of these nerves are either sensory, motor, or both if they have multiple functions. They may have autonomic function or may be voluntary. Table 15.2 lists the number and corresponding names of these nerves, its type, and key functions.

Table 15.3 has a mnemonic device to remember the names and types of each nerve.

The Autonomic Nervous System

The autonomic nervous system is divided into the **sympathetic** and **parasysmpathetic nervous systems.** Figure 15.15*A* illustrates the neuronal arrange-

Table 15.2: Name and Functions of the 12 Cranial Nerves

Number	Name	Type	Key Function
I	Olfactory	Sensory	Smell
II	Optic	Sensory	Vision
III	Oculomotor	Motor	Eye movement, pupil size, focusing lens
IV	Trochlear	Motor	Eye movement
V	Trigeminal	Both	Sensations from the head and teeth, chewing
VI	Abducens	Motor	Eye movement
VII	Facial	Both	Face and scalp movement, salivation, tears, taste
VIII	Vestibulocochlear	Sensory	Balance and equilibrium, hearing
IX	Glossopharyngeal	Both	Tongue movement, swallowing, salivation, taste
X	Vagus	Both	Sensory from GI tract and respiratory tree; motor to larynx, pharynx; parasympathetic motor to the abdominal and thoracic organs
XI	Accessory	Motor	Head movement
XII	Hypoglossal	Motor	Tongue movement

Table 15.3: Mnemonic for Remembering Cranial Nerve Names and Type

Number	Name	Name-Mnemonic Word	Type-Mnemonic Word
I	Olfactory	on	six
II	Optic	old	sailors
III	Oculomotor	Oregon's (Oklahoma's, Olympus, etc.)	made
IV	Trochlear	towering	merry
V	Trigeminal	tops	but
VI	Abducens	a	my
VII	Facial	fine	brother
VIII	Vestibulocochlear	veterinary	said
IX	Glossopharyngeal	gastroenterologist	bad
X	Vagus	viewed	business
XI	Accessory	a	my
XII	Hypoglossal	horse	man

ment of the somatic, sympathetic, and parasympathetic nerves. The parasysmpathetic nerves arise from the cranial nerves and sacral nerves. The sympathetic nerves arise from the thoracic and lumbar spinal nerves (Figure 15.15*B*).

The sympathetic nerves' presynaptic fibers synapse at the sympathetic chain, or other ganglia, and use **acetylcholine** as the neurotransmitter at this junction (see Figure 15.15*A*). The postsynaptic fibers innervate organs of the chest and abdomen and use **norepinephrine** as their neurotransmitter. The parasympathetic nerves' presynaptic fibers are long and synapse at or just below the surface of the organ they innervate. A short postsynaptic fiber innervates the organ. Both use acetylcholine as their neurotransmitter substance.

The easiest way to remember the effects of the sympathetic and parasympathetic nervous systems is to think about which system is active in a given set of circumstances. Consider a cow being chased by wolves: The cow's sympathetic nervous system is fully turned on, and the parasympathetic turned off. The cow's heart rate increases; the force of its heart contraction increases; the bronchioles dilate; the pupils dilate; its gastrointestinal motility shuts down; the diameter of the skin blood vessels decreases; the diameter of the muscle vessels increases; and the diameter of the kidney blood vessels decreases.

Conversely, a cow lying down in the field, relaxed and chewing its cud, is in the exact opposite situation: The cow's parasympathetic nervous system is active. The cow's heart rate is decreased; the bronchioles constrict slightly; the pupils constrict slightly; and the gastrointestinal blood flow, motility, and secretions (including saliva) increase. There is no significant effect on the force of heart contractions or on the diameter of skin, muscle, or kidney blood vessels. If a drug such as atropine (called a parasymptholytic), which turns off the parasympathetic nervous system, is given to an animal, the sympathetic nervous system will take over. Heart rate will increase and mucous membranes will become dry, among other effects.

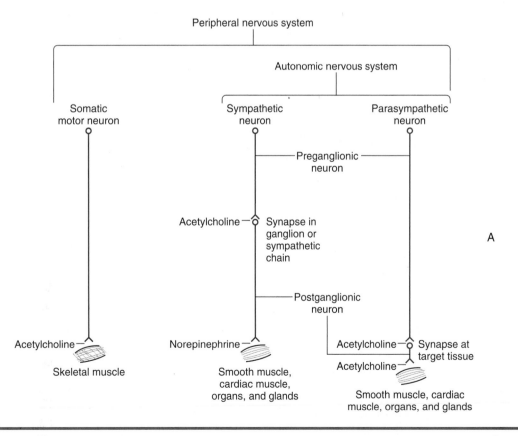

Figure 15.15: *A.* The peripheral nervous system.

Continued

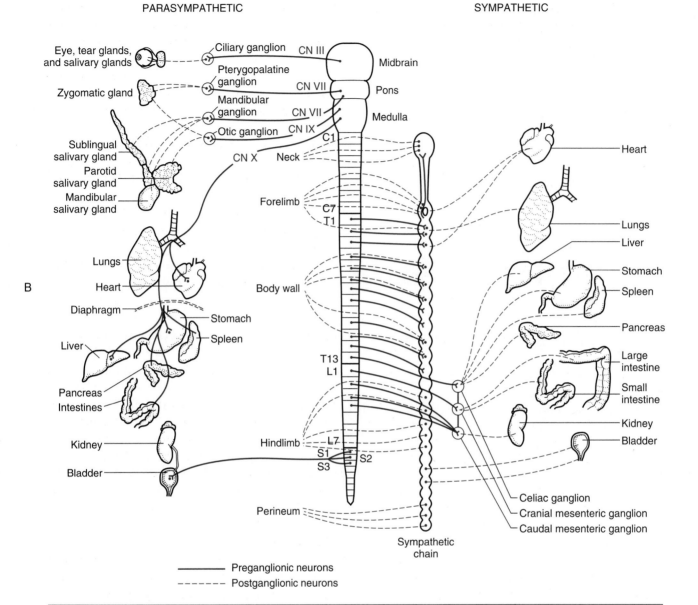

PARASYMPATHETIC

SYMPATHETIC

Figure 15.15, *cont'd: B.* Sympathetic and parasympathetic preganglionic and postganglionic innervation.

Anatomy and Physiology of the Brain

The *brain's* regions originate in the embryo. They are discussed in the following text.

The **telecephalon** is composed of the **cerebrum,** or **cerebral hemispheres.** These are divided into functional lobes, which roughly correspond to the area of the cerebrum underlying the bone of the same name.

1. The **frontal lobe** contains the part of the motor cortex associated with voluntary movement, as well as areas associated with psychomotor skills.

2. The **parietal lobe** also contains part of the motor cortex, as well as the somesthetic interpretation center, which controls conscious perception and localization of pain, touch, and temperature.

3. The **occipital lobe** is associated with visual interpretation.

4. The **temporal lobe** is associated with auditory function, behavior, and memory.

5. The **piriform lobe** is found on the ventral surface of the cerebrum and is associated with olfaction (Figure 15.16).

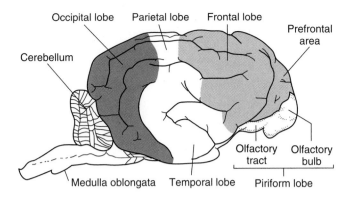

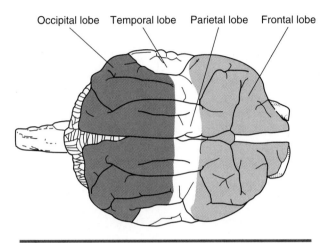

Figure 15.16: The lobes of the cerebrum of a sheep.

The major connection between the right and left cerebral hemispheres is the **corpus callosum.** Ventral to the corpus callosum and **septum pellucidum** is the body of the **fornix.** It is the white matter-connection between the hippocampal gyrus, and nearby areas, such as the *mammillary body* and the *brainstem.*

Other areas of the brain include the **diencephalon,** which consists mainly of the **thalamus,** but also includes smaller areas of the *epithalamus, subthalamus, metathalamus,* and **hypothalamus.** The *thalamus* functions as a sensory relay center; it receives general sensory impulses andtransmits them to the telencephalon. The other parts of the diencephalon also act as relay centers. The *hypothalamus* consists of the **optic chiasm, tuber cinereum, mammillary body, infundibulum** and **hypophysa.** Its functions were discussed in chapter 13 on the endocrine system. The diencephalon and telencephalon together may be referred to as the *forebrain.*

The **mesencephalon,** or **midbrain,** is located immediately caudal to the diencephalon and consists of the **corpora quadrigemina** (made up of the **rostral** and **caudal colliculi**) and the paired **cerebral peduncles.** The rostral colliculi are associated with vision and the caudal colliculi with hearing. The cerebral peduncles are essentially the continuation of the right and left halves of the spinal cord and brainstem into the respective cerebral hemispheres. They contain nerve fiber tracts and nuclei.

The embryonic **rhombencephalon (hindbrain)** develops into the **metencephalon** and **myelencephalon.** The metencephalon consists of the **cerebellum** dorsally and the **pons** ventrally. The *cerebellum* has a **cortex** and a **medulla** made up of **white matter** which, on a cut surface, has a branching appearance called the **arbor vitae.** The cerebellum is connected by the **rostral, middle,** and **caudal cerebellar peduncles** to the cerebral peduncle, pons, and brainstem, respectively. The cerebellum coordinates motor activity. The pons is visible on the ventral surface, rostral to the brainstem. It is a mixture of white and gray matter and contains the ascending reticular activating system (ARAS), which maintains alertness or awareness by way of the cerebrum, and controls the apneustic center and pneumotaxic center for respiration, the nuclei of the vestibular apparatus, and the motor nucleus of cranial nerve five (CN V). The *myelencephalon* is the **medulla oblongata,** or **brainstem.** It contains the center for the functions of the heart, as well as for respiration, swallowing, and vomiting.

EXERCISE 15.3

ANATOMY OF THE BRAIN

Procedure

1. Obtain a preserved sheep's brain and rinse it with tap water.

2. The outer, heavy, fibrous layer of the **meninges** is the **dura mater** (Figures 15.17 and 15.18). The dura mater is intimately attached to the inside of the skull, but there is a slight space between it and the vertebral canal as it covers the spinal cord. Carefully remove this covering. The dura mater contains a longitudinal fold, called the **falx cerebri,** which penetrates into the *longitudinal fissure* between the two cerebral hemispheres. The transverse fold separating the cerebrum from the cerebellum is the **tentorium cerebelli.** The pituitary gland will break off from the infundibulum during the removal of the dura mater.

The inner two layers of the meninges can now be seen covering the brain. The **arachnoid,** or middle layer, lies between the dura mater and the **pia mater,** the innermost, vascular layer of the meninges (these layers also cover the spinal cord). The arachnoid is easily distinguished from the pia mater in the region overlying the grooves on the brain surface because the pia mater dips into the grooves and the arachnoid traverses them (see Figure 15.18). The subarachnoid space is filled with **cerebrospinal fluid,** which has been removed from this specimen, and in the preserved brain, the arachnoid appears to be attached to the pia mater.

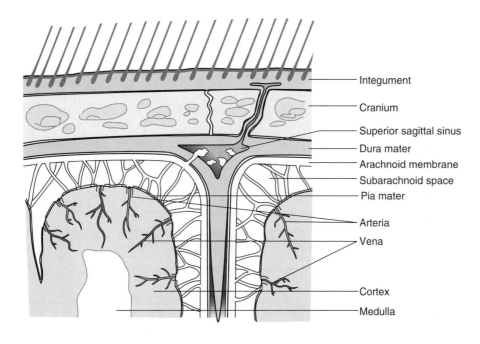

- Integument
- Cranium
- Superior sagittal sinus
- Dura mater
- Arachnoid membrane
- Subarachnoid space
- Pia mater
- Arteria
- Vena
- Cortex
- Medulla

Figure 15.17: Diagram of the meninges of the brain (all species).

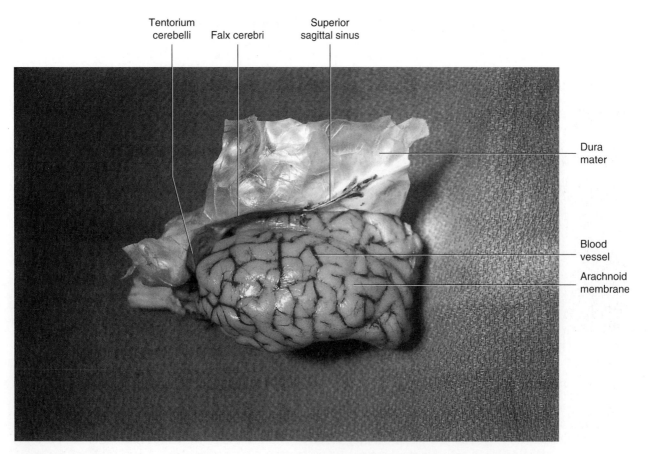

Tentorium cerebelli   Falx cerebri   Superior sagittal sinus

Dura mater

Blood vessel

Arachnoid membrane

Figure 15.18: The meninges of the sheep's brain.

3. Place the brain in the dissecting tray so the dorsal surface is visible. Observe the paired **cerebral hemispheres** and the caudal *cerebellum*. The cerebral hemispheres are separated by the **cerebral longitudinal fissure,** and the cerebellum is separated from the cerebral hemispheres by the **transverse fissure** (Figures 15.19 and 15.20).

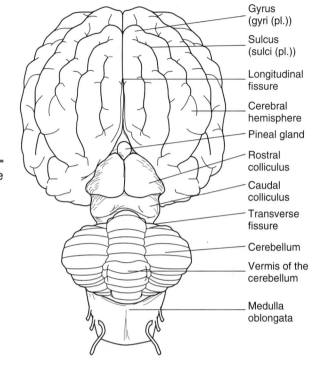

Figure 15.19: Diagram of the dorsal view of the sheep's brain.

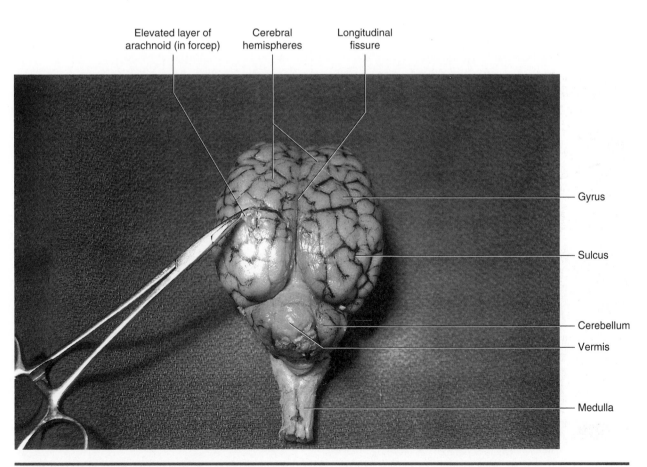

Figure 15.20: Dorsal view of the sheep's brain.

4. The surface of each hemisphere is composed of numerous ridges and grooves. The raised ridges are called the **gyri** (**gyrus** is singular), and the depressions are **sulci** (**sulcus** is singular).

5. Spread the cerebral hemispheres and cerebellum to see the roof of the midbrain, or mesencephalon (Figure 15.21). Four prominent, round swellings, the **corpora quadrigemina,** form the roof of the midbrain. The larger, rostrodorsal pair are called the **rostral colliculi** and the smaller, caudoventral pair the **caudal colliculi.** The **pineal gland** can be seen between the superior colliculi. The **trochlear nerve (cranial nerve [CN] IV)** appears as a thin, white strand, directed ventrally and slightly caudal to the caudal colliculi.

6. Caudal to the cerebral hemispheres is the **cerebellum.** The cerebellum is connected to the brainstem by three prominent fiber tracts, or peduncles. Lift the cranial edge of the cerebellum and locate the **rostral cerebellar peduncle** that connects the cerebellum to the cerebral peduncle of the midbrain. The **middle cerebellar peduncle** connects the cerebellum with the pons. Slightly caudal to this is the **caudal cerebellar peduncle,** connecting the cerebellum to the medulla.

7. Turn the brain over so the ventral surface is visible. A pair of **olfactory bulbs** can be seen beneath the cerebral hemispheres. These bulbs lie over the cribriform plate of the ethmoid bone and receive the olfactory neurons from the nose (Figures 15.22 and 15.23).

8. A white band, the **olfactory tract,** extends from each olfactory bulb along the ventral surface of the cerebral hemispheres.

9. On the ventral surface of the *diencephalon* is the **hypothalamus,** located posterior to the olfactory tracts. The **optic nerves (CN II)** undergo a partial crossing over (decussation) to form the rostral border of the hypothalamus, known as the **optic chiasm.**

10. The rest of the hypothalamus is the oval area lying caudal to the optic chiasm. The **infundibulum** can be seen connecting the pituitary to the hypothalamus. Caudal to the infundibulum are the paired, rounded **mammillary bodies.** Around the infundibulum is the **tuber cinereum** (see Figures 15.22 and 15.23).

11. Observe the **cerebral peduncles** on the ventral surface of the midbrain. The large **oculomotor nerves (CN III)** arise from the cerebral peduncle caudal to the mammillary body.

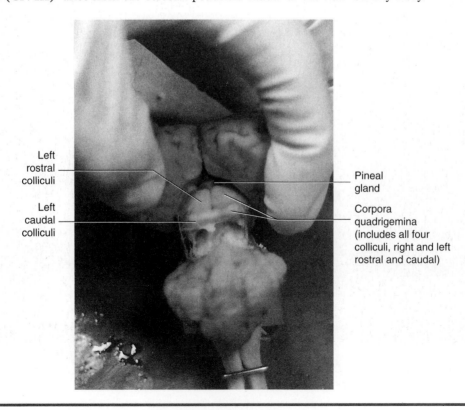

Left rostral colliculi

Left caudal colliculi

Pineal gland

Corpora quadrigemina (includes all four colliculi, right and left rostral and caudal)

Figure 15.21: Corpora quadrigemina of the sheep's brain.

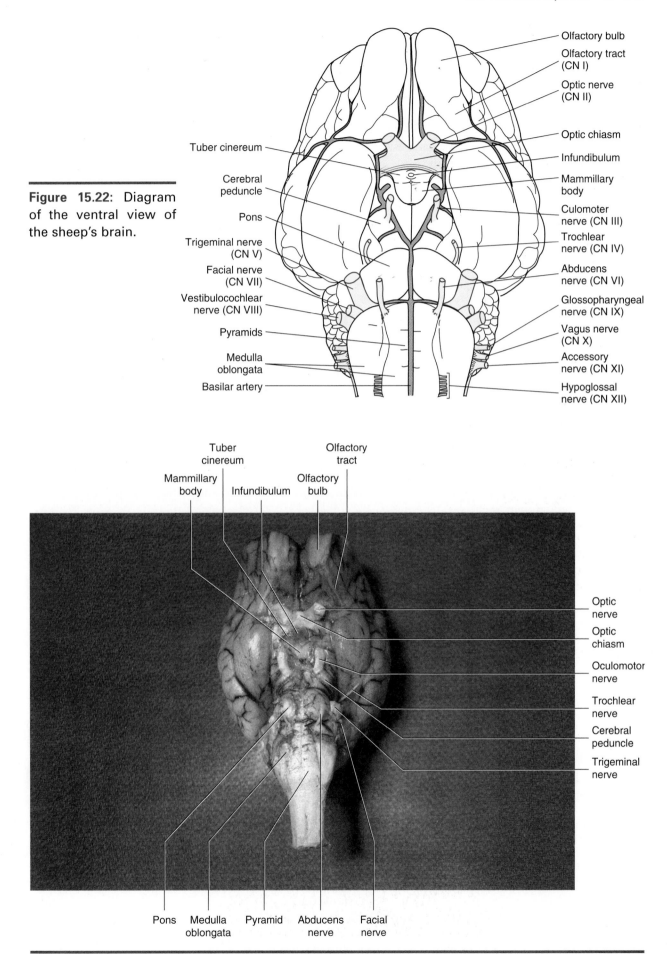

Figure 15.22: Diagram of the ventral view of the sheep's brain.

Olfactory bulb

Olfactory tract (CN I)

Optic nerve (CN II)

Optic chiasm

Infundibulum

Mammillary body

Culomoter nerve (CN III)

Trochlear nerve (CN IV)

Abducens nerve (CN VI)

Glossopharyngeal nerve (CN IX)

Vagus nerve (CN X)

Accessory nerve (CN XI)

Hypoglossal nerve (CN XII)

Tuber cinereum

Cerebral peduncle

Pons

Trigeminal nerve (CN V)

Facial nerve (CN VII)

Vestibulocochlear nerve (CN VIII)

Pyramids

Medulla oblongata

Basilar artery

Tuber cinereum

Olfactory tract

Mammillary body

Infundibulum

Olfactory bulb

Optic nerve

Optic chiasm

Oculomotor nerve

Trochlear nerve

Cerebral peduncle

Trigeminal nerve

Pons

Medulla oblongata

Pyramid

Abducens nerve

Facial nerve

Figure 15.23: Ventral view of the sheep's brain.

12. Caudal to the midbrain is the **pons.** This is composed primarily of white fibers, many of which run transversely across the pons and out to the cerebellum. Just caudolateral to the pons on both sides is the large **trigeminal nerve (CN V),** and caudolateral to this nerve are two nerves, the **facial nerve (CN VII)** and the **vestibulocochlear nerve (CN VIII).** The **abducens nerve (CN VI)** is located at the border between the pons and the medulla just off the midline on both sides. The rest of the nerves will be difficult to locate because they are often broken off as the dura mater is removed.

13. The **medulla oblongata** is caudal to the pons. The longitudinal bands of tissue on either side of the ventral median fissure on the ventral surface of the medulla are known as the **pyramids** (see Figures 15.22 and 15.23).

14. Cut the brain in half along the longitudinal sulcus to create a sagittal section of the sheep brain (Figures 15.24 and 15.25). Locate the **corpus callosum,** or the white fibers connecting the two cerebral hemispheres.

15. A thin covering of tissue separating the left and right **lateral ventricles** is the **septum pellucidum,** which is ventral to the corpus callosum. The septum should be intact on one side of the brain, and an opening to the *lateral ventricle* should be seen on the opposite side (as it is virtually impossible to split the septum pellucidum down the middle). Inside the lateral ventricles is the **choroid plexus,** a network of capillaries protruding into each ventricle and covered by a layer of ependymal cells (derived from the lining membrane of the ventricles). This produces the *cerebrospinal fluid (CSF).* It flows out of each lateral ventricle through the **foramen of Monro** to the **third ventricle** through the **mesencephalic aqueduct,** and into the **fourth ventricle.** Then the CSF either flows down the **central canal** of the spinal cord or passes out into the **subarachnoid space** via the **foramen of Luschka.** The CSF forms a cushion to protect the brain and spinal cord.

16. The **fornix,** a band of white fibers, lies ventral to the septum.

17. The **third ventricle** and **thalamus** lie ventral to the fornix. The narrow third ventricle, the walls of which are covered by a shiny layer of epithelium, is on the midline. The thalamus forms the lateral walls of the third ventricle.

18. The **massa intermedia** connects the two sides of the thalamus. The third ventricle forms a partial circle around the massa intermedia. This structure appears as a dull, circular area not covered by the epithelium.

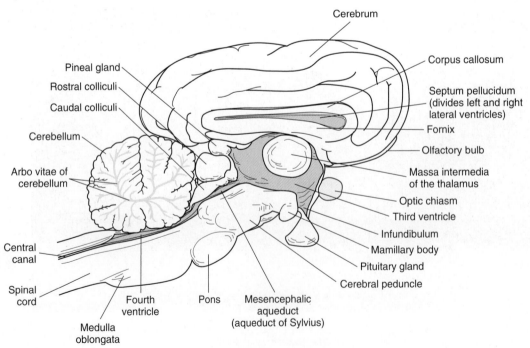

Figure 15.24: Diagram of the sagittal view of the sheep's brain.

19. The **foramen of Monro,** the opening through which each lateral ventricle communicates with the third ventricle, lies in the depression rostral to the massa intermedia. Find this connection by passing a dull probe through it.

20. Relocate the hypothalamus. This lies ventral to the third ventricle. Also note the pineal body dorsal to the midbrain, near the rostral colliculi. Observe the narrow **mesencephalic aqueduct (Aqueduct of Sylvius)** leading through the midbrain and connecting the third and fourth ventricles.

21. The **fourth ventricle** lies above the pons and medulla and below the cerebellum.

22. The beginning of the spinal cord may be seen connected to the medulla. A canal known as the **central canal,** which is connected to the fourth ventricle, is present in the center of the cord.

23. Note the tree-like arrangement of gray and white matter in the *cerebellum.* This arrangement is known as the **arbor vitae** (tree of life). The gray matter of the cerebellum is the **cortex,** and the white is the **medulla.**

24. Make a cut into the cerebral hemisphere. The outer layer of each cerebral hemisphere is its cortex, composed of **gray matter.** The white area beneath the cortex is the **white matter.**

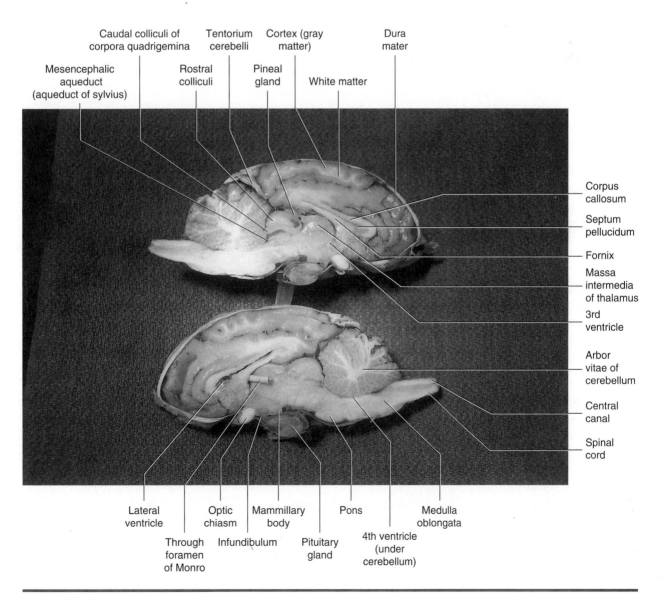

Figure 15.25: Sagittal view of the sheep's brain.

EXERCISE 15.4

THE CROSS SECTION OF THE SPINAL CORD

The spinal cord is part of the central nervous system and starts at the foramen magnum of the skull and lies within the vertebral canal of the vertebrae.

1. Obtain a slide of a cross-sectional view of mammal spinal cord (Figure 15.26).

2. The spinal cord is divided into two mirror halves, right and left. The two halves are divided dorsally by a sulcus, the **dorsal median sulcus,** and ventrally by a fissue, the **ventral median fissure.** The **gray matter** is located centrally on both sides, surrounded by **white matter** peripherally.

3. The white matter contains the white tracts, or columns, that communicate with the brainstem, cerebellum, and cerebrum. The dorsal white column is called the **dorsal funiculus;** the lateral column, the **lateral funiculus;** and ventral column, the **ventral funiculus.**

4. In the center is the **central canal,** which communicates with the fourth ventricle.

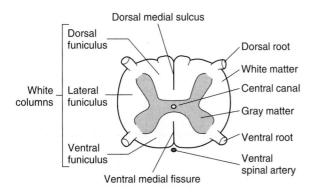

Figure 15.26: The cross section of the spinal cord (any species).

EXERCISE 15.5

SPINAL REFLEXES

A spinal reflex arc is a pathway of a nerve impulse. From a **sensory afferent neuron** to the spinal cord, the nerve impulse synapses at an **interneuron** (also called an *internuncial neuron,* or *Renshaw cell*) and synapses again at a **motor efferent neuron** (Figure 15.27).

The neurotransmitter at the *efferent motor neuron* is excitatory. It establishes an **excitatory postsynaptic potential** in the efferent motor neuron that goes to the **agonist muscle.** This neurotransmitter is *acetylcholine.* Simultaneously, another interneuron establishes an **inhibitory postsynaptic potential** to the efferent motor neuron of the **antagonist muscle** by releasing an inhibitory neurotransmitter, **glycine,** at the spinal cord level. **Gamma-amino butyric acid (GABA)** is the inhibitory neurotransmitter in the brain. This way the agonist muscle is contracting while its opposing muscle, the antagonist, is relaxing, thus allowing easy movement of a limb or other body part. The excitatory neurotransmitter is removed from the synaptic cleft by the release of an enzyme, in this case, **acetylcholinesterase.**

Procedure

Using a live dog or cat, find the following reflexes. They are used to gauge the depth of anesthesia.

1. **palpebral (blink) reflex:** Touch the hairs on the lateral side of the eyelid or the medial canthus of the eye; the response should be a blink.

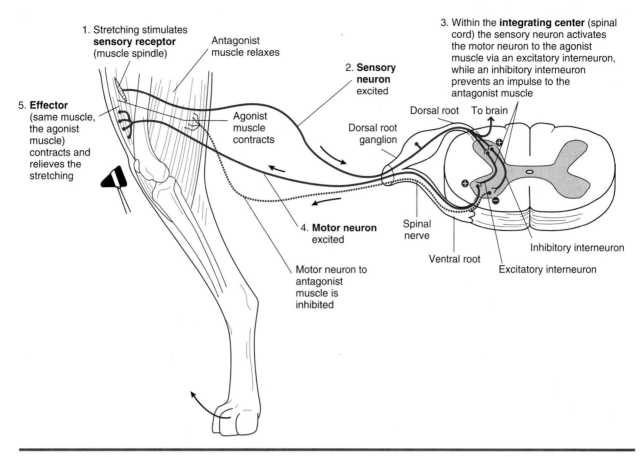

1. Stretching stimulates **sensory receptor** (muscle spindle)

Antagonist muscle relaxes

2. **Sensory neuron** excited

3. Within the **integrating center** (spinal cord) the sensory neuron activates the motor neuron to the agonist muscle via an excitatory interneuron, while an inhibitory interneuron prevents an impulse to the antagonist muscle

Dorsal root   To brain

5. **Effector** (same muscle, the agonist muscle) contracts and relieves the stretching

Agonist muscle contracts

Dorsal root ganglion

4. **Motor neuron** excited

Spinal nerve

Ventral root

Inhibitory interneuron

Excitatory interneuron

Motor neuron to antagonist muscle is inhibited

Figure 15.27: The spinal reflex arc in a dog.

2. **corneal reflex:** Touch the cornea with a wisp of cotton from a cotton ball; the response should be a blink.

3. **oral-pharyngeal reflex:** Try to open the animal's mouth, or place an object in the mouth, while touching the pharyngeal region. The normal response should be for animal to close its mouth. This reflex is often stimulated by pilling an animal.

4. **laryngeal reflex:** Touch a part of the larynx, such as the epiglottis, with a wooden tongue depressor. The normal response should be for the animal to cough or swallow; the epiglottis should close and so should the vocal folds.

5. **ear pinna reflex:** Lightly touch the lateral aspect of the upper part of the ear; the normal response is for the ear to twitch.

6. **pedal reflex (toe pinch):** Extend the rear limb while simultaneously pinching the middle toe; the normal response is a reflexive withdrawing of the limb.

7. **patellar reflex:** Strike the straight patellar tendon lightly with a percussion hammer; the normal response is reflexive jerking of the lower leg.

8. **flexors of the tarsus and extensors of the digits:** Place your index finger on the cranial tibial muscles and strike your finger semi-softly with the percussion hammer. The response should be flexion of the tarsus and extension of the digits.

9. **extension of the tarsus:** Grasp behind the stifle with your thumb, and strike it with the percussion hammer. The response should be for the animal to extend the tarsus.

10. **extension of the elbow:** Place your finger on the tendon of the long head of the triceps and strike your finger with the percussion hammer. The animal should extend the elbow joint.

QUESTIONS

1. Why is it necessary to strike your finger or thumb placed over the muscle instead of the muscle directly?

2. Why is the patellar reflex not abolished during general anesthesia?

Discussion

With the patellar reflex, the percussion hammer strikes a relatively tough tendon, whereas at the other sites it would strike the muscle directly and could damage it. Plus, by striking your finger or thumb you spread the stimulus across the entire muscle belly.

The patellar reflex, being an entirely spinal reflex, is not abolished during general anesthesia because the anesthetic is only depressing the brain and not the spinal cord.

CLINICAL SIGNIFICANCE

The nervous system is so intricate and delicately balanced that even minor problems can have catastrophic effects. Take for instance the disease hydrocephalus. The condition may be congenital or acquired. The congenital form occurs in small and brachycephalic breeds, such as the bulldog, Chihuahua, Maltese, Pomeranian, toy poodle, Yorkshire terrier, Llasa apso, Cairn terrier, Boston terrier, pug, Pekingese, and the Siamese cat. It usually appears a few weeks after birth, but may present at up to one year of age. Acquired hydrocephalus can occur in any breed at any age.

There are two types of hydrocephalus. The non-communicating type is caused by an obstruction of the CSF flow within the ventricular system of the brain. The communicating type is due to a lack of resorption into the subarachnoid space. Congenital hydrocephalus is usually obstructive, caused by a congenitally narrow cerebral aqueduct. Acquired hydrocephalus is usually due to trauma, inflammatory conditions, or neoplasia. The result of both types is an increased pressure within the lateral ventricles, which hydraulically compresses the cortex of the brain. This pressure may result in seizures, altered mental status, visual deficits, and an uncoordinated gait.

A set of diseases classified as peripheral neuropathies results in demyelination of the spinal cord and peripheral nerves. This condition may occur in many species and may be a primary condition or may be secondary to axonal degeneration. It most often affects the cervical spine and spinal nerves, causing progressive ataxia and spastic incoordination in the rear legs (paraparesis), which may progress to include the front legs (tetraparesis). The cause is unknown.

*There is also a host of viral and bacterial causes of inflammation of the brain and spinal cord. Canine distemper and rabies are two examples. The nervous system was designed to keep unwanted molecules out. The pia mater follows the vessels into the brain, enveloping them and allowing only certain-sized molecules to pass through. This is known as the **pial barrier**, or **blood-brain barrier**. It is also the reason why certain antibiotics will not enter nervous tissue. In inflammatory conditions, however, the pores in the pia mater may become bigger, allowing unwanted molecules in, but also allowing us to use certain antibiotics we could not use under normal circumstances.*

Veterinary Vignettes

I remember being told in veterinary school that boxers are considered walking tumor factories. In my practice, however, this was not the case. *My* boxers were not getting tumors! It was magic, and I was the magician. It was a nice delusion while it lasted. Then Tobey Tuscalini arrived at my practice door. I wish I could have said "Never more!" But he just kept coming back. This dog made up for all the tumors other boxers didn't get. Fortunately, his tumors were not malignant. Most were skin tumors, but he seemed to have one of every type. Every veterinarian keeps samples of unusual tumors and growths, but Tobey had his own shelf in my cabinet. If he lived to be 15, I figured he would have his own wing in Ripley's Believe It or Not!

It was on a fine, sunny day in June that Anthony Tuscalini carried Tobey in and set him on my exam table. His hind legs were paralyzed, and he was without a deep pain response. This was not a good sign. Testing for deep pain involves pinching the toe so that it is painfully stimulated. If the dog withdraws its leg without evidence of feeling pain, the peripheral nerves and spinal centers are normal, but the spinal cord is not. If the dog shows evidence of feeling pain, by turning its head and crying, the cord and thalamus are normal as well.

As I walked to the front office area to get another page for Tobey's medical record (I was already on page 23), I took a quick look outside. The sky had suddenly grayed, and I heard distant thunder. It was an ominous sign. I took Tobey to radiology and took a lumbar spine series of radiographs. The spinous process of the L3 vertebra had a moth-eaten appearance, which corresponded to the exact location where Tobey's panniculus reflex had stopped. This reflex comes from the dorsal rami of the spinal nerves to move the muscles of the skin. This was definitely not a good sign. After viewing the radiographs, I reexamined Tobey and was able to palpate two hard lumps under the skin on either side of the L3 vertebra.

"Tony, I think Tobey has a tumor around his backbone and spinal cord," I told him.

"Can you do surgery to repair it?" he asked.

"You say he was moving his legs last night?" I asked skeptically.

"He wasn't walking very well, but he was moving," he replied.

This may have been the best news yet, because that meant the spinal cord had been functional a short time ago, and was still viable. "OK, we can try to do surgery, but it's a long shot, and I can't make any promises."

"Do it!" Tony said. "We've got to give him a chance."

About an hour later, I was opening Tobey's back. After clipping the hair off, I could see the tops of two golf ball-sized lumps buried deep under the skin on either side of the vertebral column. I made one long incision above the spinous processes of vertebral bodies T13 through L5. There were two semi-fibrous masses, one on either side of the vertebra, with a connecting stalk that passed directly through the spinous process of the L3 vertebra. I excised the two masses, which left massive holes in the subcutaneous fat and muscle tissue. The ventral part of each mass was necrotic and gelatinous rather than fibrous, and getting it out was difficult.

At that time, I felt my effort to remove every neoplastic cell had not been successful. With my 3-mm bone rongeurs, I removed bits and pieces of the dorsal lamina and spinous process. This surgery is called a dorsal laminectomy. Normally, as a decompressive surgery on the spinal cord I would do a hemilaminectory, only opening one side of the lamina to view the spinal cord. In this case, I had no choice but to remove the entire lamina and much of the dorsal spinous process. I tried to leave as much of the pedicle of the vertebra on either side of the spinal cord as possible. It was necessary to remove both articular processes of the L3-L4 junction and the left cranial articular process of the L2-3 junction. This exposed the spinal cord, which had a significant dent in it where the tumor's stalk had passed above. I knew I couldn't leave the vertebra without support; otherwise it would tear apart at the L3-L4 intervertebral disc.

I had read about this technique in a veterinary journal and decided to use a thin, intramedullary pin and bend it on both ends around the spinous processes of L1 and L5. Then, I wired it to each vertebrae, L1 - L5; there was just enough of the L3 spinous process left to attach it to the pin. Had I had a spinal plate I might have used it, but attaching it to what remained of L3 would have been difficult. As it was, this worked fine.

I put Tobey on post-operative steroids and antibiotics, and 5 days later he walked out of the hospital. The pathology report said the tumor was a very hot fibrosarcoma (a malignant tumor of fibrous origin). I suggested that Tony consult a veterinary oncologist, but he decided against it. I thought for sure the tumor would return, but it never did. And, the amazing thing was, Tobey never developed any more tumors. My only explanation was that we turned on his immune system. I know I didn't get every tumor cell, so his body must have destroyed them. Either that or maybe...I am a magician?

SUMMARY

Much of this chapter was devoted to explaining the differences between the central nervous system and the peripheral nervous system. You learned about the neurons and the supporting cells of nervous tissue. The structure of neurons was discussed, and you learned to classify neurons based on number of axons. You once again looked at a giant multipolar neuron, but now with greater knowledge. Dissection of the cat's periphral nervous system and the sheep brain were done in detail. You learned the names of the cranial nerves, their type, and function. Finally, using a live dog you learned how to test the various reflexes used to assess neurological function.

CHAPTER 16

ORGANS OF SPECIAL SENSE

- differentiate between the three parts of the ear
- understand the mechanisms of hearing and balance
- name and locate parts of the ear using models and diagrams
- dissect the sheep's eye and identify its structures
- generally understand the mechanism of sight
- identify the layers of the cornea
- know the innervation of the eye and ear

MATERIALS:

- preserved sheep eye
- Mayo dissecting scissors
- probe
- 1 × 2 thumb forceps or Adson tissue forceps
- #4 scalpel handle with blade
- rubber gloves
- models of the ear
- plastic red, green, fluorescent orange, and white balls
- sheets of colored paper to match the balls
- live dog that knows how to retrieve balls

Introduction

You have already briefly studied the mechanisms of taste in the chapter on the digestive system (chapter 9) and smell in the chapter on the respiratory system (chapter 11). This chapter will cover the other two special senses: hearing and sight.

The Ear and the Mechanism of Hearing

Hearing occurs when sound waves enter the ear and stimulate its tiny bones, which in turn stimulate hair cells in the **cochlea.** This sends information up the **vestibulocochlear nerve (CN VIII),** through the *caudal colliculi,* and to the cortex of the *temporal* part of the brain.

In the exercises in this chapter the important structures are listed in colored bold print. If a structure is mentioned prior to its dissection it will be italicized. Structures discussed prior to dissection may also be in bold print for special emphasis.

ANATOMY AND PHYSIOLOGY OF THE EAR

The ear can be divided into three anatomical and functional parts: the **external ear, middle ear,** and **internal ear.**

Procedure

1. Using both the provided diagram of the canine ear (Figure 16.1) and a model of the ear (can be a human ear model), locate the **pinna,** the **vertical** and **horizontal ear canals** of the **external auditory canal,** and the **tympanic membrane (ear drum).** Human ear canals are not divided into vertical and horizontal canals. The support for the pinna of the animal ear is mostly elastic cartilage, and the attached muscles enable it to be aimed in the direction of sound. The pinna acts as a funnel to direct the sound waves into the external auditory canal. Note the vertical and horizontal components of the external ear canal in Figure 16.1. When placing the speculum of an otoscope into a dog or cat's ear, it initially must be directed down the vertical canal, then turned horizontally so you can see into the horizontal canal.

   The **tympanic membrane** is the termination of the external ear. It is oriented so the dorsal aspect is more superficial than the ventral aspect. It is a paper-thin membrane, composed of connective tissue, and is tightly stretched across the opening of the middle ear cavity. As the sound waves strike it, the membrane vibrates at the same frequency through a process called *sympathetic vibration.*

2. The **middle ear** is located within a hollow area bounded by the temporal bone and the *tympanic bullae,* and it is lined with soft tissue. The middle ear is filled with air through a communication with the middle ear via the **auditory** or **Eustachian tube** from the nasopharynx (which you studied as

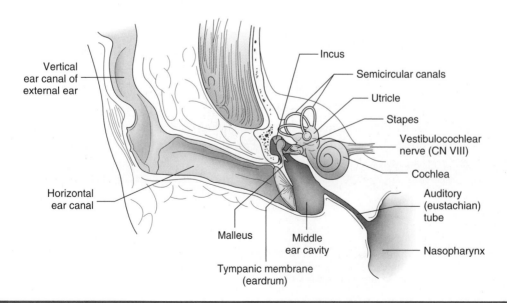

Figure 16.1: The external, middle, and internal ear of a dog or cat.

part of the respiratory system). This tube enables the pressure on the eardrum to be equal on its inside surface (via the middle ear) and on the outside surface (via the external auditory canal).

The eardrum is attached to the first of the three *ossicles* of the middle ear, which is called the **malleus,** but may also be known by its common name, the *hammer.* Next is the **incus** (or *anvil*), then the **stapes** (or *stirrup*), which is connected to and covers the **oval window** of the cochlea. An acronym to help remember these names—**m**alleus, **i**ncus, **s**tapes—is MIS (so you will not miss it on a test). The ossicles act as a system of levers that transmits the sound vibrations to the **cochlea.** In the process, they decrease the amplitude (size) of the vibrations but increase their force. This way the cochlea is not damaged. A tiny muscle attached to the malleus, the **tensor tympani,** acts to dampen vibrations. The **stapedius,** another muscle, acts to restrict movement of the stapes caused by loud sounds. Thus, it also prevents cochlear damage.

3. The **inner ear** has components that contribute to both hearing and balance. The hearing component is housed in a snail-shell-shaped spiral cavity within the temporal bone called the **cochlea** (Figure 16.2). Inside of this is the soft, multilayered, fluid-filled **organ of Corti,** which contains the receptor cells for hearing (Figure 16.3). This organ runs the length of the cochlea in a tube called the **cochlear duct,** which is filled with a liquid called **endolymph.** On either side of this duct is a tube containing **perilymph;** these ducts communicate at the tip of the cochlea.

The cochler duct starts at the **oval window,** which is attached to the stapes, goes to the tip of the cochlea, and returns to the **round window.** Nothing is attached to the round window. The organ of Corti runs the length of the cochlear duct, resting on a membrane called the **basilar membrane** (Figure 16.4). The organ's functional parts are the **hair cells, supporting cells,** and **tectorial membrane.** The hair cells are receptor cells, with tiny hair-like projections on their surfaces. The tectorial membrane is gelatin-like and lies gently on the hairs. The supporting cells provide physical support to the hair cells.

Sound waves transmitted to the cochlea's oval window via the stapes cause the membrane of the window to move back and forth. This causes the perilymph around the cochlear duct to vibrate back and forth. The round window acts as a pressure-relief mechanism by alternately bulging out and in with the vibrations of the perilymph. The perilymph vibrations cause the tectorial membrane to rub against the hair cells, which bends the sensory hairs and generates nerve impulses that travel to the brain and are interpreted as sound.

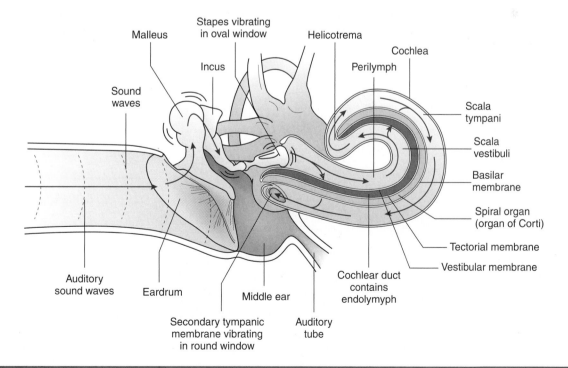

Figure 16.2: The ear of a dog, showing ossicles and a longitudinal cross section of the cochlea.

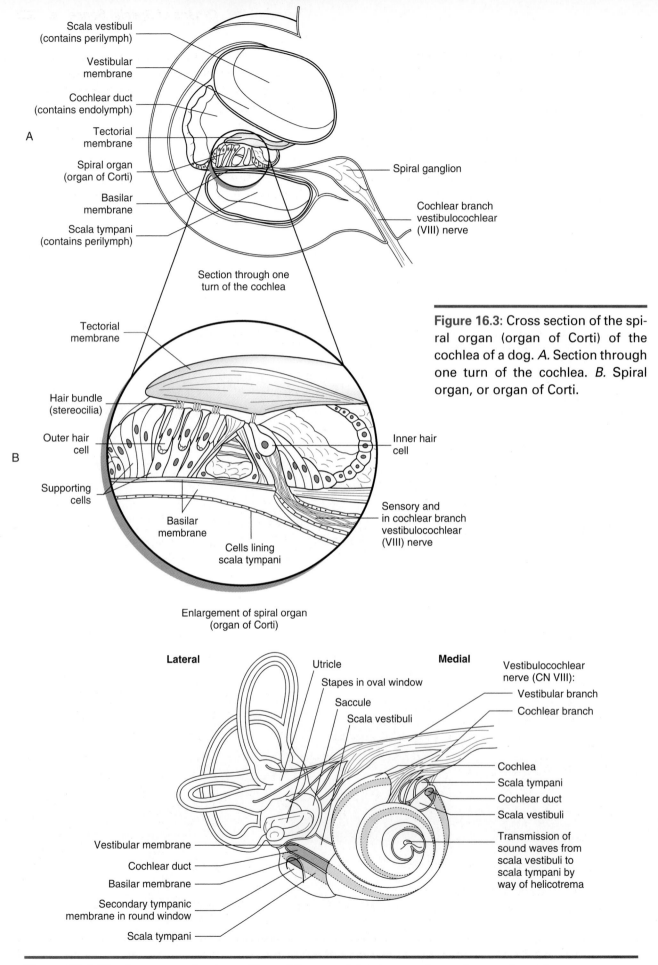

Scala vestibuli (contains perilymph)

Vestibular membrane

Cochlear duct (contains endolymph)

Tectorial membrane

Spiral organ (organ of Corti)

Basilar membrane

Scala tympani (contains perilymph)

A

Spiral ganglion

Cochlear branch vestibulocochlear (VIII) nerve

Section through one turn of the cochlea

Tectorial membrane

Hair bundle (stereocilia)

Outer hair cell

B

Supporting cells

Inner hair cell

Sensory and in cochlear branch vestibulocochlear (VIII) nerve

Basilar membrane

Cells lining scala tympani

Enlargement of spiral organ (organ of Corti)

Figure 16.3: Cross section of the spiral organ (organ of Corti) of the cochlea of a dog. *A.* Section through one turn of the cochlea. *B.* Spiral organ, or organ of Corti.

Lateral

Utricle

Stapes in oval window

Saccule

Scala vestibuli

Medial

Vestibulocochlear nerve (CN VIII):

Vestibular branch

Cochlear branch

Cochlea

Scala tympani

Cochlear duct

Scala vestibuli

Transmission of sound waves from scala vestibuli to scala tympani by way of helicotrema

Vestibular membrane

Cochlear duct

Basilar membrane

Secondary tympanic membrane in round window

Scala tympani

Figure 16.4: The inner ear of a dog with cochlea, semicircular canals, and nerve branches.

The inner ear also contains the **semicircular canals** and **vestibule.** These are important in maintaining balance and equilibrium. The vestibule is a part of the inner ear located between the cochlea and semicircular canals. It contains two sac-like spaces called the **utricle** and **saccule** (see Figure 16.4), which are continuous with the cochlear duct and are filled with endolymph. These two sacs are surrounded by perilymph. Each sac contains an area of sensory epithelium called the **macula.** Each also contains **hair cells** and **supporting cells** covered by a gel-like matrix containing tiny crystals of calcium carbonate called **otoliths** (Figure 16.5). Gravity causes the otoliths and the gelatinous matrix to put constant pressure on the hairs, as long as the head stays still. Movement of the head bends these sensory hairs, which generates nerve impulses and gives the brain information about the position of the head.

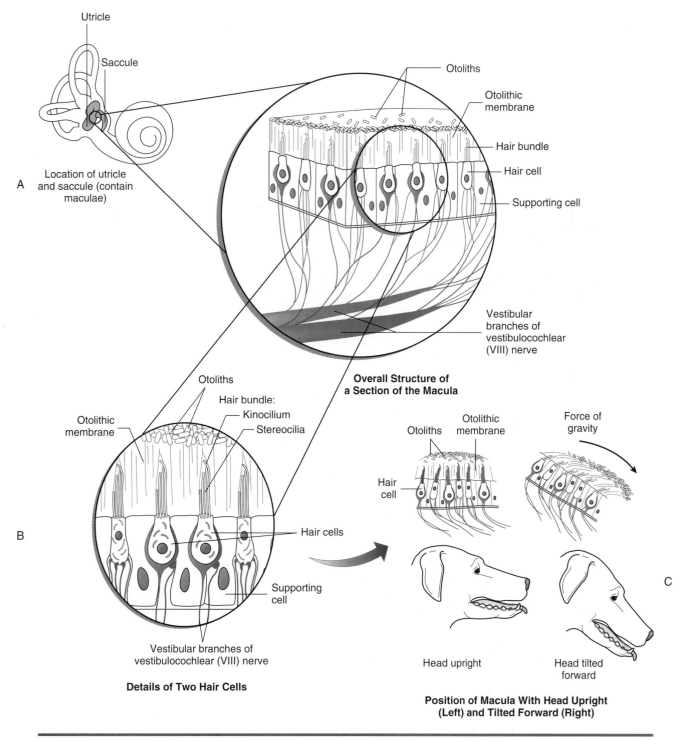

Figure 16.5: Receptors of the utricle and saccule of a dog and their mechanism of action. *A.* Section of the macula. *B.* Two hair cells. *C.* Position of macula with head upright and tilted forward.

There are three **semicircular canals,** each oriented in a different plane at right angles to one another. These also are filled with endolymph and surrounded by perilymph. Near the utricle end of each semicircular canal is an enlargement known as the **ampulla.** It contains the receptors, known as the **crista ampullaris (crista).** The ampulla functions in the same way as the macula of the vestibule. It consists of a cone-shaped area of hair cells and their supporting cells, which project into a gelatinous structure called the **cupula** (Figure 16.6). But the ampulla has no otoliths. The cupula functions as a float that

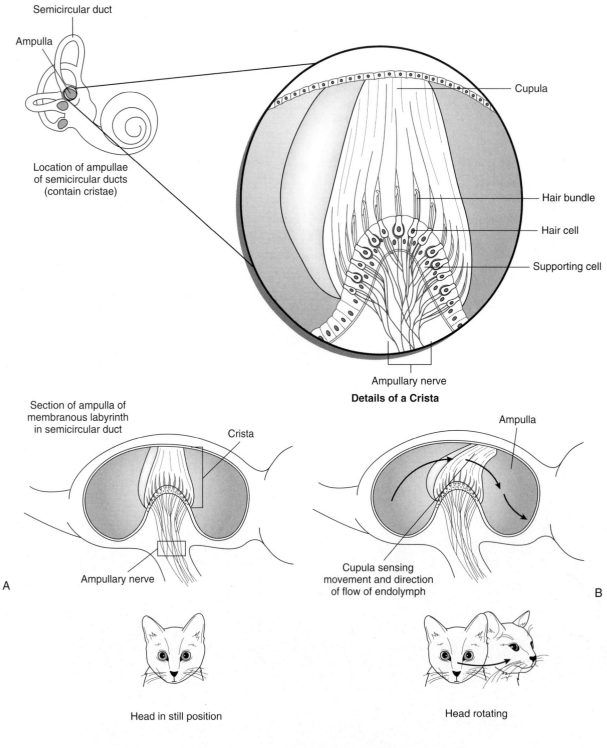

Figure 16.6: Receptors of the semicircular canal of a cat and their mechanism of action. *A.* A crista. *B.* Position of a crista when head is still and when it rotates.

moves with the endolymph as the head changes position. As the head moves in one of the planes of the semicicular canals, the inertia created causes the endolymph to lag behind the movement of the canal. The relative movement of the endolymph pulls on the cupula, bending the hairs, and generating a nerve impulse that is sent up the vestibular portion of the *vestibulocochlear nerve* to the brain.

EXERCISE 16.2

ANATOMY AND PHYSIOLOGY OF THE EYE

Vision is a process in which light waves of differing wavelengths enter the eye and are focused on the back of the eyeball, which is lined by the **retina.** The retina's nerve cells pick up the visual image and transmit it via the **optic nerve,** which crosses to the opposite side of the brain at the *optic chiasm* and is relayed through the rostral colliculi to the cortex of the occipital part of the cerebrum. This information is processed and interpreted as a picture by the brain. The image is sent to the back of the eye upside down, but the brain converts it so the world is right-side up.

DISSECTION OF THE SHEEP EYE

1. Obtain a sheep eye and note the fat surrounding the surface of the eye. This cushions the eye from shock in its **bony orbit.**

2. Identify the **sclera:** the tough, fibrous, external, white coat.

3. The **conjunctiva** is a thin, transparent membrane that covers the front portion of the sclera and lines the inside of the eyelids. It is composed of two parts: the **bulbar conjunctiva,** attached to the sclera, and the **palpebral conjunctiva** on the inner eyelids. The space between these two parts is the **conjunctival sac,** and they meet at the **fornix.**

4. The **palpebra** are the eyelids. The **dorsal** and **ventral palpebra** join at the **medial canthus,** or inner corner of the eye, and at the **lateral canthus,** or outer corner of the eye. Along the margins of the eyelids are small pores that are the openings to the **tarsal glands (Meibomian glands).** These glands produce a waxy substance that helps prevent tears from overflowing onto the face. The **eyelashes (cilia)**

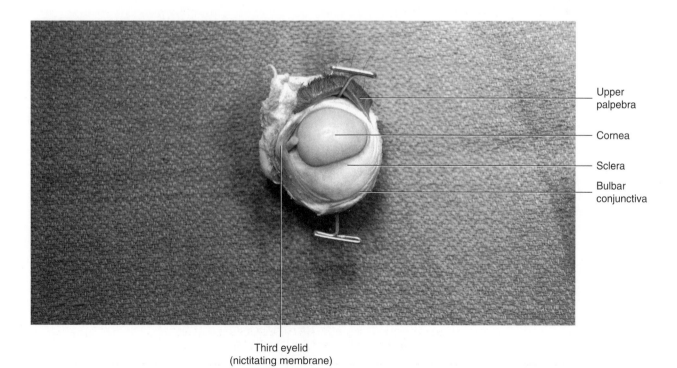

Upper palpebra

Cornea

Sclera

Bulbar conjunctiva

Third eyelid
(nictitating membrane)

Figure 16.7: External view of a sheep's eye.

are more prominent on the upper lid. Domestic animals also have a **third eyelid,** or **nictitating membrane,** which is located medially between the globe of the eye and the eyelids (Figure 16.7). It is supported by an internal T-shaped piece of cartilage and is covered by conjunctiva. Behind the third eyelid, on its ocular surface, are *lymph tissue* and the **gland of the third eyelid,** an accessory lacrimal gland (tear-producing gland).

5. The **lacrimal apparatus** is the tear drainage route for the eye (Figure 16.8). At the medial corners, just above and below the medial canthus, a small hole or **punctum** (plural is **puncta**) will be found. The **dorsal punctum** connects to the **superior canaliculus,** and the **ventral punctum** connects to the **inferior canaliculus;** these canaliculi join to form the **lacrimal sac,** which drains into the **nasolacrimal duct.** This duct exits to the nasal cavity just inside external nares on the medial surface of the wall. Patency of these ducts can be tested by placing fluorescein dye in the eyes and checking for its appearance around the nose. It is sometimes necessary to use a Wood's light (an ultraviolet light) to observe this.

6. On your specimen, the eye muscles may be attached to the globe (eyeball), but they will probably be too damaged to distinguish which is which. Four straight muscles and two oblique muscles form the **extraocular eye muscles** (Figure 16.9). There also are four **rectus muscles** (the ventral, dorsal, lateral, and medial rectus muscles) and two **oblique muscles** (the ventral and dorsal oblique muscles). Many animals also have a **retractor bulbi muscle,** which humans do not.

7. The **cornea** is the anterior, transparent tissue (though it may appear opaque in your specimen because of the preservative) that attaches to the sclera at the **limbus.** The cornea has multiple layers. The outer **epithelium** is composed of stratified squamous cells that are continuous peripherally with the conjunctiva. The outer epithelium is thicker in the middle than at the periphery and shows great power of regeneration. Beneath the epithelium is the basement membrane. The **substantia propria,** or **stroma,** is the thickest layer of the cornea and is made of transparent collagen fibrils without blood vessels. Its transparency is dependent on the exact amount of water it contains. If there is too much (corneal edema), it becomes cloudy. It contains many pain receptors, making it extremely sensitive. Below this is **Descemet's membrane,** a thin membrane just deep to the stroma. The **endothelium** is a single layer of squamous epithelial cells. Descemet's membrane is approximately four times thicker than the **endothelial layer** of cells (Figure 16.10).

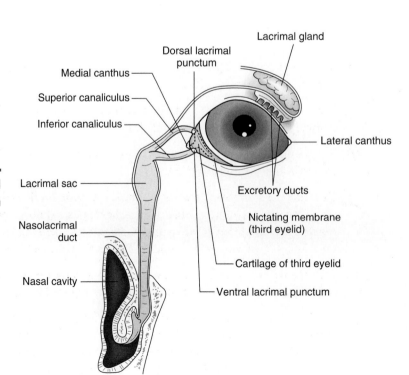

Figure 16.8: The eye and an internal view of the nasolacrimal system of a dog.

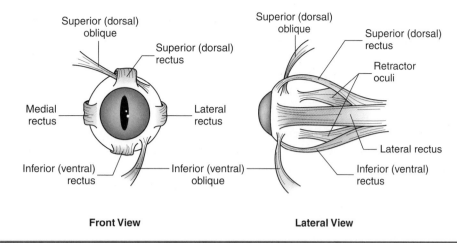

Figure 16.9: The extrinsic eye muscles of a dog.

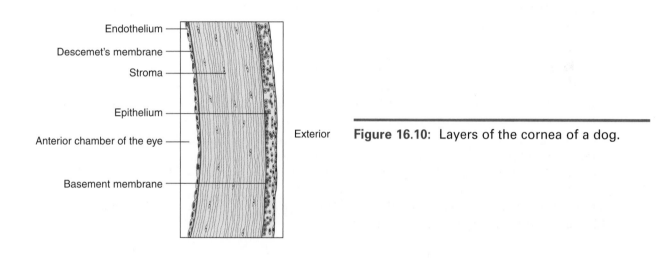

Figure 16.10: Layers of the cornea of a dog.

8. The **optic nerve** is located on the caudal surface of the eye. The nerve has a solid, white core and is approximately 3-mm thick.

9. Make a stab incision through the sclera about $^1/_2$ cm from the edge of the cornea. Using scissors, cut completely around the eye, parallel to the cornea.

10. You should now see the **vitreous humor (vitreous body)** in the caudal part of the eye and the **lens** sitting in the rostral part (Figure 16.11).

11. Examine the interior of the rostral part of the eye. Look at the black structure around and behind the lens. This structure, consisting of the ciliary muscle and its processing appear as radial folds; this is the **ciliary body** (Figures 16.12 and 16.13). Locate the **suspensory ligaments,** the delicate fibers connecting the ciliary body to the lens. They hold the lens in position.

12. Detach the **lens** from the ciliary body and remove the lens. The remnants of the suspensory ligaments can be seen attached to the lens (see Figure 16.13).

13. The **iris** is now visible anterior to the former position of the lens (see Figure 16.13). This also appears black (as did the ciliary body) on the caudal surface but it may be brownish-tan on the cranial surface. Try to distinguish between the circular and radial fibers that make up the iris. These are the muscles that control the aperture opening of the iris, known as the **pupil.**

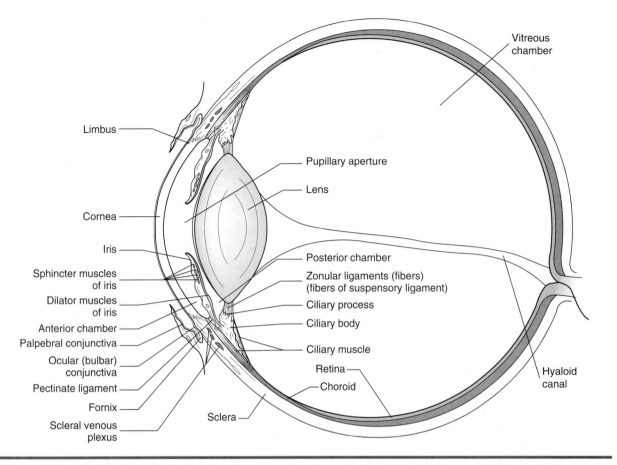

Figure 16.11: Sagittal section of a sheep's eye, showing vitreous humor, lens, and ciliary body.

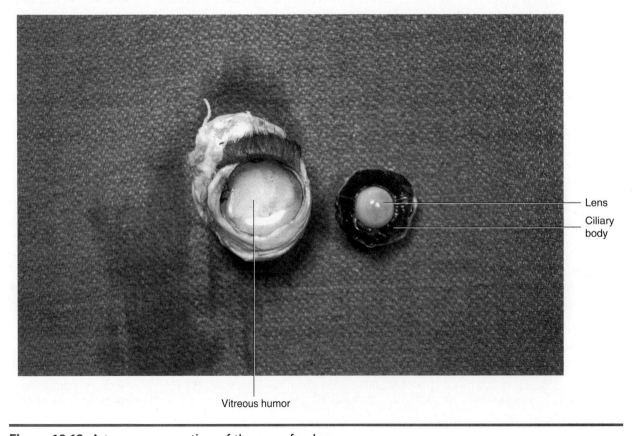

Figure 16.12: A transverse section of the eye of a dog.

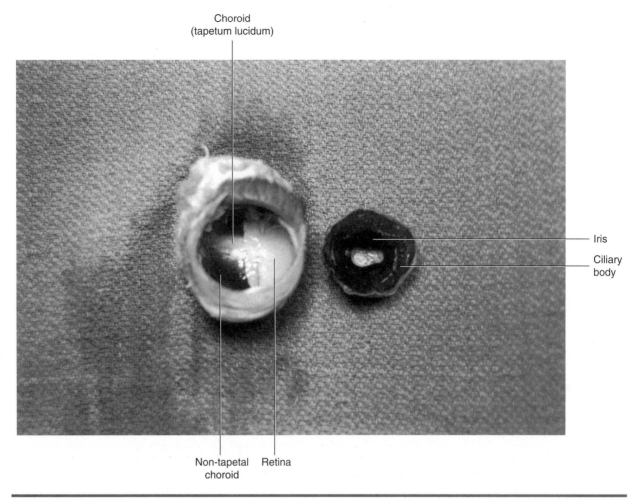

Choroid
(tapetum lucidum)

Iris

Ciliary
body

Non-tapetal   Retina
choroid

Figure 16.13: Sheep's eye, showing retina, choroid, and iris.

14. Find the space between the ciliary body and the iris, then find the space between the iris and the cornea. These are the **posterior** and **anterior chambers,** respectively. The ciliary processes produce the **aqueous humor,** which flows between the ciliary body and the iris, in front of the lens, and through the pupil into the anterior chamber. The aqueous humor flows out of the anterior chamber at the angle formed between the iris and cornea *(iridocorneal angle)* through tiny holes in the pectinate ligament (called the *spaces of Fontana*), into vessels leading to the *scleral venous plexus* (called the *canals of Schlemm*), and is absorbed back into the bloodstream. The pectinate ligament is a network of fine trabeculae connecting the iris to the inner wall of the sclera.

15. This first thing to note in the caudal part of the eye is the **vitreous humor.** During life this substance is perfectly clear (see Figure 16.13). Remove it from the eyeball.

16. The **retina** is the white inner coat that was covered by the vitreous humor (it is white because of the preservative). Determine the point at which the retina is attached caudally; this is the location of the **optic disc,** or the inner attachment of the optic nerve (see Figure 16.13).

17. The **middle vascular layer** of the eye is called the **uvea.** It consists, from anterior to posterior, of the *iris, ciliary body* and the **choroid.** It is located between the retina and sclera. The retina is easily separated from the **choroid,** the pigmented vascular layer. The choroid is continuous with the ciliary body and completely envelops the posterior hemisphere of the eyeball (the area posterior to the lens and ciliary body). It consists of six layers, which from outmost inward are: the suprachoroid, the perichoroidal lymphatic space, the vascular layer, the reflective layer **(tapetum lucidum),** the choriocapillary layer, and the basal lamina.

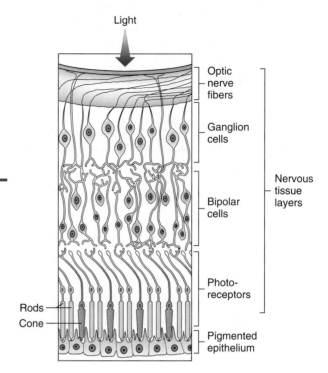

Figure 16.14: The layers of the retina of a dog.

The **tapetum lucidum** (see Figure 16.13) is a highly reflective, iridescent area. It is not present in swine or humans. The function of the tapetum lucidum is to reflect light back onto the retina. As the light passes the photoreceptors (the **rods** and **cones**) of the retina, it reflects off the tapetum and passes through the photoreceptors once again, thus amplifying the entering light. The tapetum lucidum is also what causes an animal's eyes to shine in the dark.

The *retina* is a multilayered structure that is the functional nervous tunic of the eye. It consists of five layers (Figure 16.14). Moving from the inside of the retina toward the choroid the layers are: the *optic nerve fibers,* the *ganglion cells,* the *bipolar cell layer,* the *photoreceptor layer,* and the *pigmented epithelial layer.* The inner layer consists of axonal nerve fibers that combine to form the *optic nerve* at the *optic disc* (where all the nerve fibers converge). The disc contains only a few nerve fibers and no photoreceptors, so it is essentially a blind spot at the back of the eye.

The light must pass all the way through to the photoreceptor layer. These cells, when excited, stimulate the bipolar cells, which in turn stimulate the ganglion cells (the next layer). The axons of the ganglion cells leave the retina as a tight bundle of fibers to form the optic nerve. The photoreceptors are neurons with dendrites modified to be sensory receptors for light. There are two types of photoreceptors, which are named for their shape: **rods** and **cones.** The rods are more sensitive to light and produce a somewhat coarse image in shades of gray. The cones are more sensitive to color and detail and do not function well in dim light.

Most domestic animals are described as color-blind because they have many rods and few cones. However, this description is not entirely accurate; animals can see colors to some extent, but they appear washed out or faded. They also do not perceive detail as sharply as humans. This is because humans have an area highly concentrated with cones, called the *fovea centralis,* in the center of the retina. Most animals need good vision to survive. The eyes of prey animals are located to the sides of their head, giving them a greater field of vision and a better chance to escape. Predators need good binocular vision to catch prey; therefore, their eyes are located more toward the front of their heads.

COLOR VISION IN DOGS

This is an entertaining experiment that illustrates color vision, or the lack of it, and visual acuity in dogs.

Procedure

1. Obtain a dog that will retrieve. Take red, green, fluorescent orange, and white balls and get the dog to retrieve each of them. Then, wash each of them to remove any contaminating odors.

2. Blindfold a dog. In a field of green grass, place one of the colored balls at the 12, 3, 6, and 9 o'clock positions at the perimeter of a circle 60 ft in diameter, so one ball sets at each position. (Note that one ball must be the same color as the grass.) Place the blindfolded dog in the center of the circle.

3. Remove the blindfold and release the dog. Observe which ball it retrieves first.

4. Remove that ball from the game and continue with the other three. Continue until all the balls are retrieved.

5. Place the white ball on a sheet of white paper, and each colored ball on sheets of colored paper of the same color. Repeat the experiment. This should be a test of visual acuity.

QUESTIONS

1. Which ball did you expect the dog to retrieve first?

2. Why would placing the balls on sheets of paper of the same color be a test of visual acuity?

Discussion

The ball the dog should see easiest is the white ball. Even the fluorescent orange ball will have a gray-scale appearance equal to that of the green and red balls. It is easy for a human to see, but not a dog. Fluorescent-orange boat bumpers are great to teach a retriever hand signals because you can see the bumper from a distance in high grass, but the dog cannot and needs your help to find it. Placing the balls on similarly colored sheets of paper removes the contrasting background, and the dog must then rely on its sense of visual acuity, or how well it sees the shadow cast around the ball, to find it.

CLINICAL SIGNIFICANCE

The amount of aqueous humor in the anterior chamber of the eye determines its intraocular pressure. If the animal does not have the ability to absorb this liquid, the volume within the eye will increase and so will the pressure. This is a disease called glaucoma. *It results in degenerative changes in the retina and optic nerve and eventually leads to blindness.*

When a veterinary ophthalmologist examines an animal's eye, one of the tests performed is to visualize the angle between the cornea and iris (the irido-corneal angle). This is done using a slit lamp. A narrow beam of light is passed through the cornea and viewed from an angle perpendicular to the beam. Often glaucoma is caused by a narrowed angle *between the iris and cornea, which impairs the drainage through the spaces of Fontana and into the canals of Schlemm. We can measure the pressure within the ocular chamber using a Schiotz tonometer. Pressures greater than 25 to 30 mmHg in dogs or more than 30 mmHg in cats indicate glaucoma. Increased pressures may also be secondary to other diseases, such as primary lens luxation, anterior uveitis, and hyphema.*

Most forms of glaucoma are best treated surgically. Surgery can enhance the outflow or reduce the production of aqueous humor. Unfortunately, veterinarians often are presented with cases that are already quite advanced, and treatment is unsuccessful.

Veterinary Vignettes

"**D**oc, Tom Foley is on his way in with his Rottweiler, Granite. He's just been hit by a train," my receptionist rushed in to tell me. "He'll be here in about 30 minutes," she said.

"You mean a real train? Uh . . . like a real choo-choo?" I asked. I have always liked using the most technical terms with my staff; it makes us appear professional.

"Yep."

"How bad is it?" I asked.

"Don't know; he didn't say. He just hung up in a hurry," she told me.

Oh no! I thought. The last time I had one of those, the dog had chest trauma, was throwing all sorts of abnormal electrocardiogram wave patterns, and was hemorrhaging internally. I rushed into the treatment room to tell my technician what was going on and to get the trauma cart ready. "This could be a bad one," I told her. I turned up the radio when we heard the news flash come in.

"We advise you to route around Highway 99 West at Glen Market Road. We have reports that the northbound Union Pacific express is stalled at this intersection. It seems an animal was injured and is on its way to an emergency veterinary hospital. We will keep you posted on these developments as they come in. This is Sharon Schilling, live on the scene for KPTX."

I had thought I was going to get to go home early this evening. It had been a relatively slow day; a few vaccinations, 14 cases in a row of flea allergic dermatitis, and a man wanting codeine for his non-coughing dog with kennel cough.

"He's here, Doctor," the receptionist told me.

We rushed out to help. Granite was standing in the waiting room, tongue hanging out the side of his mouth, panting and drooling on the floor, and wagging his tail! He jumped up, as he usually does, putting his front feet on the counter top to look over at the receptionist. His left eye had popped out of its socket, a condition called proptosis. Tom had had the good sense to hobble Granite by tying his front feet loosely together, and he had covered the eye with a moist cloth to protect it. He had completed a first aid course at work and was using his training on his dog. Smart man!

"I heard the report on the radio about a train on 99, is that the same one?" I asked him.

"Yeah, he ran out in front of it, got hit on the back of the head and knocked about 20 feet."

That explained the dog. What about the train? I thought.

"I don't know why it stopped," Tom added. "Maybe he derailed it." He laughed.

We checked Granite over and found nothing else wrong. We replaced his eye and sewed the lids together to protect it and hold it in place. I injected some aqueous cortisone behind the eye, and explained to Tom that it would take a while for the hemorrhage and swelling in the fat pad behind Granite's eye to go down. Because we repaired the eye quickly, he would probably not lose his sight, which he didn't. The funny thing about Granite was, he didn't even act like he had a headache.

We found out that the engineer of the train saw the dog running toward the track, then tried to look back to see if he had hit it. In the process, he banged his head and cut a huge hole in his scalp. He probably had a worse headache than Granite. But then, maybe his head wasn't as hard!

SUMMARY

The focus of this chapter was on the sense of sight, balance and equilibrium, and hearing. The sense of smell and taste were covered briefly in previous chapters. You learned the anatomy of the external, middle, and internal ears using the provided drawings and a plastic model. The physiological mechanism of hearing and the transmission of the nerve impulses to the brain also were discussed. Using the illustrations provided, the mechanism by which the inner ear—using the semicircular canals and the vestibule—recognizes changes in posture and spatial orientation was covered. You were able to dissect a sheep's eye and learn its anatomical structures. Illustrations were used to describe how the retina works to facilitate vision. Discussion also included the mechanism of aqueous humor production and resorption, the microscopic anatomy of the cornea, and the mechanism of sight. Finally, a fun exercise using a live dog was utilized to illustrate color vision, or the lack of it, and visual acuity. In this you learned that dogs have trouble viewing items of similar gray-scales, like green grass and orange balls, but objects that have substantially different contrast (black vs. white) is much easier for them to see.

BIBLIOGRAPHY

Bacha, W. J., Jr., and L. M. Bacha, *Color Atlas of Veterinary Histology.* 2nd ed. Baltimore: Lippincott, 2000.

Boyd, J. S., *Color Atlas of Clinical Anatomy of the Dog and Cat.* 2nd ed. Philadelphia: Mosby, 1999.

Cochran, P. E., *Guide to Veterinary Medical Terminology.* St. Louis: Mosby, 1991.

Coles, E. H., *Veterinary Clinical Pathology.* 4th ed. Philadelphia: W. B. Saunders, 1986.

Colville, T., and J. Bassert, *Clinical Anatomy & Physiology for Veterinary Technicians.* St. Louis: Mosby, 2002.

Cooper, E. L. and L. D. Burton, *Agriscience Fundamentals & Applications.* 3rd ed. New York: Delmar Thomson Learning, 2002.

Edwards, N. J., *ECG Manual for the Veterinary Technician.* Philadelphia: W. B. Saunders, 1993.

Evans, E. E., *Miller's Anatomy of the Dog.* 3rd ed. Philadelphia: W. B. Saunders, 1993.

Evans, H. E., and A. deLattunta, *Miller's Guide to the Dissection of the Dog.* 3rd ed. Philadelphia: W. B. Saunders, 1993.

Ettinger, S. J., *Textbook of Veterinary Internal Medicine, Diseases of the Dog and Cat.* Philadelphia: W. B. Saunders, 1975.

Frandson, R. D., and T. L. Spurgeon, *Anatomy and Physiology of Farm Animals.* 5th ed. Philadelphia: Lea & Febiger, 1992.

Gilbert, S. G., *Pictorial Anatomy of the Cat.* 7th ed. Seattle: University of Washington, 1987.

Holmstrom, Steven E., *Veterinary Dentistry for the Technician & Office Staff.* 1st ed. Philadelphia: W. B. Saunders, 2000.

Hudson, L. C., and W. P. Hamilton, *Atlas of Feline Anatomy for Veterinarians.* Philadelphia: W. B. Saunders, 1993.

Jenkins, T. W., *Functional Mammalian Neuroanatomy.* Philadelphia: Lea & Febiger, 1972.

Kealy, J. K., *Diagnostic Radiology of the Dog and Cat.* 2nd ed. Philadelphia: W. B. Saunders, 1987.

Marieb, E. N., *Human Anatomy & Physiology Laboratory Manual.* 7th ed. San Francisco: Benjamin Cummings, 2002.

McDonald, L. E., and M. H. Pineda, *Veterinary Endocrinology and Reproduction.* 4th ed. Philadelphia: Lea & Febiger, 1989.

Neal, K. G., and B. H. Kalbus, *Dissection Guide for the Cat (and Selected Sheep Organs).* Minneapolis: Burgess, 1971.

Patten, B. M., *Foundations of Embryology.* 2nd ed. New York: McGraw-Hill, 1964.

Pratt, P. W., *Principles and Practice of Veterinary Technology.* St. Louis: Mosby, 1998.

Reagan, W. J., T. G. Sanders, and D. B. DeNicola, *Veterinary Hematology Atlas of Common Domestic Species.* Ames, Iowa: Iowa State University, 1998.

Rebar, A. H., *Handbook of Veterinary Cytology.* St. Louis: Ralston Purina, 1978.

Sandman, K. M., and J. Hatari. "Cranial Cruciate Ligament Repair Techniques: Is One Best?" *Veterinary Medicine* 96(11): 850-856, 2001.

Sisson, S., and J. D. Grossman, *The Anatomy of the Domestic Animals.* 5th ed. Philadelphia: W. B. Saunders, 1975.

Stashak, T. S., *Adam's Lameness in Horses.* 4th ed. Philadelphia: Lea & Febiger, 1985.

Synbiotics Corporation. *ICG Status-Pro & ICG Status-LH.* San Diego: Synbiotics Corporation, 1998.

Tilley, L. P., and F. W. K. Smith, *The 5-Minute Veterinary Consult, Canine and Feline.* Baltimore: Williams & Wilkins, 1997.

Tortora, G. J., and S. R. Grabowski, *Principles of Anatomy and Physiology.* 9th ed. New York: John Wiley & Sons, 2000.

INDEX

Rumen, 194
Ruminant stomach, 194–195

S

S-T segment, 233
SA node, 231
Saccule, 327
Sacral, 4
Sacrospinalis, 169
Sacrotuberous ligament, 132
Sacrum, 112
Sagittal crest, 114
Sagittal plane, 5
Sagittal suture, 114
Salivary glands, 47, 175, 179
Saphenous nerve, 305
Sarcomere, 139
Sarcoplasm, 138
Sarcoplasmic reticulum, 20
Sartorius, 160, 161
Satellite cells, 297
Scapula, 114, 116
Scapular, 4
Schwann cells, 297, 298
Sciatic nerve, 164, 165, 305, 306
Sclera, 329
Scrotal ligament, 288
Scrotum, 281, 288
Sebaceous glands, 87
Sebum, 91
Second heart sound, 207, 208
Secondary bronchi, 244, 246, 247
Secondary lamina, 96
Secretory vesicles, 21
Sectioning methods, 32–35
Segmental veins, 255
Segmented neutrophils (segs), 69
Self-propagating action potential, 301
Semimembranosus, 165, 166
Seminal fluid, 282, 286
Seminal vesicles, 282
Semitendinosus, 164, 165, 166
Septum pellucidum, 311, 316
Serosa, 175, 191
Serosal surface, 32
Serratus dorsalis, 168, 170

Serratus ventralis, 149–152
Serum, 68
Sesamoid bones, 104
Sheep
    brain, 273, 311–317
    eye, 329, 332, 333
    feet, 130
    lobes of cerebrum, 311
    lungs, 246
    meninges (brain), 312
    pineal gland, 274
    spinal cord, 79
Sheet of cells, 34
Short bones, 104
Short pastern bone, 95
Silver stains, 28
Simple columnar epithelial cells, 42
Simple columnar epithelium, 38, 42–43
Simple cuboidal epithelial cells, 41
Simple cuboidal epithelium, 38, 41
Simple epithelium, 37
Simple squamous epithelial cells, 33
Simple squamous epithelium, 38–40
Simplex uterus, 280
Sino-atrial (S-A) node, 231
Skeletal muscle, 75–76
Skeletal system, 103–136
    articular projections, 105
    bone. *See* Bone
    calcium measurements, 134
    carpus, 117, 118, 127
    cat, 109–121
    chicken, 131
    comparative arthrology/dermatology, 132
    comparative osteology, 122–131
    depressions, 105
    dog, 122, 130
    femur, 118, 119
    horse, 126–129
    humerus, 116, 117
    leg/foot, 118–121, 126–130
    mandible, 115
    metacarpal bones, 117, 118
    metatarsals, 120, 121
    non-articular projections, 105
    pelvis, 118, 119

NOTES

NOTES

NOTES

NOTES

NOTES

NOTES